Lecture Notes on Multidisciplinary Industrial Engineering

Series Editor

J. Paulo Davim, Department of Mechanical Engineering, University of Aveiro, Aveiro, Portugal

"Lecture Notes on Multidisciplinary Industrial Engineering" publishes special volumes of conferences, workshops and symposia in interdisciplinary topics of interest. Disciplines such as materials science, nanosciences, sustainability science, management sciences, computational sciences, mechanical engineering, industrial engineering, manufacturing, mechatronics, electrical engineering, environmental and civil engineering, chemical engineering, systems engineering and biomedical engineering are covered. Selected and peer-reviewed papers from events in these fields can be considered for publication in this series.

Carolina Machado
Editor

Human Resource Management in a Disrupted World

Springer

Editor
Carolina Machado
School of Economics
Management and Political Science
University of Minho
Braga, Portugal

ISSN 2522-5022 ISSN 2522-5030 (electronic)
Lecture Notes on Multidisciplinary Industrial Engineering
ISBN 978-3-031-89947-8 ISBN 978-3-031-89948-5 (eBook)
https://doi.org/10.1007/978-3-031-89948-5

This Springer imprint is published by the registered company Springer Nature Switzerland AG
The registered company address is: Gewerbestrasse 11, 6330 Cham, Switzerland

Conference Committees

Organizing Committee

Ana Luísa Veloso, School of Psychology, University of Minho, Braga, Portugal
Ana Paula Ferreira, School of Economics, Management and Political Science, University of Minho, Braga, Portugal
Carolina Machado, School of Economics, Management and Political Science, University of Minho, Braga, Portugal
Emilia Fernandes, School of Economics, Management and Political Science, University of Minho, Braga, Portugal
Gina Gaio Santos, School of Economics, Management and Political Science, University of Minho, Braga, Portugal
Isabel Silva, School of Psychology, University of Minho, Braga, Portugal
João Leite Ribeiro, School of Economics, Management and Political Science, University of Minho, Braga, Portugal
Regina Leite, School of Economics, Management and Political Science, University of Minho, Braga, Portugal

Scientific Committee

Alexandra Costa, ISEP, University of Porto, Portugal
Amélia Marques, ESCE, Polytechnic Institute of Setubal, Portugal
Ana Cláudia Rodrigues, ISCAP & CEOS.PP, Polytechnic Institute of Porto, Portugal
Ana Isabel Couto, ISCAP & CEOS.PP, Polytechnic Institute of Porto, IS-UP & Falculty of Economics of Porto, Portugal
Ana Luísa Martinho, ISCAP & CEOS.PP, Polytechnic Institute of Porto & A3S, Portugal
António José Almeida, ESCE, Polytechnic Institute of Setubal, Portugal
Beatriz Martín del Rio, Universidad Miguel Hernández de Elche, Spain

Professional Committee

Preface

This book seeks to reflect a set of selected communications that were presented at the XII International Conference on Intervention in Human Resources, under the theme "HRM in a Disrupted World: at the crossroads of people management." Taking place on October 10 and 11, 2024, at the University of Minho, Braga, Portugal, this conference, now in its 12th edition, is a prominent conference in the area of Human Resources that has been organized systematically over the years. Counting on Professors Chris Brewster (from Henley Business School, University of Reading, UK), Santiago Gutiérrez (from the University of Castilla-La Mancha, Spain), and Eduardo Oliveira (from the University of Porto/Faculty of Economics, Portugal), as keynote speakers, this conference sought to bring together the best researchers who have been developing their research in the area of Human Resources Management.

Why talk about Human Resource Management in a disruptive world?

Managing human resources in a disruptive world is a significant challenge for modern organizations. Disruption refers to rapid and profound changes that affect the way companies operate, often driven by technological advances, changes in worker expectations, globalization, economic and environmental crises, among other factors.

In a disruptive context, there are many challenges facing HRM. Among the main aspects of HRM with strong expression in a disruptive context, adaptation to digital transformation; flexibility and hybrid work models; the focus on well-being and mental health; diversity, equity and inclusion; requalification and continuous development; talent management; innovation and organizational agility; ethics and corporate responsibility; between others, are highlighted here. In particular, and looking at each of these challenges, we can say that, in terms of adapting to digital transformation, it is increasingly clear that technology is transforming all aspects of work, from recruitment to performance management. Artificial intelligence, automation and data analysis tools are being used to optimize HR processes. Companies need to ensure their workforce is prepared to adopt and integrate new technologies, which may require ongoing reskilling and training programs.

Regarding flexibility and hybrid work models, the pandemic period that the world has experienced has accelerated the adoption of remote and hybrid work models. This has brought to light the need to create flexible work policies that balance productivity with employee well-being. HR management needs to be able to lead the transition to these new working models, ensuring that organizational culture and employee engagement are maintained.

When it comes to focusing on well-being and mental health, increased workplace pressures and economic uncertainty have led to greater attention to employee mental health and well-being. Support programs such as counseling and mental health resources have become essential. HR management must be aware of these aspects, promoting a healthy and sustainable work environment.

Social disruption, including global movements for racial justice and gender equity, has significantly influenced HR management. Companies are being pressured to adopt more inclusive practices and build diverse teams. This includes reviewing recruiting, compensation, and promotion practices to eliminate bias and promote equity.

With rapidly evolving job market demands, the skills needed for success are changing. Hence, the need to promote requalification and continuous development. At this level, it is observed that HR management needs to create reskilling and upskilling strategies to ensure that employees can keep up with these changes. Investing in continuous learning is crucial to the competitiveness of the workforce.

Globalization and the ability to work remotely allow companies to recruit talent from anywhere in the world. This brings challenges, such as managing geographically dispersed teams, cultural differences, and varying labor legislation. HR management needs to develop effective strategies to integrate and manage these global teams.

Innovation and organizational agility are also an increasingly emerging reality. In times of disruption, the capacity for innovation and organizational agility become competitive advantages. HR management must foster a culture that values experimentation, continuous learning, and the ability to respond quickly to market changes. This may involve changes to the organizational structure and the way performance is evaluated and rewarded.

Finally, we come across issues associated with ethics and corporate responsibility. In an ever-changing world, companies face new ethical issues, especially regarding the use of technologies and environmental impact. HR management plays a fundamental role in defining and implementing responsible practices, ensuring that the company acts in an ethical and sustainable manner.

Managing human resources in a disrupted world requires a proactive and adaptive approach. Companies need to be prepared to navigate constant change while ensuring their employees are equipped, motivated, and healthy to face future challenges.

Due to the relevance that the theme assumes, this book, looking to bring together some selected communications that were presented at the conference, has, as its main objective, to promote reflection, discussion, and dissemination of knowledge among all those who express their interest in this area of knowledge, with emphasis, on particularly, to undergraduate students, academics researchers, human resource managers, managers, engineers, and other professionals in related matters with disruptive environments.

In short, conscious of these challenges, with this book, entitled *Human Resource Management in a Disrupted World*, we look to present the different contributions that have been made for a deeper understanding of the challenges, opportunities, and best practices related to management and business in a disrupted environment. It looks to provide managers, leaders as well as all kind of professionals with the most recent and critical knowledge and abilities/competencies, to deal with the processes, tools, and mechanisms able to build dynamic, innovative, and successful organizations

Human Resource Management in a Disrupted World can be used by a variety of potential stakeholders. Contributing to stimulate the growth and development of each individual in a dynamic, competitive, global, responsible, and sustainable economy, the diversity of subjects discussed in this book reflects the knowledge and know-how of different experts from this field of research. Organized into 15 chapters, the book starts with the presentation of the conference's opening *keynote speaker*, which discusses "Continuity and Change in HRM: A (Slightly) Subversive View." Next comes *Thematic Area: Digital Transformation and Automation*, which integrates "Managing Digital Nomads: Insights from Four Continents," just followed by "Industry 4.0: Knowledge Management with Artificial Intelligence?" and "Generative Artificial Intelligence in Candidate Selection." We then come across *Thematic Area: Evolution of Work and the Workforce*, which highlights "A Performance Evaluation System in a Non-technological Portuguese SME: How Far Do We Need to Run?", and "Work-Family Conflict, Stress, and Job Satisfaction: The Mediating and Moderating Roles of Stress, Influence and Meaningful Work." *Thematic Area: Organizational Culture and Employee Engagement*, focuses on "Leadership, Organizational Culture, and Change Management: Navigating Complexity in a Disrupted World," continuing to address the challenges of a disrupted world, *Thematic Area: Talent Development and Continuous Learning* focuses on three key issues, such as "Online Recruitment: A Focus Group Study with Portuguese Recruiters," followed by "Executive Coaching as a Mental Health Intervention Instrument Integrated in Human Resources Management," and "Career Transitions: Insights from a Narrative Approach." In the *Thematic Area: Change Management and Organizational Resilience* take center stage "Telework and Well-Being: Investigating the Role of Motivation in Shaping Remote Work Experiences," and the discussion about "Expatriate Workers: Impact of Change on Adaptation to the Destination Country." The book ends with *Thematic Area: Sustainability and Corporate Social Responsibility*, where it is discussed "Work Time Reduction: A Critical Analysis of the Main Arguments for Adopting a 4-Day Workweek," and finally the *Thematic Area: The Importance of HRM for Achieving the Sustainable Development Goals (SDGs)*, which analyzes "Voices from Within: How University Workers Perceive HRM," soon followed by *"HRM in the Small Family Business: Case Studies in the Interior of Portugal."*

Human Resource Management in a Disrupted World can be used by various potential stakeholders, including academics/researchers, managers, engineers, practitioners, and other professionals in the different areas of business, management, and engineering. Students from different undergraduate/graduate levels (undergraduate, master, and doctoral/PhDs) in the areas of management, human resource management, psychology, and industrial engineering are also potential stakeholders in this book, who need to know and understand the challenges that organizations are facing nowadays, in a world characterized by continuous disruptions.

The Editor acknowledges her gratitude to Springer for this opportunity and their professional support. Finally, we would like to thank all chapter authors for their interest and availability to collaborate on this project.

Braga, Portugal Carolina Machado

Contents

Part I Keynote Speaker

1 **Continuity and Change in HRM: A (Slightly) Subversive View** 3
Chris Brewster and Carolina Feliciana Machado

Part II Thematic Area: Digital Transformation and Automation

2 **Managing Digital Nomads: Insights from Four Continents** 21
Joana Fernandes and Ana Cláudia Rodrigues

3 **Industry 4.0: Knowledge Management with Artificial Intelligence?** 47
Elsa Maria Rodrigues and Ana Veloso

4 **Generative Artificial Intelligence in Candidate Selection** 89
Joana Santareno

Part III Thematic Area: Evolution of Work and the Workforce

5 **A Performance Evaluation System in a Non-technological Portuguese SME: How Far Do We Need to Run?** 113
Bárbara Balbina Macedo Marques and Ana Paula Ferreira

6 **Work-Family Conflict, Stress, and Job Satisfaction: The Mediating and Moderating Roles of Stress, Influence and Meaningful Work** 141
Alexandra Fontes, José Carlos Pinho, Gina Gaio Santos, and Carlos Menezes

Part IV Thematic Area: Organizational Culture and Employee Engagement

7 **Leadership, Organizational Culture, and Change Management: Navigating Complexity in a Disrupted World"** 163
Paulo Viana and Carolina Feliciana Machado

Part V Thematic Area: Talent Development and Continuous Learning

8 **Online Recruitment: A Focus Group Study with Portuguese Recruiters** 195
Ana Sofia Freitas and Catarina Brandão

9 **Executive Coaching as a Mental Health Intervention Instrument Integrated in Human Resources Management** 215
Manuel Salvador Gomes de Araújo and Vanda Gisela Conde Pereira Branco Afonso

10 **Career Transitions: Insights from a Narrative Approach** 243
Carla Malta, Cassiana Damo, and Viviana Meirinhos

Part VI Thematic Area: Change Management and Organizational Resilience

11 **Telework and Well-Being: Investigating the Role of Motivation in Shaping Remote Work Experiences** 275
Rita Couto, Sílvia Lopes, Vânia Sofia Carvalho, and Maria José Chambel

12 **Expatriate Workers: Impact of Change on Adaptation to the Destination Country** 309
Ana Francisca Oliveira Ribeiro and Isabel S. Silva

Part VII Thematic Area: Sustainability and Corporate Social Responsibility

13 **Work Time Reduction: A Critical Analysis of the Main Arguments for Adopting a 4-Day Workweek** 333
Ana Oliveira, Sofia Alexandra Cruz, Ana Isabel Couto, Jorge Cerdeira, Cristina Parente, and Carlos Manuel Gonçalves

Part VIII Thematic Area: The Importance of HRM for Achieving the Sustainable Development Goals (SDGs)

14 **Voices from Within: How University Workers Perceive HRM** 359
Aléxia Ferreira Gomes, Ana Veloso, Patrício Costa, and Teresa Proença

15 HRM in the Small Family Business: Case Studies in the Interior of Portugal 389
Maria Amélia Marques and Jorge Rodrigues

Index 411

About the Editor

Carolina Machado received her Ph.D. degree in Management Sciences (Organizational and Policies Management area/Human Resources Management) from the University of Minho in 1999, her Master's degree in Management (Strategic Human Resource Management) from the Technical University of Lisbon in 1994, and Degree in Business Administration from the University of Minho in 1989. Teaching in Human Resources Management subjects since 1989 at the University of Minho; she is since 2004 Associate Professor (with Habilitation since 2022), with experience and research interest areas in the field of Human Resource Management, International Human Resource Management, Human Resource Management in SMEs, Training and Development, Emotional Intelligence, Management Change, Knowledge Management and Management/HRM in the Digital Age/Business Analytics, Sustainability and Higher Education Sustainability. She is Head of the Human Resources Management Work Group at the School of Economics, Management and Political Science at the University of Minho, Coordinator of Advanced Training Courses at the Interdisciplinary Centre of Social Sciences, Member of the Interdisciplinary Centre of Social Sciences (CICS.NOVA.UMinho), University of Minho as well as Chief Editor of the International Journal of Applied Management Sciences and Engineering (IJAMSE), Guest Editor of journals, books Editor and book Series Editor, as well as reviewer in different international prestigious journals. In addition, she has also published both as editor/co-editor and as author/co-author several books, book chapters, and articles in journals and conferences. e-mail: carolina@eeg.uminho.pt

Part I
Keynote Speaker

Chapter 1
Continuity and Change in HRM: A (Slightly) Subversive View

Chris Brewster and Carolina Feliciana Machado

Abstract Human resource management plays a fundamental role in organizations and can be seen as a strategic program or a more humanistic field of study. In the current context, marked by disruption and the VUCA world (volatility, uncertainty, complexity, and ambiguity), HRM faces increasing challenges. The Covid-19 pandemic has accelerated changes in work, making so-called "wicked problems" more evident, that is, complex issues with no single solution. The CRANET database provides insights into HRM practices at a global level, contributing to adaptation to new realities. Furthermore, technological transformation continues to reshape the relationship between work and talent management. Stakeholder HRM is gaining relevance in academia, highlighting the need to align organizational practices with social and scientific expectations. With this dynamic scenario, HRM must balance efficiency and well-being, adopting innovative approaches to managing people in times of constant change. The challenge we set ourselves with this chapter is to know whether, in disruptive times, such as the ones we are living in, we need to rethink HRM.

Keywords Continuity · Change · HRM · Subversive view · Disrupted world

C. Brewster
Henley Business School, University of Reading, Reading, UK

C. F. Machado (✉)
School of Economics, Management and Political Science, University of Minho, Braga, Portugal
e-mail: carolina@eeg.uminho.pt

Interdisciplinary Centre of Social Sciences (CICS.NOVA.UMinho), University of Minho, Braga, Portugal

C. Machado (ed.), *Human Resource Management in a Disrupted World*, Lecture Notes on Multidisciplinary Industrial Engineering,
https://doi.org/10.1007/978-3-031-89948-5_1

1.1 Introduction

Human Resource Management (HRM) in a disrupted world faces unprecedented challenges and opportunities driven by factors such as digital transformation, automation and globalization, and de-globalization. Among some of the main challenges that have been most frequently manifested, digitalization and automation are perhaps the most significant, with artificial intelligence (AI) and automation redefining tasks and functions, requiring human resource management to adapt to new requalification strategies (reskilling) and skills improvement (upskilling) (Parry & Battista, 2019). Remote and hybrid work is a growing modern reality. The traditional work and employment model is giving way to, or taking us back to, more flexible formats, requiring, for many businesses, a rethinking of organizational culture, with new approaches to leadership and employee engagement. Issues related to diversity and inclusion, and continuous adaptation and organizational agility may be needed to face these challenges within organizations. Shortage of specialized labor are pressuring governments in relation to migration and forcing organizations to adopt new technologies and innovative attraction and retention strategies, such as employer branding and well-being programs. Although they have faced a major push-back in the USA recently, diversity and inclusion may be a valuable tool for organizations, enabling them to develop more varied workforces and requiring the development of active policies and an open mindset to change (O'Higgins & Zsolnai, 2017). To develop continuous adaptation and greater agility, organizations in certain sectors at least will be required promote agile HRM, capable of responding quickly to changes in the market and society in general, promoting a mindset of innovation and continuous learning.

These challenges are a constant concern in many organizations today, but they also constitute a source of opportunities for HRM. The increasing use of technology in human resources, such as People Analytics, AI, and automation, at least in larger organizations, contributes to optimizing processes such as recruitment, onboarding, and performance management. Creating work environments that prioritize employee experience and well-being helps to increase productivity and talent retention (Parry & Battista, 2019). At the same time, it is essential that HRM promotes a culture of continuous learning and adaptation, thus helping employees remain relevant assets in the market. Leadership, from transformational leaders who inspire, communicate well, and are adaptable will play a fundamental role in leading teams in the challenging times in which we live.

In short, in a disrupted world like ours, HRM must be proactive, innovative, and people-centered to ensure that organizations thrive in a constantly changing environment. Following these challenges and opportunities, this chapter seeks, in a critical way, to address the issue of HRM in a disrupted world, so the central question that arises is whether, *in these disrupted times, do we need to rethink HRM?*

1.2 Human Resource Management: Construct Clarity

Human Resource Management is a multifaceted field that can be analyzed both as a proposed action program and as an academic and professional field of study. The aim below is to explore both aspects with a focus on strategic HRM and its connection with organizational performance.

1.2.1 HRM as Program

HRM can be seen as a set of strategic policies, processes, and practices that an organization implements to manage its employees effectively. Within this perspective, the approach that has been (somewhat redundantly) titled Strategic Human Resource Management seeks to align HRM practices with the company's strategic objectives.

Strategic Human Resource Management

The "hard" perspective of HRM, initially associated with the Michigan School, emphasizes human resource management as an instrument for achieving organizational efficiency. In this approach, workers are seen essentially as resources that must be managed rationally to ensure the company's productivity and competitiveness. Like other resources, the objective is to obtain the best possible quality at the lowest possible price, to develop them as much as possible and to get rid of them when they are no longer productive or can be replaced by some other person or technology capable of doing the work more cost-effectively. This approach notes that the labor force is, in nearly all organizations, the largest single element of variable costs–and the source of innovation and customer service. The Michigan model is based on a unitarist logic, where the interests of the organization are seen as the same as, or (where that is not possible) prevail over, labor force interests, with HRM focused on structured processes such as recruitment, performance evaluation and results control. The main objective of SHRM is to align management practices with the company's strategy, ensuring that employees contribute effectively to organizational success (Boselie et al., 2009). This perspective is characterized by a strong emphasis on quantitative indicators, such as performance targets, operational efficiency and return on investment in human capital. The training and development of workers are seen as means to maximize their usefulness within the company, while motivation and individual well-being receive less attention. Thus, the Michigan School represents a pragmatic and mechanistic view of people management, in which employees are considered productive factors to be optimized.

1.2.2 HRM as a Field of Study

On the other hand, what was initially termed the Harvard School adopts a "soft" approach to HRM, focused on people and their appreciation as strategic assets of the organization. This perspective argues that workers should not be seen simply as resources but rather as stakeholders, partners in the organizational process, whose well-being and personal development directly influence the company's performance. The Harvard model is based on a more pluralistic logic, recognizing that the interests of employees and the organization are not always perfectly aligned, and it is necessary to find a balance between business objectives and individual needs (Beer et al., 2015). Human resource management, in this view, is not limited to obtaining short-term results but is also concerned with aspects such as motivation, involvement, organizational culture, and employee satisfaction. To achieve this, organizations focus on open communication, active participation of employees in decision-making, and the continuous development of skills. This softer approach to HRM recognizes that more humane and integrated management practices contribute to better organizational performance in the long term, as satisfied and committed workers tend to be more productive, innovative, and loyal to the company.

Although they are different approaches, the perspectives can be complementary. The "hard" vision ensures a clear alignment between organizational strategy and people management, guaranteeing efficiency and measurable results, and the "soft" approach promotes a healthier and more sustainable work environment, where workers feel valued and motivated to contribute to their maximum potential (Boselie et al., 2009). Today, many organizations seek a balance between these two views, combining the discipline and structure of the Michigan model with the appreciation of human capital advocated by the Harvard model. This way, they can achieve both their strategic objectives and a more harmonious and productive work environment.

This fundamental difference of views is just one of many debates in HRM: the field has generated much discussion about its real effectiveness. Criticisms are frequently heard, for example, about the disconnect sometimes felt between theory and practice, where models developed in the research are not always applicable to the organizational context; the growing challenge in terms of measuring the impact of developed HRM (as the relationship between HRM practices and organizational performance is very often neither linear nor immediate); or even the growing influence that technology and automation have been exerting in challenging the traditional role of HRM, while at the same time raising ethical questions about the future of work.

In short, HRM can be seen both as an organizational program applied mainly in large private sector organizations (with a strategic focus on talent management and creating value for the company) and as a field of study (which seeks to understand and optimize the relationship between people and organizations in every context). However, its effectiveness depends on the ability to overcome challenges such as the lack of a clear connection between HRM practices and business results, as well as the need to continually adapt to new market and technology demands (Boselie et al., 2009).

1.3 A Disrupted World

The future of work and human resource management (HRM) is being shaped by emerging technologies, globalization, and changing workforce demographics. Artificial intelligence, robotics, and digitization are automating tasks and enabling more flexible work arrangements (Parry & Battista, 2019). This technological disruption presents increasing, and increasingly complex, challenges for HRM professionals, who must help businesses to decide on the impact of the technology on the amount of human labor that will be required–whether redundancies will occur–and help those employees who will retain their jobs to update skills. They will also have to address potential negative effects on wellbeing (Parry & Battista, 2019). The evolving landscape requires new HRM strategies to manage global workforces, balancing flexibility with employee well-being, and developing the needed skills for the digital age (Brewster et al., 2000; Burke & Cooper, 2006; Holland et al., 2024). Organizations are exploring innovative approaches to employee engagement, performance management, and workplace ethics in technology-driven environments (Gupta & Sharma, 2016; Hough et al., 2015). While these changes offer opportunities for efficiency and new occupations, they also raise concerns about job obsolescence, job loss, and working conditions in the ever expanding gig economy (Wilkinson & Barry, 2020).

1.3.1 The VUCA World

The concept of VUCA emerged at the end of the Cold War and was popularized in the 1990s to describe a *Volatile, Uncertain, Complex, and Ambiguous* world. This model is currently widely used in the business and management contexts to describe a constantly changing and highly unpredictable environment.

Taking a closer look at the dimensions of the VUCA world (Taskan et al., 2022), we can highlight that each of them presents specific challenges for organizations and for HRM (Hamid, 2019). When we talk about *volatility*, we are referring to changes that occur quickly and unexpectedly (such as pandemics, financial crises, or disruptive technological advances). In terms of HRM, this volatility implies a growing need for rapid adaptation with regard to labor strategies, as well as the promotion of greater flexibility in ways of working. The difficulty in predicting the future, given the few historical references that help anticipate trends, is at the root of *uncertainty*. Often resulting from the development of new digital business models that challenge traditional industries, and traditional ways of working, this uncertainty leads to the need to promote greater investment in continuous learning and organizational resilience. *Complexity* results from the interaction of forces at a global level, which brings multiple, inter-connected, challenges that are difficult to understand and resolve. The impact of social networks on organizations, as well as the use of global supply chains, are examples of this complexity, which leads to a greater need for knowledge and specialization, as well as the existence of leaders capable of developing systemic thinking. Finally, we are faced with the *ambiguity* resulting from the difficulty in obtaining clear and/or objective answers to the different situations that organizations face. The interpretation of data itself becomes increasingly subjective.

To deal with this disrupted world, organizations need to develop new approaches to people management. Specifically, it is urgent to promote the development of agile and resilient leaders capable of making quick and effective decisions (Raja, 2021). Agile organizations (with less hierarchical and more flexible structures to respond to environmental changes), the strategic use of technologies (using, for example, artificial intelligence to make data-based decisions), continuous training (based on promoting lifelong learning and flexibility of skills), as well as a focus on well-being and culture (through the construction of a strong organizational culture that supports employees in times of change), are some of the possible strategies that can be developed by HRM as a way of responding to the VUCA world (Hamid, 2019).

It is important to highlight here that more recently some researchers (Koman et al., 2024; Vinutha et al., 2024) have begun to emphasize the need to replace the VUCA concept with the BANI (Brittle, Anxious, Nonlinear, Incomprehensible) model, arguing that the world, in our days, is not only volatile but also fragile, anxious, nonlinear and incomprehensible. However, whether VUCA or BANI, the fundamental thing that is emphasized is that organizations and HRM professionals need to adopt a mindset of continuous adaptation and innovation, contributing to organizations being able to prosper, even in uncertain scenarios.

1.3.2 But is that Really the Case?

Whether we are living in a VUCA or a BANI world where HRM is more difficult to manage is a moot point–people have been saying this for centuries (Grint, 2022). In his critical article, "Wicked Problems in the Age of Uncertainty", Grint (2022) explores the concept of "wicked problems" in the context of the contemporary era, marked by significant uncertainties. He argues that we are living in unprecedented times, where complex and difficult-to-solve problems, such as the COVID-19 pandemic, challenge traditional approaches to leadership and management. Originally introduced by Rittel and Webber in (1973), the term "wicked problems" refers to issues that are difficult to define and lack clear or definitive solutions. These problems are characterized by multiple interconnected causes, stakeholders with divergent perspectives, and unpredictable consequences. The COVID-19 pandemic is a contemporary example of a wicked problem requiring adaptive and collaborative responses. Grint (2022) suggests that tackling wicked problems requires a form of leadership that goes beyond traditional command and control approaches. In this context, he advocates for leadership that involves multiple stakeholders (thus recognizing the diversity of perspectives and experiences needed to address complex problems), that promotes experimentation and continuous learning (since there are no predefined solutions, it is crucial to test different approaches and learn from the results), and that focuses on building organizational resilience (preparing organizations to adapt to rapid and unexpected changes).

In short, Grint's (2022) article highlights the importance of recognizing the complex nature of contemporary problems and the need for leadership and management approaches that are flexible, inclusive, and oriented toward continuous learning. The COVID-19 pandemic provided an exemplary case.

1.3.3 The COVID-19 Pandemic

The COVID-19 pandemic was an extreme example of the VUCA reality, accelerating transformations in work, the economy, and society. Organizations had to respond quickly to unforeseen challenges, and, especially in larger organizations, HRM was central to that. As one of the most disruptive events of the modern era, the COVID-19 pandemic accelerated social, technological, and organizational transformations, forcing organizations and professionals to adapt quickly (Caligiuri et al., 2020; Machado & Davim, 2022a). HRM took on a central role in this adaptation by promoting new ways of working, employee well-being, and organizational resilience strategies. The pandemic triggered changes that were already underway, but which ended up being accelerated by the crisis. As an example, remote work, and hybrid remote/office work, became the norm in a significant number of organizations, requiring new forms of collaboration and leadership. At this level, digital acceleration, based on technologies such as artificial intelligence, automation, and People Analytics, has assumed a crucial role in the management of remote teams (Amankwah-Amoah et al., 2021). As issues related to mental health and well-being begin to take on particular prominence, some organizations have begun to invest more in the emotional and psychological support of employees (Henke, 2022). It has been argued that reskilling and upskilling and the promotion of adaptive leadership have become more of a reality (Hüsing et al., 2021). Indeed, if, on the one hand, the increased need for new skills led to greater investment in ongoing training, on the other hand, in terms of leadership, managers needed to develop their skills in terms of empathy, resilience, and agile decision-making capacity.

As a result, even of there may be evidence that the pandemic effects o management are rec3ding and in many cases things have reverted to pre-pandemic approaches, the crisis emphasized that HRM must be more strategic and agile in order to face future disruptive scenarios. The pandemic was a resilience test for organizations and employees, accelerating trends that were already underway. In the context of a disrupted world, HRM plays a fundamental role in creating flexible, innovative and people-centered companies, capable of facing future uncertainties with greater capacity for success.

1.4 HRM in a Disrupted World

As we have observed, HRM in a disrupted world requires agility, innovation, and a strategic focus on human capital driven by technology, socioeconomic changes, and new ways of working (Minbaeva, 2021). Leaders must prioritize continuous adaptation, fostering flexible organizational cultures, digital skills development, and employee well-being. Artificial intelligence, automation, and hybrid work models are redefining careers, requiring skills-based recruitment, continuous learning, and empathetic leadership (Machado & Davim, 2023; Machado & Davim, 2022b). Furthermore, despite what has been happening in the USA, diversity, inclusion, and sustainability become fundamental pillars for engaging talent and maintaining competitiveness in a dynamic and uncertain environment. But if it is true that many challenges are faced, it is also true that there are many questions that are raised. Let's look at some of them.

1.4.1 The Cranet Database (https://cranet.la.psu.edu/)

The Cranet database is an international initiative that aims to collect and analyze data on HRM policies and practices in organizations in different countries. Established in 1989, Cranet has become the largest and most consolidated scientific research network in the field of HRM, currently covering more than 40 universities and business schools worldwide (CRANET Research Network, 2025). The network conducts regular surveys, usually every four or five years, to collect comparative data on topics such as HRM structure and policies, recruitment and selection, staff development, compensation and benefits, and labor relations. The results of these surveys provide a valuable basis for comparing the activities of different organizations, both nationally and internationally. This allows for a systematic analysis of trends in HRM, with the results disseminated through academic publications, conferences and seminars. Aware of all the challenges and opportunities faced by HRM specialists, the latest executive report (CRANET Executive Report, 2023:1) highlighted that "the rise of disruptive trends, such as automation and the algorithmization of HRM, often promoted through HRM or people analytics tools, has gained momentum. These trends have significantly affected the daily work of HRM professionals worldwide, prompting questions about the enduring changes in core HRM activities and responsibilities. Additionally, there is growing interest in comparing how different countries manage the impact of these challenges." Nevertheless, in many of the basic areas of HRM, the surveys show that change is limited and slow. Despite the scholarly and consultancy rhetoric, most of the basic aspects of HRM have continued to operate with only limited change.

1.4.2 Technological Change

Technological change has been a constant throughout history, but it has not always been linear or homogeneous. Since the Industrial Revolution, through electrification and the digital age, innovation has brought disruptive leaps that have transformed societies and economies. However, adoption of these changes varied significantly across countries, sectors, and companies. While some nations and industries have thrived on new technologies, others have lagged behind, stuck in outdated infrastructure or business models that resist transformation.

The technological acceleration of recent years is not unprecedented. Since the 1950s, we have envisioned automation and artificial intelligence as forces that could replace human labor and reconfigure markets. What changes now is the depth and scale of the transformation. Generative Artificial Intelligence (GAI), with models like ChatGPT, marks a possible turning point: it is no longer just about machines that automate repetitive tasks but about systems that create, analyze, and interact in an increasingly sophisticated way (Budhwar, 2023).

This new paradigm raises critical questions about the concentration of power in the technology sector. Large companies in WEIRD (Western, Educated, Industrialized, Rich, and Democratic) countries (Wooliscroft & Ko, 2023) have the capital and access to talent to dominate this revolution, reinforcing oligopolies and making it more difficult for new companies or emerging economies to compete. The asymmetry in the development and control of these technologies can intensify global inequalities, both between nations and within countries themselves.

In the job market, the impact will be profound and ambivalent. While automation and AI can increase productivity and create new opportunities, many roles will become obsolete or reconfigured. Creative, legal, and even technical professions are under threat as a new balance between humans and machines is sought. The question is not just "Which jobs will disappear?" but rather "How will the concept of work itself be restructured?".

This latest technological change is different. In previous industrial revolutions jobs that disappeared were (eventually, and usually after many years of misery and stress) replaced by other jobs. In the AI revolution jobs that were done by human beings will be replaced by tasks that the AI can complete itself: there may be no alternative roles for humans to go into. It is too early to be able to say what the new AI world will look like and how it will affect work, employment, and our societies more generally, but never before have we faced such broad and rapid change, driven by machines that learn and create. The challenge is not just technological, but political and ethical: how to regulate, distribute benefits and mitigate risks, without stifling innovation? The answer will determine whether this inflection point results in a future of greater prosperity or a new era of inequality and instability.

1.4.3 HRM and Work

HRM has traditionally focused its studies on employment–a formal contract for work to be done between an employer and an employee. This is a pattern that historically has only existed for a couple of hundred years, that sectorally only applied to certain industries and people, and that geographically does not apply to most people in the world who do their work in the informal economy. Only a relatively narrow range of jobs fit into formal models of wage employment and standardized employment relationships. This analytical limitation has become even more evident in a context of accelerated disruption, where the concept of traditional employment is being eroded.

In the developed economies, however, employment has dominated the debate amongst governments, businesses, and individual workers. Employment has always played a central role for multiple stakeholders: employers, investors, retirees, employees, employees' families, governments, and supranational bodies such as the European Union. The stability and predictability of formal employment have supported social protection systems, pension financing, tax collection, and the functioning of national economies. However, this centrality is decreasing. Automation, the platform economy, temporary work, and the growth of the informal sector question the viability of employment as a structuring axis of the economic and social system (Janampa et al., 2022).

In the WEIRD countries, there has always been a substantial amount of work that does not occur through formal employment. Domestic, voluntary, care work, creative work, and especially the unpaid work of (mainly women) carers and charity workers often escape the metrics of HRM and traditional economics. In the so-called "majority world", the situation is even more extreme: a large part of the population has never been part of the classic model of wage employment, depending on informal economies, subsistence, or community dynamics to survive (ILO, 2018). The current disruption is merely making this reality more evident, rather than being an entirely new phenomenon.

The proportion of work outside formal employment is increasing substantially, both in WEIRD countries and in the majority world. The precariousness of work, the rise of freelancers and platform workers, as well as the growth of unpaid work, indicate that the classic employment model is losing strength (Pathiranage, 2024; Sutherland et al., 2020). HRM, by remaining focused on the traditional employer-employee relationship, risks becoming obsolete and irrelevant to understanding and managing new work realities.

Stakeholders appear uncertain about the response to this structural change. Governments hesitate between trying to regulate new forms of work or letting them flourish without interference. Companies benefit from flexibility but face challenges in retaining talent and creating organizational culture. Social protection systems remain designed for an era of stable employment, while citizens increasingly find themselves responsible for their own livelihoods without conventional safety nets (Vandevenne & Vanroelen, 2024). The lack of a strategic vision worsens uncertainty and increases inequalities.

Perhaps it is time to change the focus of HRM studies. Instead of studying jobs, we should study work in all its forms. This would imply a more comprehensive and interdisciplinary approach, which recognizes the different types of work that sustain society and the economy. By expanding the analysis beyond the boundaries of formal employment, we can better understand how people work, how they create value, and how they can be supported in a world where employment is no longer the norm.

1.4.4 The Rise of Stakeholder HRM (in Academia)

The stakeholder perspective has always been part of the tradition of human resource management as a field of study, being central to the Harvard model proposed by Beer et al. (1984). This innovative model highlighted that HRM should not be seen solely as a business function focused on the interests of employers but rather as a system that balances the needs of multiple stakeholders, including employees, investors, governments, and society at large. However, despite this initial vision, mainstream HRM practice has been captured by a more instrumental and one-dimensional approach, focused on organizational performance and maximizing productivity, often at the expense of the well-being of workers and other stakeholders.

Three decades later, Beer et al. (2015) reassessed this multi-stakeholder perspective, arguing that HRM should return to its roots to address contemporary work challenges. Their critique points to the field's failure to effectively integrate social and ethical concerns at a time when job insecurity and economic inequalities are on the rise. Although stakeholder rhetoric has been revived in recent discourses, in practice, HRM remains largely dominated by short-term metrics and an emphasis on organizational performance, to the detriment of a genuine commitment to balancing interests. Thus, the rise of stakeholder HRM in academia seems more like a belated return than a true innovation, highlighting the need for a renewed effort to make this approach an effective reality in work management.

In recent years, academia has seen a proliferation of 'adjectival' HRM, with subfields such as Green HRM (Alcaraz et al, 2017; Harris & Tregidga, 2012; Renwick et al., 2016), Socially Responsible HRM (Aust et al., 2020; Raubenheimer & Rasmussen, 2013; Sarvaiya et al., 2018; Shen & Benson, 2016), Sustainable HRM (Brewster & Brookes, 2024; Guerci & Carollo, 2016; Ren & Jackson, 2020; Stahl et al., 2020) and Common Good HRM (Aust et al., 2020). These approaches seek to expand the role of HRM beyond traditional business interests, incorporating environmental, social, and ethical concerns. However, this fragmentation raises critical questions about the coherence of the field. Rather than representing a solid conceptual advance, this multiplication of terminologies can dilute the impact of HRM by dispersing research efforts and hindering its practical implementation.

Furthermore, despite the increasing focus on social responsibility and sustainability, the actual influence of these approaches on labor management remains limited. Many of these variants are superficially adopted by companies as part of branding and institutional legitimacy strategies, without causing significant structural changes in the relationship between labor and capital. Models such as Triple Bottom Line HRM (Bush, 2020; Elkington, 2018) and Quadruple Bottom Line HRM (O'Higgins & Zsolnai, 2017) promise to balance economic, social, and environmental objectives, but they often end up being instrumentalized to reinforce existing business practices. Thus, the rise of stakeholder HRM in academia reflects more of a rebranding trend than an effective transformation of working practices in a disrupted world.

Finally, it is important to highlight that the rise of stakeholder HRM in academia reflects different perspectives on the role of employment in a disrupted world. Some scholars take an interest congruence view, suggesting that it is possible to align the needs of employers with those of workers and other stakeholders without deeply questioning the foundations of the capitalist system (Buchanan et al., 2025). This approach often emphasizes strategic talent management, worker well-being, and corporate social responsibility, but remains rooted in a market logic where employment remains the main vehicle for wealth distribution and social stability.

On the other hand, some perspectives adopt a broader view, recognizing that technological transformations, the precariousness of work and the growth of the informal sector challenge the centrality of traditional employment (Song, 2024). These scholars argue that HRM needs to expand its focus beyond formal employment relationships and consider new forms of work organization and value redistribution (Lynn et al., 2023). At the more radical end of the spectrum, some advocate a complete overhaul of existing capitalist systems, instigated by HRM experts themselves. They argue that structural inequalities, exacerbated by the erosion of stable employment, require a redefinition of economic and social institutions, including the redistribution of work and wealth in a more equitable way (ILO, 2021). However, this view faces significant resistance, both in academia and in business practice, as it directly challenges the interests of the economic and political elites that traditionally shape labor policies.

1.5 Final Remarks

… So, in these disrupted times, do we need to rethink HRM?

In a disrupted world, HRM faces the challenge of reinventing itself. For decades, the field has focused primarily on the relationship between employers and employees in the context of formal wage labor, ignoring the multiple ways in which work is organized and performed. To remain relevant, HRM needs to expand its scope and examine work in all its manifestations–not just traditional employment but also informal, voluntary, domestic, and digital work. This expansion of the field of study does not mean abandoning organizational analysis, but recognizing that work management transcends the limits of the conventional company and extends to multiple sectors and forms of organization.

Furthermore, a critical rethinking of HRM requires the inclusion of all stakeholders. For a long time, the discipline has predominantly served the interests of employers and managers, treating workers as resources to be optimized for productivity and profit. However, a more holistic approach would recognize that the impacts of HRM extend to families, communities, governments and society as a whole. The current crisis of precarious work and labor inequalities highlights the need for a model that not only optimizes organizational efficiency, but also promotes social justice and sustainability.

HRM must also go beyond its geographical boundaries and recognize the diversity of work contexts. Many of the relevant theories and practices were shaped in the big private sector businesses of Western industrialized countries (WEIRD), disregarding realities not only among most businesses in those countries but also of most businesses, large and small, in the "majority world", where formal employment is often the exception rather than the rule. Applying the same models to such different contexts without recognizing their specificities results in a limited vision and, sometimes, in the imposition of values that do not reflect local needs. By adopting a global and inclusive perspective, HRM can become more relevant and applicable to different economic, social and cultural realities.

Finally, we must recognize the tensions and power relations inherent in HRM, without illusions about its transformative potential. The field of academic study alone will not change the world, but it can provide analytical tools for better understanding of the inequalities and contradictions of contemporary work. Accepting this limitation does not mean giving up criticism, but adopting a realistic stance on the role of academia in constructing a more informed and conscious discourse about work. Rethinking HRM is not just a theoretical necessity–it is an essential step towards understanding and navigating the complex dynamics of work in a changing world.

References

Alcaraz, et al. (2017). The human resources management contribution to social responsibility and environmental sustainability: Explorations from Ibero-America. *The International Journal of Human Resource Management, 30*(22), 3166–3189. https://doi.org/10.1080/09585192.2017.1350732

Amankwah-Amoah, J., et al. (2021). COVID-19 and digitalization: The great acceleration. *Journal of Business Research, 136*, 602–611. https://doi.org/10.1016/j.jbusres.2021.08.011

Aust, Matthews, & Muller-Camen. (2020). Common good HRM: A paradigm shift in sustainable HRM?. *Human Resource Management Review, 30*, 100705. https://doi.org/10.1016/j.hrmr.2019.100705

Beer, M., et al. (1984). *Managing human assets*. The Free Press.

Beer, M., Boselie, P., & Brewster, C. (2015). Back to the future: Implications for the field of HRM of the multi-stakeholder perspective proposed 30 years ago. *Human Resource Management, 54*(3), 427–438.

Boselie, P., Brewster, C., & Paauwe, J. (2009). In search of balance–managing the dualities of HRM: An overview of the issues. *Personnel Review, 38*(5), 461–471. https://doi.org/10.1108/00483480910977992

Brewster, C., & Brookes, M. (2024). Sustainable development goals and new approaches to HRM: Why HRM specialists will not reach the sustainable development goals and why it matters. *German Journal of Human Resource Management, 38*(2), 183–201. https://doi.org/10.1177/23970022241229037

Brewster, C., Mayrhofer, W., & Morley, M. (2000). New challenges for European resource management. *Palgrave Macmillan.* https://doi.org/10.1057/9780230597952

Buchanan, S., Shourkaei, M. M., & Dyck, B. (2025). 'Vote the assholes out': How value congruence work aligns stakeholders for corporate activism. *Journal of Management Studies.* https://doi.org/10.1111/joms.13068

Budhwar, P., et al. (2023). Human resource management in the age of generative artificial intelligence: Perspectives and research directions on ChatGPT. *Human Resource Management Journal, 33*(3), 606–659. https://doi.org/10.1111/1748-8583.12524

Burke, R. J., & Cooper, C. (2006). The new world of work and organizations: Implications for human resource management. *Human Resource Management Review, 16*(2), 83–85. https://doi.org/10.1016/j.hrmr.2006.03.013

Bush, J. T. (2020). Win-win-lose? sustainable HRM and the promotion of unsustainable employee outcomes. *Human Resource Management Review, 30*(3), 100676. https://doi.org/10.1016/j.hrmr.2018.11.004

Caligiuri, P., et al. (2020). International HRM insights for navigating the COVID-19 pandemic: Implications for future research and practice. *Journal of International Business Studies, 51*, 697–713. https://doi.org/10.1057/s41267-020-00335-9

CRANET Executive Report 2023. (2023). CRANET executive report on international human resource management, summary and analysis of 2021–2022 survey data. Retrieved February 12, 2025, from https://www.unilu.ch/fileadmin/fakultaeten/wf/institute/hrm/dok/Dateien/SSRN-id4647015.pdf

CRANET Research Network. (2025). Retrieved February 12, 2025, from https://cranet.la.psu.edu/

Elkington, J. (2018). 25 years ago, I coined the phrase "triple bottom line." Here's why it's time to rethink it. *Harvard Business Review.* Retrieved February 12, 2025. https://hbr.org/2018/06/25-years-ago-i-coined-the-phrase-triple-bottom-line-heres-why-im-giving-up-on-it

Grint, K. (2022). Critical essay: Wicked problems in the age of uncertainty. *Human Relations, 75*(8), 1518–1532. https://doi.org/10.1177/00187267211070770

Guerci, M., & Carollo, L. (2016). A paradox view on green human resource management: Insights from the Italian context. *The International Journal of Human Resource Management, 27*, 212–238. https://doi.org/10.1080/09585192.2015.1033641

Gupta, N., & Sharma, V. (2016). Exploring employee engagement—a way to better business performance. *Global Business Review, 17*(3). https://doi.org/10.1177/0972150916631082

Hamid, S. (2019). The strategic position of human resource management for creating sustainable competitive advantage in the VUCA world. *Journal of Human Resources Management and Labor Studies, 7*(2), 1–4. https://doi.org/10.15640/jhrmls.v7n2a1

Harris, C., & Tregidga, H. (2012). HR managers and environmental sustainability: Strategic leaders or passive observers? *The International Journal of Human Resource Management, 23*(2), 236–254. https://doi.org/10.1080/09585192.2011.561221

Henke, R. M. (2022). Knowing well, being well: Well-being born of understanding: Supporting workforce mental health during the pandemic. *American Journal of Health Promotion, 36*(7), 1213–1244. https://doi.org/10.1177/08901171221112488

Holland, P., Brewster, C., & Kougiannou, N. (Eds.). (2024). *Work, employment and flexibility: Innovation.* Routledge.

Hough, C., Green, K., & Plumlee, G. (2015). Impact of ethics environment and organizational trust on employee engagement. *Journal of Legal Ethical and Regulatory Issues, 18*(3), 45.

Hüsing, T., et al. (2021). *Upskilling and reskilling in the post-Covid era: Fostering new services and jobs creation – three scenarios for 2030.* European Commission. European Innovation Council and SMEs Executive Agency (EISMEA).

ILO–International Labour Organization. (2018). *More than 60 percent of the world's employed population are in the informal economy*. Retrieved February 18, 2025, from https://www.ilo.org/resource/news/more-60-cent-world%E2%80%99s-employed-population-are-informal-economy

ILO–International Labour Organization. (2021). *Inequalities and the world of work*. International Labour Conference, 109th Session, Report IV. International Labour Office.

Janampa, J. C., Bonnet, F., & Garcia, J. M. (2022). *Growth, economic structure and informality*. International Labour Organization, Working papers. https://doi.org/10.54394/UQOF2851

Koman, G., Boršoš, P., & Kubina, M. (2024). Sustainable human resource management with a focus on corporate employee recruitment. *Sustainability, 16*, 6059. https://doi.org/10.3390/su16146059

Lynn, T., et al. (Eds.). (2023). *The future of work*. Palgrave Macmillan.

Machado, C., & Davim, J. P. (Eds.). (2022). *Organizational management in post pandemic crisis*. Springer.

Machado, C., & Davim, J. P. (Eds.). (2022). *Organizational innovation in the digital age*. Springer.

Machado, C., & Davim, J. P. (Eds.). (2023). *Industry 5.0: Creative and innovative organizations*. Springer.

Minbaeva, D. (2021). Disrupted HR? *Human Resource Management Review, 31*(4), 100820. https://doi.org/10.1016/j.hrmr.2020.100820

O'Higgins, E., & Zsolnai, L. (Eds.). (2017). *Progressive business models: Creating sustainable and pro-social enterprise*. Springer.

Parry, E., & Battista, V. (2019). The impact of emerging technologies on work: A review of the evidence and implications for the human resource function. *Emerald Open Research, 1*(5). https://doi.org/10.12688/emeraldopenres.12907.1

Pathiranage, H. S. (2024). Precarious employment in the gig economy: Understanding the roles of employed poor. *Open Journal of Business and Management*, 12, 2288–2307. https://doi.org/10.4236/ojbm.2024.124117

Raja, A. S. (2021). Business research in the VUCA world (volatility, uncertainty, complexity and ambiguity). *Ushus-Journal of Business Management, 20*(1), 5–16. https://doi.org/10.12725/ujbm.54.0

Raubenheimer, K., & Rasmussen, E. (2013). Employee-focused corporate social responsibility in practice: Insights from managers in the New Zealand and Australian financial sector. *New Zealand Journal of Employment Relations, 38*(3), 35–56.

Ren, S., & Jackson, S. E. (2020). HRM institutional entrepreneurship for sustainable business organizations. *Human Resource Management Review, 30*(3), 100691. https://doi.org/10.1016/j.hrmr.2019.100691

Renwick, D. W. S., et al. (2016). Contemporary developments in green (environmental) HRM scholarship. *The International Journal of Human Resource Management, 27*(2), 114–128. https://doi.org/10.1080/09585192.2015.1105844

Rittel, H. W. J., & Webber, M. M. (1973). Dilemmas in a general theory of planning. *Policy Sciences, 4*(2), 155–169.

Sarvaiya, H., Eweje, G., & Arrowsmith, J. (2018). The roles of HRM in CSR: Strategic partnership or operational support? *Journal of Business Ethics, 153*, 825–837. https://doi.org/10.1007/s10551-016-3402-5

Shen, J., & Benson, J. (2016). When CSR is a social norm: How socially responsible human resource management affects employee work behavior. *Journal of Management, 42*(6), 1723–1746. https://doi.org/10.1177/0149206314522300

Song, Y. (2024). Impact of technological advancements on work and employment patterns. *Journal of Advanced Sociology, 5*(2), 1–14. https://doi.org/10.47941/jas.1856

Stahl, G., et al. (2020). Towards more relevant HRM research: Understanding the role of HRM in corporate sustainability and social responsibility. *Human Resource Management Review, 30*(3). https://doi.org/10.1016/j.hrmr.2019.100708

Sutherland, W., et al. (2020). Work precarity and gig literacies in online freelancing. *Work, Employment and Society, 34*(3), 457–475. https://doi.org/10.1177/0950017019886511

Taskan, B., Junça-Silva, A., & Caetano, A. (2022). Clarifying the conceptual map of VUCA: A systematic review. *International Journal of Organizational Analysis, 30*(7), 196–217. https://doi.org/10.1108/IJOA-02-2022-3136

Vandevenne, E., & Vanroelen, C. (2024). Measuring employment precariousness in gig jobs: A pilot study among food couriers in Brussels. *Work, 77*(2), 487–510. https://doi.org/10.3233/WOR-220691

Vinutha, K., Samaddar, A., & Elijala. S. (2024). Navigating the bani landscape for organisational success. *European Economic Letters (EEL), 14*(2), 979–986. https://doi.org/10.52783/eel.v14i2.1429

Wilkinson, A., & Barry, M. (2020). *The future of work and employment*. Edward Elgar Publishing.

Wooliscroft, B., & Ko, E. (2023). WEIRD is not enough: Sustainability insights from Non-WEIRD countries. *Journal of Macromarketing, 43*(2), 171–174. https://doi.org/10.1177/02761467231169880

Part II
Thematic Area: Digital Transformation and Automation

Chapter 2
Managing Digital Nomads: Insights from Four Continents

Joana Fernandes and Ana Cláudia Rodrigues

Abstract Digital nomadism has gained popularity recently, mainly due to the widespread use of technology. However, only some studies examine this disrupted work arrangement from an organisational perspective. We employed grounded theory methodology to explore the key differences in management strategies for digital nomads compared to fully remote workers. We conducted three rounds of semi-structured interviews with managers from companies hiring digital nomads, totalling eight interviews across eight countries and four continents. We identified three major categories regarding the level of intervention: organisational, human resources management, and team. While we anticipated that managing this type of worker would require disrupted strategies, the results showed a significant similarity between the management approaches for digital nomads and those for fully remote workers. Based on these findings, we propose a Remote Employee Lifecycle Operational Framework highlighting digital nomads' specificities based on the employee life cycle approach. Our diverse sample supports the conclusion that remote people management practices strongly converge despite geographical differences.

Keywords Digital nomads · Human resources · People management strategy · Operational framework · Remote employee lifecycle · Remote working

2.1 Introduction

Business leaders have recognised that flexible working arrangements have enhanced company productivity (International Workplace Group, 2019). Digital nomadism represents the most extreme and mobile form of flexible work (Schlagwein & Hossein Jarrahi, 2020).

J. Fernandes (✉)
ISCAP, Polytechnic Institute of Porto, São Mamede de Infesta, Portugal
e-mail: joana.silvafernandes28@gmail.com

A. C. Rodrigues
ISCAP, CEOS.PP, Polytechnic Institute of Porto, São Mamede de Infesta, Portugal

C. Machado (ed.), *Human Resource Management in a Disrupted World*, Lecture Notes on Multidisciplinary Industrial Engineering,
https://doi.org/10.1007/978-3-031-89948-5_2

Socioeconomic shifts, technological advancements, and improvements in transportation systems are disrupting the labour market and making travel more accessible (Müller, 2016; Orel, 2019). As a result, the number of digital nomads continues to grow steadily.

The relevance of this study stems from extensive bibliographic research, which has revealed a gap between the number of articles on digital nomadism and those that thoroughly explore its organisational perspective. Several authors suggest that further investment is needed to analyse this theme.

Wang et al. (2018) point out that while previous studies have concentrated on the lifestyles of digital nomads, there has been little focus on their integration into companies' strategic positioning. Moreover, despite the discussions about the values and norms of digital nomads about organisations, no studies were dedicated to this analysis until 2019 (Kong et al., 2019).

Considering the labour market's continued disruption, Richter and Richter (2020) pose important questions about how digital nomadism influences work within organisations and the potential opportunities and challenges of integrating such workers into an organisational framework.

It is crucial to study this topic comprehensively—not just regarding its integration into organisations but also as an economic activity and a career management style (Khatri & Shukla, 2022). Furthermore, examining its impact on company productivity is essential (Fikriy et al., 2023).

The primary goal of this study is to comprehend the differences between managing remote workers and digital nomads and to outline the human resources management strategy tailored explicitly for digital nomads.

2.2 Literature Review

2.2.1 Concept of Digital Nomad

The first definition of digital nomad appeared in 1997 in the book "Digital Nomad" by Makimoto and Manners, where it is mentioned that the desire to travel and the ability to do so are essential for digital nomadism. However, it was only in 2006 that the first scientific article appeared in the "Web Of Science" database on this phenomenon (Šímová, 2023).

Digital nomads seek to eliminate dissatisfaction with structured work and the lack of work-life balance, assuming a lifestyle that sees travel not as an occasional situation restricted to tourist activities but as a prominent and possibly constant part of life (Reichenberger, 2017).

These nomads work digitally through mobile phones, computers, or other portable computer means. Most of these workers do not engage in local work activities, relying on remote temporary jobs that provide flexibility to work while travelling (Sutherland & Jarrahi, 2017).

Digital nomadism can be seen as a set of four concepts: short-term autonomous projects, digital work, nomadic work, and global travel (Nash et al., 2018). Although several jobs include one of these concepts, they are only considered digital nomad workers if they meet all four characteristics. For example, while a nomadic worker travels to different locations due to their work, the digital nomad can choose the place they want to travel to (Nash et al., 2018). Whether you work entirely from home or occasionally from the homes of friends and family should be excluded from the definition of a digital nomad (Cook, 2023). Mobility is not only a part of what it is to be a digital nomad but a necessity to be considered so (Hemsley et al., 2020).

Digital nomadism is not just a new lifestyle but a new way of working and organising personal life (Reichenberger, 2017; Wang et al., 2018). This is a cultural phenomenon and a new economic activity driven by technology. The development of information and communication technologies allows them greater flexibility and adaptability to different activities (Hermann & Paris, 2020).

The balance between work and leisure while the search for a better work environment proved to be one of the most significant motivating factors for digital nomads. These workers value the freedom that commuting allows them and actively pursue their goals, focusing primarily on leisure rather than just working in a different geographical location (Orel, 2019).

Richter and Richter (2020) support the above notions, characterising digital nomads as strongly independent individuals who build a lifestyle that allows them to live the way they wish. They combine the realisation of these short-term autonomous projects through digital platforms that will enable digital nomads to work in independent and unrestricted locations. Digital nomadism can be considered an exemplary way of working that allows the full range of digital work mobilities to be explored (Schlagwein & Hossein Jarrahi, 2020).

Digital nomadism can be viewed as a subset of remote work that merges it with a sense of wanderlust. This desire for travel is the second most common interest among remote workers, leading to an increase in the number of digital nomads following the COVID-19 pandemic. As a result of the unique circumstances created by the pandemic, these workers have been developing and enhancing their skills, making them even more capable (Vagianos & Engel, 2023).

Considering the definitions described above, this study defines a digital nomad as any individual whose work activity is carried out digitally in any location globally with access to the internet and who travels frequently based only on their desire and motivation to do so.

2.2.2 *Digital Nomadism in Society*

Digital nomadism impacts society on both an organisational and migratory level, as well as from an ecological and sustainability perspective (Šímová, 2023). However, the rising number of digital nomads in certain areas can lead to imbalances that affect local communities (Bonneau et al., 2023).

Ecologically, remote work can reduce traffic congestion and lower CO^2 emissions. On the other hand, the increased consumption of technological devices contributes to a larger environmental footprint due to the high demand for rare metals and the energy consumption associated with these devices (Baliga et al., 2009; Hensellek & Puchala, 2021; Resende & Morais, 2010).

The spread of digital technology has provided digital nomads greater professional flexibility and adaptability (Hermann & Paris, 2020) in choosing their workplace. Work plays a pivotal role in the lives of digital nomads (Nash et al., 2018), not only in terms of work but also in how they socialise, find housing, and organise their trips (Mancinelli, 2020). Directories like nomadlist.com share information about the best places to work remotely by analysing the cost of living, internet speed, weather, fun, and safety (Cook, 2022).

The consumption of physical goods by digital nomads is small. However, they maintain the need to consume different services, such as transport, accommodation, community places, or workspaces (Wang et al., 2019).

Workspaces or coworking centres are widely used services by digital nomads. They encourage the establishment of collaborations and support between digital nomads (Orel, 2019), as they offer the necessary infrastructures to carry out their work (Lee et al., 2019; Šímová, 2023), contribute to efficiency, enhance access to new job opportunities and help improve the social life of workers by promoting interaction (Orel, 2019). Thus, these spaces cannot be seen only as a workplace. Users seek new experiences and a community in which to socialise (Putra & Agirachman, 2016).

For example, Selina is a hospitality network focused on remote workers. It has spaces in more than 80 destinations worldwide, where it offers different types of accommodation, coworking spaces, and activities during the stay. Selina's concept is to unite digital nomads in a place where they can live, work, interact with locals, and discover local nature and culture (Getman, 2021).

An example of a beneficial initiative is the establishment of a "Nomad Village" in Ponta do Sol, located in Madeira, Portugal. Launched in February 2021, this initiative aims to attract digital nomads to the area. Collaborating with local institutions, businesses, and residents, Ponta do Sol has been transformed into a vibrant nomad community, significantly enhancing the village's social and economic landscape (Startup Madeira, 2023).

2.2.3 *Digital Nomads' Characteristics and Challenges*

Studies indicate that digital nomads mainly consist of young individuals from the Millennial and Generation Z demographics, predominantly of Caucasian ethnicity, who belong to the middle class, possess higher education, and hold strong passports. They typically travel alone or in pairs, although there is a growing trend of them travelling with their families (Mancinelli, 2018; Reichenberger, 2017; Šímová, 2023). Digital nomadism allows for diverse lifestyles and adaptability to different stages of life, enabling individuals to shape their work around personal preferences (Frick & Marx, 2021).

Digital nomads strive for an ideal balance between work and leisure, with personal or family life often merging with their professional life (Richter & Richter, 2020). They enjoy working in community-oriented spaces and value the freedom to move frequently (Orel, 2019; Šímová, 2023). A key characteristic of digital nomads is the extended length of their travels and their choice not to maintain a residential base (Nash et al., 2018; Sutherland & Jarrahi, 2017).

Aspiring digital nomads must be open to personal change, as their achievements are seen as a key factor in their success rather than the number of hours worked (Frick & Marx, 2021). Establishing a solid foundation of knowledge, skills, and experiences is essential for beginning a journey in digital nomadism (Reczuch, 2020).

To sustain their productivity levels, digital nomads are described as autonomous individuals with a strong sense of self-determination (Cook, 2020, 2023), alongside organisational skills (Mancinelli, 2020) and proactivity (Frick & Marx, 2021). They tend to be creative, innovative, and willing to take (Bartosik-Purgat & Bartosik-Purgat, 2018). A high degree of self-discipline is crucial, as communication and collaboration with colleagues occur virtually (Frick & Marx, 2021).

Numerous studies indicate that digital nomads generally report higher happiness and productivity (Haking, 2017). Some benefits of digital nomadism include improved concentration and time management, reduced distractions and stress levels, enhanced creativity and stimulation, and opportunities for socialisation with people from diverse cultures and fellow digital nomads (Haking, 2017).

The choice of location for digital nomads is significantly influenced by the availability of lower-cost options for work and leisure logistics (Iyer, 2023; Schlagwein & Hossein Jarrahi, 2020). Other factors include leisure opportunities, climate, and cultural diversity (Ehn et al., 2022; Sánchez-Vergara et al., 2023).

Additionally, digital nomads experience lower expenses, a better quality of life, no time wasted commuting, more control over their schedules, and increased freedom to engage in leisure activities (Demaj et al., 2021; Vagena, 2021). While digital nomads often describe themselves as apolitical, they expect the countries they engage with to uphold labour laws, border controls, and regulations, allowing more flexible entry visas (Cook, 2022).

The ideal of freedom for digital nomads is often constrained by the bureaucracies arising from entering and staying in different countries (Cook, 2022). They must navigate tax systems, residence rules, and worker protection regulations. The more experienced they are, the easier it becomes to understand each country's labour laws, tax regimes, and visa regulations (Cook, 2022).

Digital nomads dislike being labelled as tourists; however, they often travel on tourist visas, benefiting from the power of their passports to enter specific locations quickly (Cook, 2022; Wang et al., 2019). Despite this convenience, they frequently face the "visa run" dilemma, which involves leaving a country to renew their visa (Mancinelli, 2020).

Border controls can place these workers in vulnerable positions concerning immigration laws due to their ambiguous status, as they work remotely rather than within the country (Mancinelli, 2020). This situation can lead to challenges in enjoying the tax benefits that regular employees receive, as their tax residence is often unclear (Tyutyuryukov & Guseva, 2021), putting them in a grey area regarding tax policy (Mancinelli & Germann Molz, 2023).

Several countries have begun to introduce new visas, residence permits, and policies aimed at facilitating the entry of digital nomads to attract new workers, support business development, and position themselves as ideal locations (Getman, 2021; Sánchez-Vergara et al., 2023). These countries aim to streamline bureaucratic procedures and reduce barriers imposed by public administration (Sánchez-Vergara et al., 2023). A notable example is Estonia, the first country to create a digital nomad visa in June 2020 (Hannonen, 2020; Khatri & Shukla, 2022; Newman, 2020).

2.2.4 Implications for Organisations: Benefits and Drawbacks

Given the increasing presence of digital nomads in the labour market, companies must adapt their strategies, including adjusting their recruitment processes for various projects (Bartosik-Purgat & Bartosik-Purgat, 2018).

Reczuch (2020) conducted a qualitative study to explore the potential of digital nomadism as a form of career management. He found that some participants actively manage their careers by taking the necessary steps to transition from traditional jobs. In contrast, other digital nomads embrace the opportunity to change careers, leveraging their previous professional experiences to align with new projects.

Digital nomads can be classified into three categories. First, some work for international companies, completing tasks remotely while travelling. In this case, the company's objective is to ensure that tasks or projects are accomplished, regardless of the location. Second, some individuals work as freelancers, performing their tasks from anywhere worldwide, as long as they have internet access. Lastly, there are startup creators who develop their businesses using internet connectivity.

For example, many American workers no longer seek long-term contracts with a single employer or plan to advance in a specific career path. They often believe that developing their professional skills will not fulfil their expectations, which could negatively impact their future performance. Instead, they are frequently hired through flexible arrangements, independent contracts, or consultants who work on specific tasks or for a set period (Friedman, 2014; Thompson, 2020).

2.2.4.1 Benefits

Organisations are increasingly employing digital nomads, as they are often more cost-effective than hiring full-time employees (Greenwood et al., 2017; Matsushita, 2023; Sutherland & Jarrahi, 2017). They do not require physical office space or supplies, allowing companies to save on overhead costs (Fikriy et al., 2023). Additionally, they avoid expenses related to health insurance, sick pay, and vacation benefits (Hensellek & Puchala, 2021; Richter & Richter, 2020; Thompson, 2019).

For organisations, digital nomads mean providing 24/7 customer service due to their presence in different time zones. They are often highly motivated and self-disciplined, contributing to increased productivity (Fikriy et al., 2023; Frick & Marx, 2021; Stickel, 2020).

Recruiting digital nomads enhances global recruitment opportunities by broadening the pool of available workers. This approach allows organisations to integrate higher-quality and better-skilled employees (Fikriy et al., 2023; Stickel, 2020). Digital nomadism can serve as an effective strategy to attract talent and motivate individuals, as it promotes trust and reduces organisational control, leading to increased employee satisfaction, autonomy, and improved quality of life (Frick & Marx, 2021). Additionally, online platforms streamline job searches for roles such as web designers, and performance evaluations on these platforms can significantly impact future job opportunities (Thompson, 2019).

While some companies require physical presence, limiting the viability of digital nomadism (Frick & Marx, 2021), the rise of short-term digital projects offers more opportunities. Workers are often hired without regard to their professional past and commitments for future employment, allowing companies to adjust salaries more easily (Friedman, 2014; Thompson, 2020).

2.2.4.2 Drawbacks

Integrating digital nomads into companies comes with several benefits and challenges. While digital nomads typically have reliable internet connectivity, they often face difficulties establishing authentic communication and interactions with coworkers and supervisors. This can lead to decreased trust and challenges in cultural alignment and adapting to new work models (Frick & Marx, 2021; Orel, 2019).

Incorporating digital nomads can enhance organisational culture and improve work processes. However, companies that adhere to more traditional work models may encounter obstacles when adopting this flexible approach. Concerns about potential decreases in performance, communication issues stemming from distance, and management difficulties can complicate the hiring of digital nomads (Fikriy et al., 2023; Frick & Marx, 2021). As a result, organisations aiming to integrate digital nomads must embrace a more open and collaborative management style characterised by reduced control and increased flexibility (Richter & Richter, 2020). Furthermore, organisations must rethink career planning, performance indicators, and communication strategies to attract and retain talent while preventing misunderstandings (Khatri & Shukla, 2022; Kong et al., 2019).

Organisations often struggle to hire digital nomads due to a lack of understanding regarding their preferences, logistical needs, and misalignment of visions. Improving this understanding can help in developing more flexible work models (Frick & Marx,

2021). Digital nomads value flexibility; clearly defined expectations from both sides can enhance trust and transparency (Kong et al., 2019).

Kong et al. (2019) also point out that communication issues frequently arise between digital nomads and their clients, mainly due to inflexible communication tools constrained by internal organisational processes.

2.2.5 Human Resources Management Applied to Digital Nomadism

There is currently no specific literature on human resource management within the context of digital nomadism. Therefore, this study examines the management of remote workers, which is the topic closest to this. The data presented here focus on the management of remote employees, and a specific analysis will be conducted regarding the management of individuals in the context of digital nomadism as part of this study.

Technology has transformed Human Resources (HR) processes and the nature of jobs, labour relations, and supervision (Stone et al., 2015). In the digital age, remote work management has evolved, highlighting the importance of transnational relationships between workers and organisations (Donnelly & Johns, 2021).

Managers' behaviours significantly impact remote employees' stress and access to office-related opportunities, such as social activities (Donnelly, 2015; Donnelly & Johns, 2021). HR teams should urge managers to encourage both formal and informal interactions through video calls, emphasising camera use (Straus et al., 2023; Wang et al., 2021).

To integrate teleworking effectively, managers and HR professionals must establish workflows and schedules prioritising employee satisfaction. Leadership development and training for remote workers are essential for fostering commitment and productivity (Kazak & Obelets, 2022). Joint training can enhance colleague trust, improving productivity and self-efficacy (Straus et al., 2023).

Effective leadership involves setting clear expectations, monitoring performance, and providing tailored training (Bell, 2012). Regular communication between remote workers and managers is crucial for identifying needs and interests (Bell, 2012).

Developing personal goal-setting skills can enhance engagement, supported by self-management and leadership training that promotes autonomy (Straus et al., 2023). Finally, monitoring remote workers' behaviours is essential to prevent overwork, and HR professionals should educate managers on its adverse effects on health and productivity (Bell, 2012).

Therefore, it is evident that the management of digital nomads involves collaboration between direct managers (see Table 2.1) and Human Resources professionals (see Table 2.2), with existing literature highlighting the importance of managerial roles.

Table 2.1 Managers' role in the management of remote workers

Managers' role	
Facilitating formal and informal interactions through video calls between teams, promoting camera activation during them	Straus et al. (2023) Wang et al. (2021)
New workflow and schedules	Kazak and Obelets (2022)
Setting employee's own goals	Straus et al. (2023)
Setting expectations and objectives	Bell (2012)
Performance monitoring	
Provision of remote mentors	
Frequent communication with management	
Monitoring the behaviour of remote workers to prevent overwork	

Table 2.2 Human resources' role in the management of remote workers

Human resources role	
Joint training sessions for all the organisation's employees	Straus et al. (2023)
Training in developing personal goal-setting skills	
Training in self-management	
Training in teamwork and e-leadership	
Training according to the needs of remote workers	Bell (2012)
Educating managers and companies about the impact of overwork on workers' health and productivity	

2.3 Methodology

Despite numerous suggestions in the literature, there is a significant gap in academic research regarding digital nomadism from an organisational perspective. The main objective of this study is to investigate the management strategies companies employ regarding digital nomads. A qualitative research approach was chosen to address this gap using the flexible and comprehensive Grounded Theory methodology.

Since the 1960s, grounded theory has become one of the most widely used research methodologies, mainly qualitative studies (Birks et al., 2019). While it shares specific characteristics with other qualitative approaches, such as coding, categorisation, and inductive analysis, grounded theory is distinct in its goal of developing a theory grounded in the data collected (Stough & Lee, 2021). This cyclical process involves data guiding interpretations, which inform further data collection. This cycle continues until researchers determine that no new data needs to be collected, achieving a level of theoretical saturation.

Since its beginning, grounded theory has been analysed and developed by various researchers, leading to three distinct perspectives: the classical approach by Glaser and Strauss (1967), the "Straussian" approach by Strauss and Corbin (1990), and the constructivist approach by Charmaz (2006).

Strauss and Corbin's perspective is the most frequently used in management research, with Glaser's perspective being the least used. However, and taking into account the objective of this research, it was considered that Charmaz's constructivist approach would be the one that would best apply to it, given the process of interaction with the interviewees and allowing a creative analysis of the data collected, as well as the creation of new data according to the categories created, which will lead to the final theoretical construction.

2.3.1 *Sample and Data Collection*

Identifying companies to participate in this study involved outreach through remote work and digital nomad job portals, LinkedIn, and the researchers' existing network. Contact was established via email, and forms for contact were completed on company websites or via LinkedIn messaging. The communication outlined the research's objectives, assured the confidentiality of all collected data, and included a request for an interview with either a human resources professional or a member of the company's management team.

We contacted 163 companies and conducted eight interviews. These included four with management members, three with human resources professionals, and one with a team leader. The sample comprises companies from eight countries across four continents: Europe, South America, North America, and Asia (Table 2.3).

Table 2.3 Sample characterisation

Interview number	Interviewee's function	The economic sector of the companies' activity	Location	Continent
1	Manager	Insurance	Germany	Europe
2	Human resources	Human resources benefits	Portugal	Europe
3	Manager	Software development	Brazil	South America
4	Manager	Services for travellers	Greece	Europe
5	Manager	Design and marketing	Canada	North America
6	Team leader	Electronic games distributor	China	Asia
7	Human resources	Insurance	Norway	Europe
8	Human resources	Software development	Romania	Europe

Legend The interview rounds are stressed with a stronger line

2.3.2 *Data Collection Instrument and Procedures*

Data for this study were gathered through semi-structured interviews conducted via video calls. Participants were briefed on the study's objectives, and informed consent was obtained to record the interviews. The interview questions were designed in alignment with the literature review and the overarching aims of the study. The topics explored included organisational characteristics, the benefits of employing digital nomads, the integration of digital nomads within organisations, perceptions of productivity, and the challenges associated with managing digital nomads. As a semi-structured format, the questions were flexible and adjusted as necessary during the interview process.

Initial coding was conducted using the grounded theory methodology. The first three interviews revealed new thematic dimensions. In response to these findings, we incorporated additional questions into the second round of interviews to further explore the emerging sub-dimensions.

The second round consisted of five interviews, which aimed to validate and, where necessary, expand the initial identified dimensions. By the end of the second round, we better understood the previously identified dimensions and reorganised the Team Management macro dimension, which ultimately included seven distinct dimensions.

To answer this study's goal, we developed a People Management model validated through a third round, consisting of an eighth interview with a Human Resources Director with a Ph.D in Human Resources.

2.4 Data Presentation

While the first three interviews indicated a high consensus regarding several identified dimensions, discrepancies on specific issues highlighted the need for further data collection to ensure a more comprehensive understanding (Table 2.4).

During the initial coding process of the first three interviews, seventeen distinct categories were identified. These were subsequently grouped into three overarching macro-categories representing the different levels within the organisation: Organizational (comprising six categories), Human Resources Management (comprising nine categories), and Team (comprising three categories) (Fig. 2.1). Additional data were gathered during the remaining five interviews to validate and, where necessary, refine the identified categories.

Following the second round of interviews, a more comprehensive understanding of the previously identified categories was achieved, leading to the reorganisation of the Human Resources Management macro-category. By the conclusion of this round, Human Resources Management encompassed eight categories.

Subsequently, a Remote Employee Lifecycle Operational Framework model was developed. An eighth interview with a Human Resources Director validated this model to ensure its accuracy and relevance.

Table 2.4 First round of three interview categories

Concordant categories	Discordant categories
1. Existence of a budget for remote workers	11. Types of contracts
2. Socialisation	12. Onboarding process
3. Creating compulsory moments of contact with the company	13. Salary progression
4. Legal management of digital nomads	14. Employee monitoring
5. Company organisation as a way of retention	15. Projects for digital nomads
6. Time management	16. Productivity perception
7. Training for employees	17. Benefits of digital nomadism for companies
8. Use of communication ICT channels and platforms	
9. Remote working as a way of attraction	
10. Financial costs of remote working	

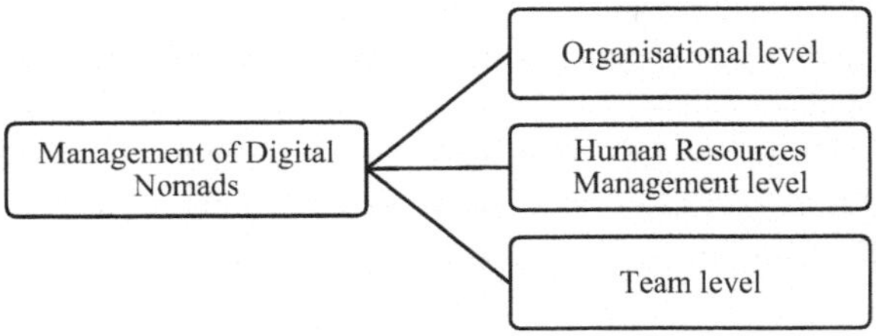

Fig. 2.1 Three overarching categories found regarding managing digital nomads at different organisational levels

Many concordant topics were identified in the first round of interviews, and the initial three macro-categories were kept. However, considering that we had only two echelons of categories in the final phase of coding, we reclassified the macro-categories into "categories" and the categories into "sub-categories" (Fig. 2.1). The following sections present each category and its sub-categories in detail.

2.4.1 Organisational Level

The organisational level involves planning and coordinating various elements within an organisation, including defining objectives, formulating strategies, allocating resources, making decisions, and creating a productive work environment that aims to achieve the established goals and objectives (University of Arizona Global Campus, 2023).

This category includes three general and three specific sub-categories related to digital nomads (Fig. 2.2).

Organisational level sub-categories transversal to all types of remote workers are (a) the utilisation of information and communication technology platforms and

Fig. 2.2 The six sub-categories of the organizational level category. *Legend* In grey–sub-categories transversal to all types of remote workers. In white–sub-categories specific to digital nomads

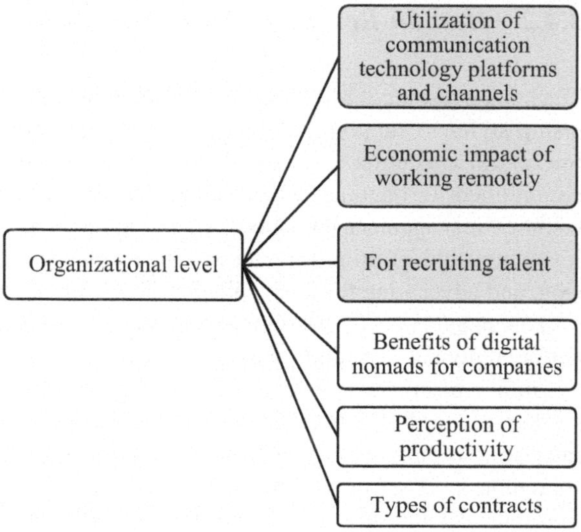

channels to interact with staff members, such as Slack, Email, Notion, Zoom, Trello, WhatsApp, and Google Meet; (b) the economic impact of working remotely is similar for remote workers and digital nomads; however, they may be more advantageous when employing digital nomads; (c) Remote work is an appealing option for companies for recruiting talent both digital nomads and a diverse range of employees seeking flexibility outside the office.

The specific sub-categories of digital nomads include (a) Perception of productivity, (b) Benefits of Digital nomads, and (c) Types of contracts.

The productivity of digital nomads was described as equal to or lower than that of other remote workers. However, the interviewees also mentioned that this productivity decreases when adapting to a new location and stabilises once the workers have adapted.

Several companies noted that integrating digital nomads enhances their competitive position. The interviews noted several benefits, such as the company's option to be 100% remote working with digital nomads, which comes naturally; since the company is working for digital nomads, employing digital nomads leads to a better understanding of customer needs; having people working in different time zones brings speed and competitive advantage to the market demands; and also employing digital nomads enables access to a broader talent pool.

The contracts of digital nomads are divided into five sub-dimensions (permanent contract, one-person company, employer of record, freelancer, professional employer organisation), and, according to the interviews, these are usually the worker's own choice.

2.4.2 Human Resources Management Level

The human resources level regards implementing human resource practices and leadership styles to support and enhance employee development (Knies et al., 2020). Successful execution involves planning, continuous application, assessment, and commitment to ongoing enhancement. Necessary actions include creating human resource management guidelines, encouraging clear communication, enabling and involving staff, offering training and development chances, supporting team cohesion, and advocating for team collaboration (Oakland & Oakland, 2001).

This category encompasses the seven sub-categories that are not explicitly targeted at digital nomads, in addition to one specifically aimed at their needs (Fig. 2.3).

Human resources management sub-categories transversal to all types of remote workers are (a) Extra financial support for remote employees is found in some of the interviewed companies trying to meet the needs of working remotely, such as budgets for remote work set-up, contingent needs, and personal development; (b) The integration stage is vital for organizations, serving as the initial point for employees and an ideal opportunity to synchronize foundational values and goals; (c) it is a Retention strategy, and companies are restructuring their operations to enhance working conditions and accommodating the unique needs and preferences of their remote staff; (d) Companies are increasingly investing in training and development initiatives aimed at improving the professional and personal growth of their remote employees, with a strong emphasis on addressing their individual needs and expectations; (e) companies are providing remote staff assistance by enhancing their working conditions and actively addressing their needs and preferences; (f) Income growth perspectives: Allowing remote work can attract a diverse workforce, making competitive salaries

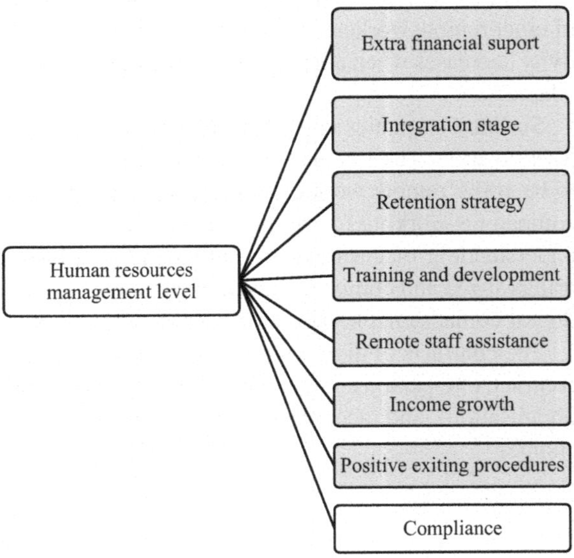

Fig. 2.3 The seven sub-categories of the human resources management level. *Legend* In grey–sub-categories transversal to all types of remote workers. In white–sub-categories specific to digital nomads

essential. Companies can utilise benchmarking, experience progression, and bonuses to stay competitive; (g) Positive exiting procedures are essential to ensure that the departing employee leaves satisfied and becomes an ambassador.

The specific sub-category of digital nomads regarding people management is Compliance. The legal framework governing digital nomads refers to the taxation regulations regarding Pay As You Earn (PAYE), social security contributions, and any other country-specific taxes applicable legally in their locations.

2.4.3 *Team Level*

Groups are compact, interconnected entities that reduce traditional hierarchical structures, increase adaptability, and incorporate insights from different functional sectors. Managing a team involves creating and applying methods and resources to effectively structure and coordinate individuals to reach a common objective. Strong leadership abilities are essential in this scenario, as they promote teamwork among members and help them devise solutions to particular challenges and issues (Gutterman, 2023).

This category includes three sub-categories, with only one targeted explicitly to digital nomads (Fig. 2.4).

Two team-level sub-categories can be used transversally across all remote workers: (a) Work-hour management includes remote workers' flexibility to choose their hours as long as they adhere to specific company guidelines. However, there are instances when companies may require remote workers to engage in synchronous work for effective collaboration and (b) staff engagement, maintained through social relationships while working remotely through various means that companies can facilitate for all remote workers, not just digital nomads.

Data from the interviews shows that digital nomads' assignment content can be differentiated. Because digital nomads frequently move from one place to another, they may engage in distinct projects compared to other remote employees within the company, as various countries have differing data processing and protection regulations.

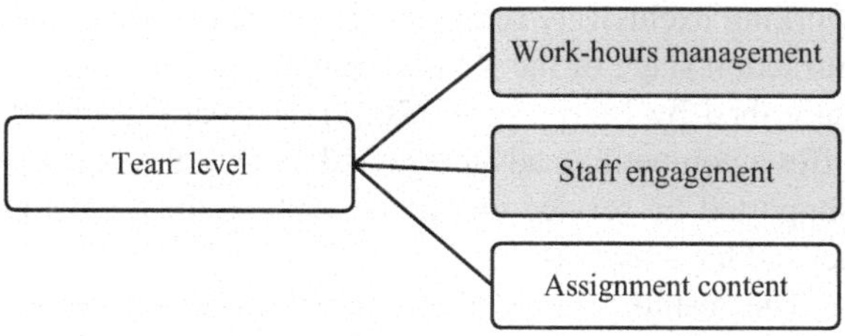

Fig. 2.4 The three sub-categories of the team level category. *Legend* In grey–sub-categories transversal to all types of remote workers. In white–sub-categories specific to digital nomads

2.5 Discussion

Based on the results obtained, it can be observed that the management of digital nomads largely mirrors the management of employees working exclusively remotely across nearly all identified categories. Notable distinctions, however, emerge in productivity perception, contract types, assigned projects, and legal management. Additionally, the analysis highlights certain advantages for companies in hiring digital nomads, differentiating them from other fully remote workers. These advantages include the competitive edge of varying time zones and the ability to access a broader talent pool.

Although productivity is characterised as either lower or equivalent, it is influenced by individual personality traits and workers' ability to prepare for and adapt to new environments. It decreases during the initial adaptation phase but stabilises once employees fully adjust to their new surroundings.

The contracts for digital nomads are distinguished by introducing two new options that, while also applicable to remote workers, hold particular significance for the legal management of digital nomads: the Employer of Record (EOR) and the Professional Employer Organization (PEO).

A Professional Employer Organisation is a human resources outsourcing service that assumes responsibility for administrative tasks on behalf of a company, effectively acting as a co-employer. In contrast, an Employer of Record is a service that facilitates hiring employees from outside the company's home country, assuming the role of the legal employer and managing compliance-related matters, such as salary, taxes, legal benefits, and employment contracts.

In this context, Human Resources teams are not required to invest additional time or specialised knowledge beyond what is necessary to manage remote workers, as these two models—PEO and EOR—provide companies with streamlined solutions for managing digital nomads.

Regarding the projects, the primary distinction identified was the need to adapt projects to the country where the worker is located due to the variations in data protection and privacy regulations across different countries.

Thus, based on the interviews conducted, it can be concluded that the management of digital nomads does not differ significantly from the management of employees working exclusively remotely. They are mainly in the hiring phase, included in the attraction stage of the Remote Employee Life Cycle. In addition to the advantages described by Fikriy et al. (2023), companies' openness to this possibility would offer a competitive advantage and expand the talent pool available for recruitment, supported by service providers, such as those mentioned above, for managing all related legal aspects.

The results showed that companies increasingly attempt to maintain physical contact among employees by organising meetings at the company's office, if available, or by gathering employees in specific countries for in-person events.

The management of work schedules for digital nomads is handled in the same way as for remote workers, with the expectation that digital nomads adapt to the meeting times established by the company.

As highlighted by Bell (2012) and Straus et al. (2023), the interviews underscored the significant emphasis placed on the monitoring of employees, as well as on understanding their expectations and establishing goals, not only in terms of professional development but also personal growth. It was also observed that companies are increasingly prioritising the individual development of their employees rather than solely focusing on professional advancement. In this context, employees are encouraged, in addition to participating in online training programs, to pursue autonomous development with the company's support. Creating a personal development budget often provides this support, allowing employees to engage in activities they believe will offer the most benefit.

2.5.1 Remote Employee Lifecycle Operational Framework

The primary objective of this dissertation was to define the human resource management strategy applied to digital nomads. In this regard, creating a model companies can use to facilitate this management is essential. In alignment with the reviewed literature (Bell, 2012; Kazak & Obelets, 2022; Straus et al., 2023; Wang et al., 2021) and drawing a parallel with the employee lifecycle, it is evident that recruitment and hiring are managed at the Human Resources level.

Examining the life cycle of employees to understand better the amount of scattered data and lack of theory has already been done regarding the engagement topic (Kwon & Park, 2019).

To make sense of the amount of data obtained and to begin and propose a new approach to digital workforce management, we use the employee life cycle stages identified by Gladka et al. (2022), who developed a conceptual base for employee life cycle (ELC), after reviewing and integrating various definitions systemically.

Gladka et al. (2022) divided the cycle into two main employee life phases: before entering the company attraction and after retention. We aim to present an operational management framework for the digital workforce. Specifically, the attraction stages are merged into one, and we present the details of the stages after the employees begin to work in the company, the retention phase.

We also incorporate the activities and organisational responsible stakeholders into this framework to help it be used operationally (Fig. 2.5).

Attraction is integrated within the organisational level, emphasising that the marketing and communication departments should address employer branding at a strategic level. Working hours management, communication, and socialisation are handled at the team management level. At the same time, onboarding, employee support, employee monitoring, income growth (including both professional and personal development) and the offboarding process are managed at both levels.

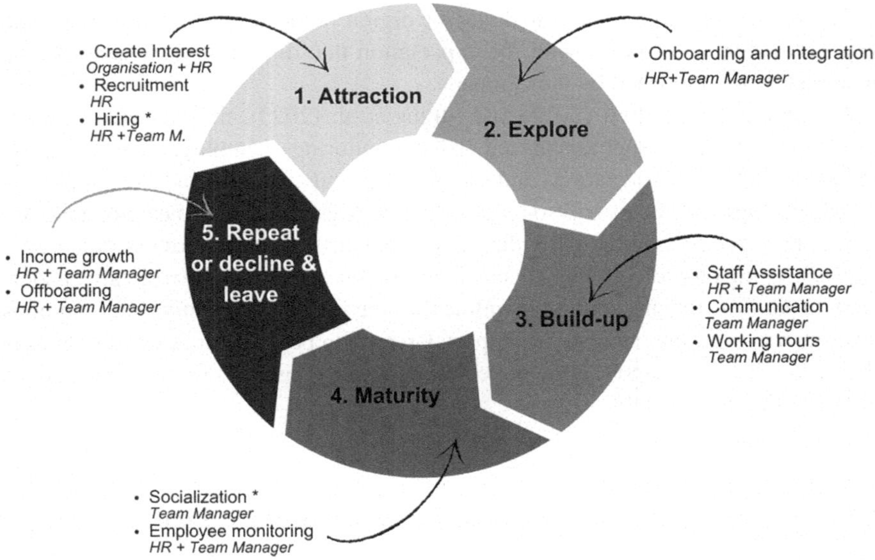

Fig. 2.5 Remote employee lifecycle operational framework. *Legend* 1. The stages are the Employee Life Cycle stages, according to Gladka et al. (2022). 2. The outer boxes include the associated organisational activities and the stakeholders implementing them (in italics). 3. The "*" symbol regards the digital nomads' management specificities

This model was constructed using data collected from seven interviews. An eighth interview was conducted with a human resources director with specialised academic training in human resources that validated the model.

2.5.1.1 Attraction

The attraction stage includes creating interest, recruitment, and hiring activities.

Creating Interest

Attraction is managed at the organisational level and encompasses all subsequent phases of the employee lifecycle. The company's description as 100% remote or "remote-friendly" is a key attraction for digital nomads and all employees seeking greater flexibility in their working hours and the opportunity to work from anywhere.

Recruitment

Recruitment is managed at the Human Resources level and can be done through the company's website, remote work platforms, or LinkedIn.

Hiring

Team managers have a say in hiring, as they usually participate in the selection phase. They also assign projects to new hires, which are fully disclosed in the selection and hiring stages.

Similarly, hiring is managed at the Human Resources level and involves all compliance-related aspects. Companies or services such as Employer of Record (EOR) or Professional Employer Organizations (PEO) may facilitate this process, especially when hiring digital nomads.

The type of contract for each employee should be determined based on the expectations of both parties and the nature of the work to be performed. Projects assigned to each employee will be managed according to their location and the relevant legal regulations that apply in those jurisdictions.

Due to their constant movement between countries, the project assignment and the type of contract differentiate digital nomads from fully remote workers.

2.5.1.2 Onboarding and Integration

The Human Resources team and the team leader carry out the onboarding process. Depending on the company's organisation, it can be implemented in various ways, including an initial administrative process for introducing content and communication platforms, online training and an introduction to the company's activities, and establishing a probationary period.

Additionally, creating a budget for remote workers and their setup improves working conditions, ensuring employees have the necessary tools and resources to perform effectively.

2.5.1.3 Build-Up

The building-up stage includes staff assistance, working-hours management and communication.

Staff Assistance

Employee support involves maintaining clear expectations and objectives while providing the means to help employees achieve them. This can be facilitated by creating a budget to address potential needs, ensuring workers have the necessary professional and personal development resources.

Working-Hours Management

The absence of strict time tracking allows employees greater autonomy and freedom. However, specific structured time slots, such as "team sync hours," may be introduced, during which all team members must be available. Outside of these periods, employees are free to choose their working hours. Other structured events, such as mandatory meetings, may be scheduled to ensure all employees are present at key moments.

Communication

Using communication platforms enhances work organisation and serves as a means of socialising among employees. During the interviews, the following platforms were cited as examples: Slack, Email, Telephone calls, Notion, Zoom, Trello, WhatsApp, Google Meet, Microsoft Teams, Loom, and Team Flow. These tools facilitate day-to-day communication and collaboration, helping remote teams stay connected and organised.

2.5.1.4 Maturity

The attraction stage consists of socialisation and employee monitoring.

Socialisation

Socialisation can be promoted through in-person meetings, either company-wide or within specific teams, which help foster relationships among employees. These events create opportunities for bonding, collaboration, and strengthening team dynamics, essential for maintaining a positive work culture in remote and hybrid environments.

Organising in-person meetings is a more significant challenge for managing digital nomads, although it is also challenging for remote workers.

Employee Monitoring

Employee monitoring aims to foster development not only in professional terms but also on a personal level. In this regard, online or in-person training sessions may be organised based on individual employee goals, or personal development can be encouraged through autonomous learning supported by a dedicated budget. This budget can also be used for personal growth in physical or emotional aspects, promoting a holistic approach to employee well-being.

2.5.1.5 Repeat or Decline and Leave

Income Growth

Recognition and remuneration are considered essential factors for employee retention. In this regard, income growth can be based on experience, with benchmarking against the continent or country where the employee operates, or it may be linked to performance bonuses tied to specific indicators and objectives.

Positive Exiting Procedures

Like the onboarding and integration processes, positive exiting procedures ensure employees leave the company on good terms and become ambassadors. Key elements of offboarding include conducting an exit interview, settling any outstanding payments, and ensuring the return of all company documents, projects, and contacts. This process helps maintain a positive reputation for the company and strengthens relationships with former employees.

2.6 Conclusions

Digital nomadism has increasingly become an everyday reality in businesses, driven by the growth of technological platforms for information and communication and the ease of movement between countries. Although various researchers have studied this new lifestyle, there has been limited focus on digital nomads from an organisational perspective. Given their growing significance for companies and their increasing prevalence in the labour market, it is crucial to understand how organisations should structure themselves to integrate, manage, and develop these workers effectively.

The primary objective of this dissertation was to "define the human resource management strategy applied to digital nomads." To achieve this, a Grounded Theory approach was employed. Three rounds of interviews were conducted, totalling eight in-depth interviews. Data analysis revealed three macro dimensions of digital nomad management: organisational, people, and team management, subdivided into seventeen specific dimensions. Throughout the interviews, it became evident that the strategy for managing digital nomads does not significantly differ from that applied to 100% remote workers, with the main distinctions being in the legal management and the nature of the projects assigned.

This study concludes that offering employees the option of digital nomadism will not significantly impact the company if objectives, values, and expectations are aligned from the beginning. Additionally, it is crucial to maintain open communication channels to adjust these expectations when necessary, providing support and guidance without resorting to control or excessive workplace pressure. The findings indicate a high degree of convergence in results despite geographical diversity in the sample. Furthermore, the study revealed that contrary to the initial literature review's preconceived notion that digital nomads primarily engage in freelance work, many companies offer stable employment contracts while still providing the freedom and flexibility that digital nomads highly value.

Organisations can enhance their talent attraction capacity by hiring digital nomads without the concern that it will require significant adjustments to their people management processes, especially if they already have employees working remotely permanently.

This study is grounded in a scattered global sample and found the similarities that enabled the building of an operational people management framework for digital nomads and full remote workers. The study contributes to academia and companies, not only those employing remote workers and digital nomads. On the one hand, this study presents results that enlighten the specificities of digital nomads and what should be regarded to employ and successfully manage them. On the other hand, this study offers a conceptual and practical framework summarising all the findings, covering the remote worker lifecycle and highlighting the digital nomad differences: Remote Employee Lifecycle Operational Framework.

The study's main limitation was obtaining the companies' participation to conduct the interviews (participation rate 5%). However, considering the data collected and the widespread consensus, we conclude that theoretical saturation has been achieved.

In future research, it would be interesting to complete the attraction stage of the original model with the employer's activities. It would also be interesting to triangulate the framework we presented with the digital staff's perspective, continuing the qualitative approach. Applying this model using a quantitative approach would also be of great interest, although challenging, considering our difficulty obtaining data for the interviews. This approach would provide empirical evidence to assess the model's generalizability, offering insights into the measurable impact of such practices on organisational outcomes, employee well-being, and performance in remote work, specifically for digital nomads.

Despite the similarities in managing digital nomads and fully remote workers, deepening the differences, based on these findings, would help build better management practices for both types of employees in the digital workforce.

References

Baliga, J., Hinton, K., Ayre, R., & Tucker, R. S. (2009). Carbon footprint of the Internet. *Telecommunications Journal of Australia, 59*(1), 5.1–5.14. https://doi.org/10.2104/tja09005

Bartosik-Purgat, M., & Bartosik-Purgat, M. (2018). *Digital nomads–entrepreneurship or new lifestyle?*.

Bell, B. S. (2012). Remote work: Examining current trends and organizational practices. *International HR Adviser, 49*.

Birks, M., Hoare, K., & Mills, J. (2019). Grounded theory: The FAQs. *International Journal of Qualitative Methods, 18*, 160940691988253. https://doi.org/10.1177/1609406919882535

Bonneau, C., Aroles, J., & Estagnasié, C. (2023). Romanticisation and monetisation of the digital nomad lifestyle: The role played by online narratives in shaping professional identity work. *Organization, 30*(1), 65–88. https://doi.org/10.1177/13505084221131638

Charmaz, K. (2006). *Constructing grounded theory.* Sage Publications. ISBN 978-0-7619-7352-2.

Cook, D. (2020). The freedom trap: Digital nomads and the use of disciplining practices to manage work/leisure boundaries. *Information Technology & Tourism, 22*(3), 355–390. https://doi.org/10.1007/s40558-020-00172-4

Cook, D. (2022). Breaking the contract: Digital nomads and the state. *Critique of Anthropology, 42*(3), 304–323. https://doi.org/10.1177/0308275X221120172

Cook, D. (2023). What is a digital nomad? definition and taxonomy in the era of mainstream remote work. *World Leisure Journal, 65*(2), 256–275. https://doi.org/10.1080/16078055.2023.2190608

Demaj, E., Hasimja, A., & Rahimi, A. (2021). Digital nomadism as a new flexible working approach: making tirana the next european hotspot for digital nomads. In M. Orel, O. Dvouletý, & V. Ratten

(Eds.), *The flexible workplace* (pp. 231–257). Springer International Publishing. https://doi.org/10.1007/978-3-030-62167-4_13

Donnelly, R. (2015). Gender, careers and flexibility in consultancies in the UK and the USA: A multi-level relational analysis. *The International Journal of Human Resource Management, 26*(1), 80–99. https://doi.org/10.1080/09585192.2014.934889

Donnelly, R., & Johns, J. (2021). Recontextualising remote working and its HRM in the digital economy: An integrated framework for theory and practice. *The International Journal of Human Resource Management, 32*(1), 84–105. https://doi.org/10.1080/09585192.2020.1737834

Ehn, K., Jorge, A., & Marques-Pita, M. (2022). Digital nomads and the Covid-19 pandemic: Narratives about relocation in a time of lockdowns and reduced mobility. *Social Media.*

Fikriy, A., Sani, A., Syed Omar, S. A., Mohamed Thaheer, A. S., Fikri Ishak, M., & Shuib, F. H. (2023). Digital nomads: Who are they? *International Journal of Academic Research in Economics and Management Sciences, 12*(2), 631–641. https://doi.org/10.6007/IJAREMS/v12-i2/17516

Frick, N., & Marx, J. (2021). Integrating digital nomads in corporate structures: Managerial contemplations. In *Hawaii International Conference on System Sciences.* https://doi.org/10.24251/HICSS.2021.260

Friedman, G. (2014). Workers without employers: Shadow corporations and the rise of the gig economy. *Review of Keynesian Economics, 2*(2), 171–188. https://doi.org/10.4337/roke.2014.02.03

Getman, N. (2021). *Developing a digital nomads destination from the ground.* Iscte-Instituto Universitário de Lisboa.

Gladka, O., Fedorova, V., & Dohadailo, Y. (2022). Development of conceptual bases of the employee life cycle within an organization. *Business: Theory and Practice, 23*(1), 39–52. https://doi.org/10.3846/btp.2022.13510

Glaser, B. G., & Strauss, A. L. (1967). *The discovery of grounded theory: Strategies for qualitative research.* Transaction.

Greenwood, B., Burtch, G., & Carnahan, S. (2017). Unknowns of the gig-economy. *Communications of the ACM, 60*(7), 27–29. https://doi.org/10.1145/3097349

Gutterman, A. (2023). Team management. *SSRN Electronic Journal.* https://doi.org/10.2139/ssrn.4527071

Haking, J. (2017). *Digital nomad lifestyle: A field study in Bali.* KTH Royal Institute of Technology.

Hannonen, O. (2020). In search of a digital nomad: Defining the phenomenon. *Information Technology & Tourism, 22*(3), 335–353. https://doi.org/10.1007/s40558-020-00177-z

Hemsley, J., Erickson, I., Jarrahi, M. H., & Karami, A. (2020). Digital nomads, coworking, and other expressions of mobile work on Twitter. *First Monday.* https://doi.org/10.5210/fm.v25i3.10246

Hensellek, S., & Puchala, N. (2021). The emergence of the digital nomad: A review and analysis of the opportunities and risks of digital nomadism. In M. Orel, O. Dvouletý, & V. Ratten (Eds.), *The flexible workplace* (pp. 195–214). Springer International Publishing. https://doi.org/10.1007/978-3-030-62167-4_11

Hermann, I., & Paris, C. M. (2020). Digital nomadism: The nexus of remote working and travel mobility. *Information Technology & Tourism, 22*(3), 329–334. https://doi.org/10.1007/s40558-020-00188-w

International Workplace Group. (2019). *The IWG global workspace survey.* https://assets.regus.com/pdfs/iwg-workplace-survey/iwg-workplace-survey-2019.pdf

Iyer, D. K. (2023). Digital nomad. *Society for the Anthropology of Work.* https://doi.org/10.21428/1d6be30e.9b08c1c9

Kazak, O., & Obelets, T. (2022). Remote employment: A short-term strategy for adapting to the crisis or the new reality of human resource management. *Baltic Journal of Economic Studies, 8*(2), 98–104. https://doi.org/10.30525/2256-0742/2022-8-2-98-104

Khatri, P., & Shukla, S. (2022). *Digital nomadism: A systematic review and research agendas.*

Knies, E., Leisink, P., & Van De Schoot, R. (2020). People management: Developing and testing a measurement scale. *The International Journal of Human Resource Management, 31*(6), 705–737. https://doi.org/10.1080/09585192.2017.1375963

Kong, D., Schlagwein, D., & Cecez-Kecmanovic, D. (2019). Issues in digital nomad-corporate work: An institutional theory perspective. In *Proceedings of the 27th European Conference on Information Systems*. ECIS, Suécia.

Kwon, K., & Park, J. (2019). The life cycle of employee engagement theory in HRD research. *Advances in Developing Human Resources, 21*(3), 352–370. https://doi.org/10.1177/1523422319851443

Lee, A., Toombs, A. L., Erickson, I., Nemer, D., Ho, Y., Jo, E., & Guo, Z. (2019). The social infrastructure of co-spaces: Home, work, and sociable places for digital nomads. *Proceedings of the ACM on Human-Computer Interaction, 3*(CSCW), 1–23. https://doi.org/10.1145/3359244

Mancinelli, F. (2018). A practice of togetherness: Home imaginings in the life of location-independent families. *International Journal of Tourism Anthropology, 6*(4), 307. https://doi.org/10.1504/IJTA.2018.096359

Mancinelli, F. (2020). Digital nomads: Freedom, responsibility and the neoliberal order. *Information Technology & Tourism, 22*(3), 417–437. https://doi.org/10.1007/s40558-020-00174-2

Mancinelli, F., & Germann Molz, J. (2023). Moving with and against the state: Digital nomads and frictional mobility regimes. *Mobilities*, 1–19. https://doi.org/10.1080/17450101.2023.2209825

Matsushita, K. (2023). How the Japanese workcation embraces digital nomadic work style employees. *World Leisure Journal, 65*(2), 218–235. https://doi.org/10.1080/16078055.2022.2156594

Müller, A. (2016). The digital nomad: Buzzword or research category? *Transnational Social Review, 6*(3), 344–348. https://doi.org/10.1080/21931674.2016.1229930

Nash, C., Jarrahi, M. H., Sutherland, W., & Phillips, G. (2018). Digital nomads beyond the buzzword: Defining digital nomadic work and use of digital technologies. In G. Chowdhury, J. McLeod, V. Gillet, & P. Willett (Eds.), *Transforming digital worlds* (Vol. 10766, pp. 207–217). Springer International Publishing. https://doi.org/10.1007/978-3-319-78105-1_25

Newman, C. (2020). *Estonia launches the world's first 'Digital Nomad Visa'*. CGTN. https://newseu.cgtn.com/news/2020-07-17/Estonia-launches-the-world-s-first-Digital-Nomad-Visa--Sb42I30bN6/index.html

Oakland, S., & Oakland, J. S. (2001). Current people management activities in world-class organizations. *Total Quality Management, 12*(6), 773–788. https://doi.org/10.1080/09544120120075370

Orel, M. (2019). Coworking environments and digital nomadism: Balancing work and leisure whilst on the move. *World Leisure Journal, 61*(3), 215–227. https://doi.org/10.1080/16078055.2019.1639275

Putra, G. B., & Agirachman, F. A. (2016). *Urban coworking space: Creative tourism in digital nomads perspective.*

Reczuch, D. (2020). Can digital nomadism be considered a career management style and a lifestyle? *Zarządzanie Mediami*, 137–148.

Reichenberger, I. (2017). Digital nomads–a quest for holistic freedom in work and leisure. *Annals of Leisure Research, 21*(3), 364–380. https://doi.org/10.1080/11745398.2017.1358098

Resende, L. V., & Morais, C. A. (2010). Study of the recovery of rare earth elements from computer monitor scraps–leaching experiments. *Minerals Engineering, 23*(3), 277–280. https://doi.org/10.1016/j.mineng.2009.12.012

Richter, S., & Richter, A. (2020). Digital nomads. *Business & Information Systems Engineering, 62*(1), 77–81. https://doi.org/10.1007/s12599-019-00615-1

Sánchez-Vergara, J. I., Orel, M., & Capdevila, I. (2023). Home office is the here and now. Digital nomad visa systems and remote work-focused leisure policies. *World Leisure Journal, 65*(2), 236–255. https://doi.org/10.1080/16078055.2023.2165142

Schlagwein, D., & Hossein Jarrahi, M. (2020). The mobilities of digital work.pdf. *Research-in-Progress Papers, 89.*

Šímová, T. (2023). A research framework for digital nomadism: A bibliometric study. *World Leisure Journal, 65*(2), 175–191. https://doi.org/10.1080/16078055.2022.2134200

Startup Madeira. (2023). *Digital nomadas madeira Islands*. https://digitalnomads.startupmadeira.eu/about-us/

Stickel, M. (2020). *Challenges and opportunities of digital nomadism and its implications for tomorrow's workforce*. Universidade Nova de Lisboa.

Stone, D. L., Deadrick, D. L., Lukaszewski, K. M., & Johnson, R. (2015). The influence of technology on the future of human resource management. *Human Resource Management Review, 25*(2), 216–231. https://doi.org/10.1016/j.hrmr.2015.01.002

Stough, L. M., & Lee, S. (2021). Grounded theory approaches used in educational research journals. *International Journal of Qualitative Methods, 20*, 160940692110522. https://doi.org/10.1177/16094069211052203

Straus, E., Uhlig, L., Kühnel, J., & Korunka, C. (2023). Remote workers' well-being, perceived productivity, and engagement: which resources should HRM improve during COVID-19? a longitudinal diary study. *The International Journal of Human Resource Management, 34*(15), 2960–2990. https://doi.org/10.1080/09585192.2022.2075235

Strauss, A. L., & Corbin, J. (1990). *Basic of qualitative research: Grounded theory procedures and techniques*. Sage Publications.

Sutherland, W., & Jarrahi, M. H. (2017). The gig economy and information infrastructure: The case of the digital nomad community. *Proceedings of the ACM on Human-Computer Interaction, 1*(CSCW), 1–24. https://doi.org/10.1145/3134732

Thompson, B. Y. (2019). The digital nomad lifestyle: (Remote) work/leisure balance, privilege, and constructed community. *International Journal of the Sociology of Leisure, 2*(1–2), 27–42. https://doi.org/10.1007/s41978-018-00030-y

Thompson, B. Y. (2020). Digital nomadism: Mobility, millennials and the future of work in the online gig economy. In G. Hearn (Ed.), *The future of creative work*. Edward Elgar Publishing. https://doi.org/10.4337/9781839101106.00019

Tyutyuryukov, V., & Guseva, N. (2021). From remote work to digital nomads: Tax issues and tax opportunities of digital lifestyle. *IFAC-PapersOnLine, 54*(13), 188–193. https://doi.org/10.1016/j.ifacol.2021.10.443

University of Arizona Global Campus. (2023). *What is organizational management*. University of Arizona Global Campus. https://www.uagc.edu/blog/what-organizational-management

Vagena, A. (2021). Digital nomads and tourism industry. *Academia Letters*. https://doi.org/10.20935/AL765

Vagianos, D., & Engel, S. (2023). The COVID-19 pandemic impact on society: The case of digital nomadism. *Communications of International Proceedings, 10*(5171/2023), 4121823.

Wang, B., Liu, Y., Qian, J., & Parker, S. K. (2021). Achieving effective remote working during the COVID-19 pandemic: A work design perspective. *Applied Psychology, 70*(1), 16–59. https://doi.org/10.1111/apps.12290

Wang, B., Schlagwein, D., The University of Sydney Business School, AU, Cecez-Kecmanovic, D., UNSW Business School, AU, Cahalane, M. C., & UNSW Business School, AU. (2018). Digital work and high-tech wanderers: Three theoretical framings and a research agenda for digital nomadism. In *Australasian Conference on Information Systems 2018*. University of Technology. https://doi.org/10.5130/acis2018.bl

Wang, B., Cecez-Kecmanovic, D., Schlagwein, D., & Cahalane, M. (2019). *Digital nomadism and the market economy: Resistance and compliance*.

Chapter 3
Industry 4.0: Knowledge Management with Artificial Intelligence?

Elsa Maria Rodrigues and Ana Veloso

Abstract The technological era in which we live reflects the profound transformation that the world is undergoing and the constant need for organisations to adapt to remain competitive. This context brings together different technological scenarios that require a multidisciplinary and attentive approach on the part of organisational psychology, since the process of Knowledge Management goes hand in hand with Industry 4.0 and Artificial Intelligence. In the current chapter, entitled 'Industry 4.0: Knowledge Management with Artificial Intelligence?' we propose to analyse the role of Knowledge Management in Industry 4.0 and interrogate the influence of Artificial Intelligence on the Knowledge Management process. A literature review was carried out using PRISMA (Preferred Reporting Items for Systematic Reviews and Meta-Analyses), as well as an analysis of the data collected in a case study, using interviews, focus group and a questionnaire survey on Knowledge Management in Industry 4.0. The era of Industry 4.0 has specific characteristics, directly associated with the incorporation of information technologies into the production process and is seen as a new industrial reality with production models in which equipment, instruments, tools and processes of different kinds are connected and coexist almost autonomously (Junior, G. T., & Saltorato, P. (2018). Impactos da indústria 4.0 na organização do trabalho: uma revisão sistemática da literatura impacts of the industry 4.0 on work organization: a systematic review of the literature, 18(2), 743–769). Our interest was to problematise this technological environment in its connection to the phenomenon of Knowledge Management in an industrial environment.

Keywords Knowledge management industry 4.0 · Artificial intelligence

---

This research was approved by the Ethics Committee of the University of Minho (CEICSH 067/2019).

E. M. Rodrigues (✉) · A. Veloso
School of Psychology, University of Minho, Braga, Portugal
e-mail: elsamariarodrigues.phd@gmail.com

C. Machado (ed.), *Human Resource Management in a Disrupted World*, Lecture Notes on Multidisciplinary Industrial Engineering,
https://doi.org/10.1007/978-3-031-89948-5_3

3.1 Introduction

This study is part of a research project which aims to contribute to understanding the processes associated with the phenomenon of Knowledge Management in the context of Industry 4.0, more specifically about the role of Artificial Intelligence in Knowledge Management in a manufacturing organisation that is currently living in the era of Industry 4.0.

With this work, we intend to add essential information that will enable us to find answers to the research questions of ‘How do the processes associated with Knowledge Management take place in organisations in the era of Industry 4.0?’ and ‘How does Knowledge Retention and Transfer occur in the Knowledge Management process in the era of Industry 4.0?’. The processes named before are fundamental to understanding the phenomenon and designing new organisational realities.

It should be noted that by carrying out the aforementioned research project we concluded that the phenomenon of Knowledge Management is transversal to the organisation and is complex to manage, and that the individual plays a crucial role as an agent for generating, retaining and transferring Knowledge, being a leveraging element for Knowledge and Knowledge Management in the era of Industry 4.0. Organisations today are challenged to live in a context of constant change and unpredictability, in a VUCA world-Volatility, Uncertainty, Complexity, Ambiguity (Bennett & Lemoinhe, 2014), where everything is changing quickly, there is unpredictability, and it is often not possible to have a history that is applicable to new situations (Balog, 2020, p. 16).

We believe that the economic, technological, and social changes we are currently experiencing are posing challenges for human resource managers and researchers in the social field with regard to the demands of Knowledge Management in a digital and technological society and in an environment of Artificial Intelligence.

3.2 Methodology

In order to reflect on the phenomenon of Knowledge Management in the era of Industry 4.0 and the influence of Artificial Intelligence on the Knowledge Management process, we conducted a literature review, exploratory interviews in different organizations (service, industry and research) and a case study of an Industrial organization.

In the literature review, we used PRISMA (Preferred Reporting Items for Systematic Reviews and Meta-Analyses) (Moher, 2009) as a guideline for the eligibility of the selection criteria in the review conducted, considering the following items from the PRISMA checklist as the basis for selecting the articles: the information within the abstract, the methodology followed, the discussion of the results and the presentation of conclusions.

For the qualitative data, collected through interviews and focus group, a template analysis (King, 2012) was used to analyze the data. The questionnaire data were analyzed using descriptive statistical procedures.

To perform the intended literature review, a search was carried out on the *Web of Science* search engine, using keywords in English with the aim of expanding our search as much as possible. The words considered were Knowledge Management *and* Industry 4.0 *and* Artificial Intelligence, on 11 October 2023. The criteria used to search for articles in the database were the existence of all three keywords or the presence of at least one of the three keywords, with a date from 2013 onwards and articles in Portuguese, French, Spanish or English. The search returned 130 articles in English and 1 article in Portuguese.

The exclusion decisions were based (in general), on the one hand, on the fact that the respective abstracts did not allude to the defined search keywords or other directly related keywords, and, on the other hand, on the fact that the abstracts showed that the topic was out of touch with the phenomenon to be analysed, e.g., if they were very focused on technical issues with little insight into organisational psychology, or very focused on digitalisation, or even applied to very specific situations such as a study oriented towards political impacts. The basic decision criterion for deciding which articles to include was not only the reference to the keywords of the research, but also the article's connection to the topic under study, for example, if the article showed that it related the different concepts and pointed out paths at the level of organisational psychology, or as an applied example that allowed us to understand the interactions of the phenomenon under study.

Literature reviews procedures

In the first phase, the researcher analysed the abstract of each article. Of the 131 articles, 77 were excluded (as the respective abstracts did not allude to the defined search keywords or other directly related keywords) and 54 were identified as potential for analysis. In a second phase, and as a way of controlling the quality of the decision, the abstracts of 57 of the 131 articles were randomly analysed by a second researcher.

Of these 57 articles, 33 were considered potential and 24 were excluded (on the basis that the abstracts showed that the topic was not related to the phenomenon being analysed). The decisions to include or exclude articles made by the two reseracher during the reading of the 57 articles coincided in 18 decisions.

In the third and final phase of this process, and after considering the reasons for the decision to include or exclude (when they were very focused on technical issues with little insight into organisational psychology, or very focused on digitalisation, or even applied to very specific situations such as a study focused on political impacts), the researcher decided to fully analyse 56 of the 131 articles resulting from the research carried out, focusing on the abstract, the methodology and the discussion of the results and the presentation of the conclusions. The decision criterion adopted by the researcher for the full analysis of the 56 articles was that the articles excluded by the researcher but considered by the second researcher would be analysed, and those excluded by the supervisor would be excluded, even though they were initially considered by the researcher. There were four exceptions regarding 4 articles, the first being that one of the articles was excluded by the researcher but was considered as a potential with doubts by the second researcher and the researcher's final decision was to exclude it because it was too focused on the organisation's policies (a topic that did not meet the research objective), the second exception was an article that was excluded by the second researcher because it was totally oriented towards the production area the researcher decided to analyse it because it touched on the topic of artificial intelligence in a shopfloor context. It should also be mentioned that this was the only article that mentioned psychology as one of the research areas, so even though it wasn't an initially defined criterion, we felt that this article should be considered. Finally, two articles were excluded because they were not available for free download.

Below we present the the systematisation of the bibliographic references considered in the literature review (Table 3.1).

It should also be noted that in our research, and in the articles selected for analysis, we were able to observe a high degree of comprehensiveness of the phenomenon in the different fields of knowledge, so we present some examples of the keywords identified by the authors and the areas of research involved. In fact, as can be seen in the table bellow, it was possible to get an overview of how these concepts are intertwined between the different fields of knowledge, and in this review, the area that is closest to Psychology, and that is mentioned very frequenctly (in more than 40 per cent of the articles), is the area of Business & Economics. In this review, Psychology is identified as a research field, alongside Engineering (Table 3.2).

3.3 Literature Review, Discussion of the Results

Knowledge Management is a vital process in organisations, so everyone must ensure that knowledge management practices are in place, which is why it is a shared responsibility (Ouédraogo & Rinfret, 2019).

As mentioned throughout this project, there are growing research trends that point to the importance of combining 'hard' Knowledge Management, which focused on organisational communication, and is based on technology, with 'soft' Knowledge Management, which is relational and organisational (Rizzi et al., 2017). These 'soft' and 'hard' factors characterise knowledge dynamics through (1) knowledge creation, (2) knowledge storage, (3) knowledge transfer and (4) knowledge sharing and promote the adoption of circular processes in inter-organisational strategies. The dynamics of Knowledge Management are crucial because enables (1) the alignment of technological, organisational, and social factors, (2) the involvement and cooperation of members and (3) the management of knowledge dynamics that include citizens, customors, institutions, increasing knowledge creation and contributing to the development of new business opportunities (Rizzi et al., 2017).

Thus, among the challenges posed to the Knowledge Management process, there is a new phase in the organisation of the industrial system, embodied in the concept

Table 3.1 Systematization of literature review references

Authors	Article title	Year published	Title source	Keywords
Lenart-Gansiniec, R	Organizational learning in industry 4.0	2019	Problemy zarzadzania-management issues	Industry 4.0; industry 4.0 implementation; learning organization; industry 4.0 challenges
Malik, N; Tripathi, SN; Kar, AK; Gupta, S	Impact of artificial intelligence on employees working in industry 4.0 led organizations	2022	International journal of manpower	Artificial intelligence; industry 4; 0; employee experiences; technostress; human resources management
De Azambuja, AJG; Plesker, C; Schützer, K; Anderl, R; Schleich, B; Almeida, VR	Artificial intelligence-based cyber security in the context of industry 4.0-a survey	2023	Electronics	Artificial intelligence; cyber security; industry 4; 0; machine learning; deep learning
Radanliev, P; De Roure, D; Nicolescu, R; Huth, M; Santos, O	Digital twins: artificial intelligence and the IoT cyber-physical systems in industry 4.0	2022	International journal of intelligent robotics and applications	Digital twin; industrial internet of things; cyber physical systems; human and robot interactions; industry 4.0; bibliometric analysis
Zeba, G; Dabic, M; Cicak, M; Daim, T; Yalcin, H	Technology mining: artificial intelligence in manufacturing	2021	Technological forecasting and social change	Artificial intelligence; bibliometric; content analysis; industry 4; 0; manufacturing
Chiarini, A; Cherrafi, A	Integrating ISO 9001 and industry 4.0. An implementation guideline and PDCA model for manufacturing sector	2023	Total quality management and business excellence	ISO 9001; Industry 4.0; quality management system; integrated model; PDCA
Noreen, U; Shafique, A; Ahmed, Z; Ashfaq, M	Banking 4.0: artificial intelligence (AI) in banking industry and consumer's perspective	2023	Sustainability	Banking sector; awareness; knowledge of AI technology; consumers
Xu, LD	Emerging enabling technologies for industry 4.0 and beyond	2022	Information systems frontiers	Industry 4; 0; artificial intelligence; 5G; 6G; quantum computing; emerging technology; enabling technology

(continued)

Table 3.1 (continued)

Authors	Article title	Year published	Title source	Keywords
Sassmannshausen, T; Burggräf, P; Wagner, J; Hassenzahl, M; Heupel, T; Steinberg, F	Trust in artificial intelligence within production management-an exploration of antecedents	2021	Ergonomics	Production management; cyber production management; trust; artificial intelligence; machine learning
Amjad, MS; Rafique, MZ; Khan, MA; Khan, A; Bokhari, SF	Blue ocean 4.0 for sustainability-harnessing blue ocean strategy through industry 4.0	2022	Technology analysis and strategic management	Blue ocean strategy; industry 4.0; six paths framework; operational excellence; business sustainability
Bag, S; Pretorius, JHC; Gupta, S; Dwivedi, YK	Role of institutional pressures and resources in the adoption of big data analytics powered artificial intelligence, sustainable manufacturing practices and circular economy capabilities	2021	Technological forecasting and social change	Big data; artificial intelligence; industry 4.0; circular economy; sustainable manufacturing
Rana, J; Daultani, Y	Mapping the role and impact of artificial intelligence and machine learning applications in supply chain digital transformation: a bibliometric analysis	2022	Operations management research	Supply chain digital transformation; industry 4.0; artificial intelligence; machine learning; bibliometric analysis; intelligent supply chains
Tseng, ML; Tran, TPT; Ha, HM; Bui, TD; Lim, MK	Sustainable industrial and operation engineering trends and challenges toward Industry 4.0: a data driven analysis	2021	Journal of industrial and production engineering	Sustainable industrial and operation engineering; industry 4.0; data driven analysis; fuzzy delphi method; bibliometric analysis
Hossain, S; Hassan, S; Karim, R	Assessment of critical barriers to industry 4.0 adoption in manufacturing industries of Bangladesh: an ISM-based study	2023	Brazilian journal of operations and production management	Industry 4.0 (I4.0); interpretive structural modelling (ISM); MICMAC; internet of things (IoT); artificial intelligence (AI)

(continued)

Table 3.1 (continued)

Authors	Article title	Year published	Title source	Keywords
Eslami, Y; Ashouri, S; Lezoche, M	Semantic interoperability and sustainability an industry 4.0 product life cycle issue	2021	Advances in production management systems: artificial intelligence for sustainable and resilient production systems, apms 2021, Pt Iv	Product life cycle; semantic interoperability; knowledge formalization; big data; data mining; sustainability models; industry 4.0
Abbate, S; Centobelli, P; Cerchione, R; Oropallo, E; Riccio, E	Investigating healthcare 4.0 transition through a knowledge management perspective	2023	Ieee transactions on engineering management	Medical services; systematics; knowledge management; solid modeling; medical diagnostic imaging; blockchains; big data; digital transformation; digitalization; enabling technologies; healthcare 4; 0; knowledge creation; knowledge management; SECI model; technological innovation
Ruiz-Real, JL; Uribe-Toril, J; Torres, JA; De Pablo, J	Artificial intelligence in business and economics research: trends and future	2021	Journal of business economics and management	Artificial intelligence; business; economics; bibliometrics; research trends; decision-making
Emmert-Streib, F	From the digital data revolution toward a digital society: pervasiveness of Artificial Intelligence	2021	Machine learning and knowledge extraction	Artificial intelligence; machine learning; data science; social data; natural language processing; industry 4; 0; ethical AI
Krzywdzinski, M; Butollo, F	Combining experiential knowledge and artificial intelligence. The digital transformation of a traditional machine-building company	2022	Management revue	Industry 4.0; manufacturing; innovation; ambidexterity; skill formation (JEL: J24, L21, L22, L64, O32)

(continued)

Table 3.1 (continued)

Authors	Article title	Year published	Title source	Keywords
Chen, J; Lim, CP; Tan, KH; Govindan, K; Kumar, A	Artificial intelligence-based human-centric decision support framework: an application to predictive maintenance in asset management under pandemic environments	2021	Annals of operations research	Artificial Intelligence; decision support; small and medium enterprises; predictive maintenance; asset management; pandemic preparedness
Dey, PK; Chowdhury, S; Abadie, A; Yaroson, EV; Sarkar, S	Artificial intelligence-driven supply chain resilience in Vietnamese manufacturing small- and medium-sized enterprises	2023	International journal of production research	Artificial intelligence; supply chain resilience; organisational factors; circular economy; agility; risk management
Lee, CS; Tajudeen, FP	Usage and impact of artificial intelligence on accounting: evidence from Malaysian organisations	2020	Asian journal of business and accounting	Artificial intelligence; accounting; AI; Malaysia; payable; qualitative
da Silva, CWB; Nascimento, JS; Ferraz, MA; Roberto, JCA; Soares, MC	Quality 4.0: emerging technologies and applications	2023	Revista De Gestao E Secretariado-Gesec	Quality 4; 0; emerging technologies; industry 4
Yaqub, MZ; Alsabban, A	Industry-4.0-enabled digital transformation: prospects, instruments, challenges, and implications for business strategies	2023	Sustainability	Digital transformation; digital technologies; industry 4; 0; business strategy; competitiveness; blockchain technology; big data analytics
Battaglia, D; Galati, F; Molinaro, M; Pessot, E	Full, hybrid and platform complementarity: exploring the industry 4.0 technology-performance link	2023	International journal of production economics	Industry 4; 0 technologies; complementarity; technology adoption; performance effect; expert interviews

(continued)

Table 3.1 (continued)

Authors	Article title	Year published	Title source	Keywords
Wolniak, R	The concept of operation and production control	2021	Production engineering archives	Operation management; industrial enterprise; industrial management; production control; operation manager; Industry 4.0; quality management
Ruggero, SM; Dos Santos, NA; Estender, AC; Da Silva, MT	Industry 4.0: expectations, impediments and facilitators	2021	Advances in production management systems: artificial intelligence for sustainable and resilient production systems, apms 2021, Pt Iv	Lean manufacturing; technology; automotive segment
Gupta, S; Justy, T; Kamboj, S; Kumar, A; Kristoffersen, E	Big data and firm marketing performance: findings from knowledge-based view	2021	Technological forecasting and social change	Big data analytics; artificial intelligence; marketing performance; knowledge-based view
Lin, LH; Wang, KJ	Talent retention of new generations for sustainable employment relationships in work 4.0 era-assessment by fuzzy delphi method	2022	Sustainability	Fuzzy delphi method; industry 4; 0; talent management; talent retention; work 4; 0
Aquilani, B; Piccarozzi, M; Abbate, T; Codini, A	The role of open innovation and value co-creation in the challenging transition from industry 4.0 to society 5.0: toward a theoretical framework	2020	Sustainability	Industry 4; 0; society 5; 0; open innovation; value co-creation; big data; artificial intelligence

(continued)

Table 3.1 (continued)

Authors	Article title	Year published	Title source	Keywords
Narwane, VS; Raut, RD; Yadav, VS; Cheikhrouhou, N; Narkhede, BE; Priyadarshinee, P	The role of big data for supply chain 4.0 in manufacturing organisations of developing countries	2021	Journal of enterprise information management	Big data analytics; digital supply chain; supply chain 4; 0; business performance; artificial neural network; structural equation modelling
Farbiz, F; Habibullah, MS; Hamadicharef, B; Maszczyk, T; Aggarwal, S	Knowledge-embedded machine learning and its applications in smart manufacturing	2023	Journal of intelligent manufacturing	Artificial intelligence; smart manufacturing; knowledge-embedded machine learning; physics-based model; data-driven model; prognostics and health management; digital twin; inductive learning; deductive learning; abductive learning
Arias-Pérez, J; Vélez-Jaramillo, J	Ignoring the three-way interaction of digital orientation, not-invented-here syndrome, and employee's artificial intelligence awareness in digital innovation performance: a recipe for failure	2022	Technological forecasting and social change	Not-invented-here syndrome; inhibitors of digital transformation; intelligent process automation; transaction cost theory; knowledge-based view; Employee's artificial intelligence awareness
Wang, R; Milisavljevic-Syed, J; Guo, L; Huang, Y; Wang, GX	Knowledge-based design guidance system for cloud-based decision support in the design of complex engineered systems	2021	Journal of mechanical design	Complex engineered system design; decision support; design guidance; complexity management; uncertainty management; knowledge management; computer-aided engineering; design for manufacturing; design representation; uncertainty modeling

(continued)

Table 3.1 (continued)

Authors	Article title	Year published	Title source	Keywords
Bajic, B; Suzic, N; Moraca, S; Stefanovic, M; Jovicic, M; Rikalovic, A	Edge computing data optimization for smart quality management: industry 5.0 perspective	2023	Sustainability	Human-cyber-physical systems (HCPS); big data analytics (BDA); industrial internet of things (IIoT); smart quality management; digital sustainability; data optimization
Ghobakhloo, M; Iranmanesh, M; Tseng, ML; Grybauskas, A; Stefanini, A; Amran, A	Behind the definition of industry 5.0: a systematic review of technologies, principles, components, and values	2023	Journal of industrial and production engineering	Digitalization; industry 5; 0; industry 4; 0 technologies; digital society; human centric; artificial intelligence
Verma, S; Singh, V	The employees intention to work in artificial intelligence-based hybrid environments	2022	Ieee transactions on engineering management	Creativity; employment; behavioral sciences; task analysis; artificial intelligence; collaboration; technological innovation; collaboration; componential theory of individual creativity; employee's behavior; human-cobot; hybrid workplace; valence theory
Li, PG; Bastone, A; Mohamad, TA; Schiavone, F	How does artificial intelligence impact human resources performance. evidence from a healthcare institution in the United Arab Emirates	2023	Journal of innovation and knowledge	HRM; AI; healthcare organization; performance measurement
Upadhyay, A; Balodi, KC; Naz, F; Di Nardo, M; Jraisat, L	Implementing industry 4.0 in the manufacturing sector: circular economy as a societal solution	2023	Computers and industrial engineering	I4; 0; circular economy; manufacturing sector; sustainable development; societal challenges
Turner, CJ; Emmanouilidis, C; Tomiyama, T; Tiwari, A; Roy, R	Intelligent decision support for maintenance: an overview and future trends	2019	International journal of computer integrated manufacturing	Machine learning; industry 4.0; E-maintenance; intelligent maintenance

(continued)

Table 3.1 (continued)

Authors	Article title	Year published	Title source	Keywords
Zeebaree, M; Ismael, GY; Nakshabandi, OA; Sattar, S; Aqel, M	Impact of innovation technology in enhancing organizational management	2020	Estudios de economia aplicada	Technology; innovation-technology; organizational management
Balog, K	The concept and competitiveness of agile organization in the fourth industrial revolution's drift	2020	Strategic management	Agile; Industry 4.0; VUCA; sharing economy; competitiveness
Carayannis, EG; Christodoulou, K; Christodoulou, P; Chatzichristofis, SA; Zinonos, Z	Known unknowns in an era of technological and viral disruptions-implications for theory, policy, and practice	2022	Journal of the knowledge economy	Industry 4; 0; industry 5; 0; Society 5; 0; risk management 5; 0; decentralized web; blockchain; internet of things
Lerch, CM; Heimberger, H; Jäger, A; Horvat, D; Schultmann, F	AI-readiness and production resilience: empirical evidence from German manufacturing in times of the Covid-19 pandemic	2022	International journal of production research	Artificial Intelligence; resilience; COVID-19 pandemic; industrial production; manufacturing
Madreiter, T; Kohl, L; Ansari, F	A text understandability approach for improving reliability-centered maintenance in manufacturing enterprises	2021	Advances in production management systems: artificial intelligence for sustainable and resilient production systems, Apms 2021, Pt I	Reliability; maintenance; human failure; text mining; Industry 4.0
Vuković, M; Thalmann, S	Causal discovery in manufacturing: a structured literature review	2022	Journal of manufacturing and materials processing	Manufacturing; causal discovery; machine learning; industry 4; 0; artificial intelligence

(continued)

Table 3.1 (continued)

Authors	Article title	Year published	Title source	Keywords
BAĞIŞ, M; Karagüzel, ES; Kryeziu, L; Ardiç, K	A longitudinal analysis on intellectual structure of human resources management: theoretical foundations and research trends	2019	Journal of mehmet akif ersoy university economics and administrative sciences faculty	Human resource management; theoretical foundations; research trends; longitudinal analysis
Biclesanu, I; Savastano, M; Chinie, C; Anagnoste, S	The role of business students' entrepreneurial intention and technology preparedness in the digital age	2023	Administrative sciences	Industry 4.0; entrepreneur; student; university; education; skill; personality; business; digital; technology
Arena, S; Manca, G; Murru, S; Orrù, PF; Perna, R; Recupero, DR	Data science application for failure data management and failure prediction in the oil and gas industry: a case study	2022	Applied sciences-basel	Predictive maintenance; data analysis; failure prediction; machine learning; data cleaning; time series
Presti, C; De Santis, F; Bernini, F	Value co-creation via machine learning from a configuration theory perspective	2023	European journal of innovation management	Digitalization; machine learning; industry 4; 0; servitization; configuration theory
Nourmohammadi, A; Fathi, M; Ng, AHC	Balancing and scheduling assembly lines with human–robot collaboration tasks	2022	Computers and operations research	Assembly line balancing; human–robot collaboration; multiple humans and robots; joint tasks; mathematical model; meta-heuristic
Kleinknecht, A	The (negative) impact of supply-side labour market reforms on productivity: an overview of the evidence	2020	Cambridge journal of economics	Varieties of capitalism; structural reforms of labour markets; innovation; productivity crisis
Adhiatma, A; Hakim, A; Fachrunnisa, O; Hussain, FK	The role of social media business and organizational resources for successful digital transformation	2023	Journal of media business studies	Knowledge absorptive capability; social media engagement; digital resources; organisational agility; business transformation

(continued)

Table 3.1 (continued)

Authors	Article title	Year published	Title source	Keywords
Lee, CY; Chien, CF	Pitfalls and protocols of data science in manufacturing practice	2022	Journal of intelligent manufacturing	Machine learning; data science; smart manufacturing; big data; prescriptive analytics; automating analytics
Caruso, L	Digital innovation and the fourth industrial revolution: epochal social changes?	2018	Ai and society	Smart factory; digital innovation; crowdworking; corporate narratives; digital economy; digital work
Queiroz, J; Leitão, P; Pontes, J; Chaves, A; Parra-Domínguez, J; Perez-Pons, ME	A quality innovation strategy for an inter-regional digital innovation hub	2020	Adcaij-advances in distributed computing and artificial intelligence journal	Digital innovation hub; innovation management; quality innovation; digital technologies

Table 3.2 Authors' keywords and research areas

Authors' keywords	Research areas
• Artificial intelligence; industry 4; 0; employee experiences; technostress; human resources management • Artificial intelligence; cyber security; industry 4; 0; machine learning; deep learning • Artificial intelligence; bibliometric; content analysis; industry 4; 0; manufacturing • Artificial intelligence; business; economics; bibliometrics; research trends; decision-making • Artificial intelligence; machine learning; data science; social data; natural language processing; industry 4; 0; ethical AI • Artificial intelligence; decision support; Small and medium enterprises; predictive maintenance; asset management; pandemic preparedness • Artificial intelligence; supply chain resilience; organisational factors; circular economy; agility; risk management • Artificial intelligence; accounting; AI; Malaysia; payable; qualitative • Artificial intelligence; smart manufacturing; knowledge-embedded machine learning; physics-based model; data-driven model; prognostics and health management; digital twin; inductive learning; deductive learning; abductive learning • Artificial intelligence; resilience; COVID-19 pandemic; industrial production; manufacturing • Production management; cyber production management; trust; artificial intelligence; machine learning • Banking sector; awareness; knowledge of AI technology; consumers • Product life cycle; semantic interoperability; Knowledge formalization; big data; data mining; sustainability models; industry 4.0 • Medical services; systematics; knowledge management; solid modeling; medical diagnostic imaging; blockchains; big data; digital transformation; digitalization; enabling technologies; healthcare 4; 0; knowledge creation; knowledge management; SECI model; technological innovation • Industry 4.0; economic complexity; patent data; knowledge accumulation strategy; relatedness; I4 technologies • Big data analytics; artificial intelligence; marketing performance; knowledge-based view • Artificial intelligence; smart manufacturing; knowledge-embedded machine learning; physics-based model; data-driven model; prognostics and health management; digital twin; inductive learning; deductive learning; abductive learning • Not-invented-here syndrome; inhibitors of digital transformation; intelligent process automation; transaction cost theory; knowledge-based view; employee's artificial intelligence awareness • Complex engineered system design; decision support; design guidance; complexity management; uncertainty management; knowledge management; computer-aided engineering; design for manufacturing; design representation; uncertainty modeling • Knowledge absorptive capability; social media engagement; digital resources; organisational agility; business transformation	• Business and economics • Computer science; • Engineering; physics • Business and economics; public administration • Business and economics • Computer science; engineering • Operations research and management science • Engineering; operations research and management science • Business and economics • Computer science; engineering • Engineering; operations research and management science • Engineering; psychology • Green and sustainable science and technology; environmental sciences; environmental studies • Computer science, information systems; telecommunications • Business; engineering, industrial; management • Computer science, information systems; engineering, electrical and electronic; telecommunications • Business; regional and urban planning • Computer science, artificial intelligence; engineering, manufacturing • Business; regional and urban planning • Engineering, mechanical • Business

of Industry 4.0 where we have the fusion of physical, digital and biological technologies (Da Silva et al., 2023, p. 12121). The fourth industrial revolution arises from the unsustainability of the production system, the technological synergies resulting from new hardware and software and the need to provide to developed countries with sufficient labour (Balog, 2020, p. 15). The concept of Industry 4.0, a term that emerged in Germany in 2011 due to the introduction of cutting-edge technologies as part of a government initiative to modernise the country's manufacturing sector, represents the ability of industrial components to communicate with each other (Pan et al., 2015, p. 1537). It represents a significant change in the way production is carried out and introduces intelligent technologies and systems. Industry 4.0 brings about significant change in the way industrial automation is produced (Da Silva, 2023).

In this way, technological changes and the need to constantly and immediately adapt to new realities and/or highly automated multifaceted contexts have greatly contributed to Industry 4.0 acquiring its status, 'promoting substantial gains in productivity and flexibility and transforming the nature of industrial work' (Junior & Saltorato, 2018, p. 743).

Specifically, Industry 4.0 includes a variety of technological solutions such as industrial automation, computerisation, digitalisation, robotisation, Big Data, Internet of Things (IoT), combining and aggregating various data in IT systems, which increase the efficiency and productivity of organisations (Lenart-Gansiniec, 2019, p. 97). In line with the author's analysis, Industry 4.0 can only become a reality if it is supported by the organisation's own guidelines, which should encourage individuals to look for new solutions and approaches, and thus for innovative, aligned behaviour that can lead to a common meaning and sense for everyone in the organisation. In fact, one of the ways of adapting to the new technological demands that is most evident in organisations is the promotion and development of training, which is essential for acquiring the skills and knowledge to keep up with the changes. There is a close relationship between learning processes and Industry 4.0, as it speeds up responses to emerging situations in everyday life and in the future and also helps in decision-making. Technological interventions are changing existing professional profiles and there is a need to acquire new skills and competences and to pay close attention to individuals in organisations who will certainly experience different situations and need to be monitored by managers (Malik et al., 2021).

Notwithstanding the identified items that can potentially benefit the Knowledge Management process due to their agility and increased decision-making capacity, there is another issue related to the development of technology that puts human action in organisations into perspective: **Artificial Intelligence**.

The essence of Artificial Intelligence isthe development of systems capable of performing actions that require human intelligence, automating repetitive tasks and processing large amounts of data, learning and adapting over time to improve performance (Da Silva et al., 2023, p. 12124). We can define Artificial Intelligence as a disruptive technology developed in the twentieth century and which encompasses concepts such as neural networks, machine learning or deep learning and which are concepts currently associated with terms such as digital marketing, decision-making, Industry 4.0, and digital business (Ruiz-Real, 2021, p. 14). In this sense, there is an intrinsic link with Industry 4.0 and the decision-making processes of managers who are challenged to live alongside the technological and human worlds. Artificial Intelligence could play an important role in supporting the management of organisations by providing interactive decision-making mechanisms shared between artificial and human intelligence (McAfee & Brynjolfsson, 2017). Artificial Intelligence can help improve decision-making by individuals in organisations, as it can increase understanding of certain events and overcome complex situations. There are positive impacts of Artificial Intelligence related to flexibility, creativity, innovation and performance.

On the other hand, we must be aware that this technology (AI) may cause some anxiety to workers, for example, due to constant connectivity, the variety of new applications, information overload and even a feeling of job insecurity due to the power of this new technological world (Malik et al., 2021). Indeed, whatever path we take in the field of Artificial Intelligence will be a multidisciplinary one, encompassing various sciences, as this technology raises questions in the fields of engineering, technology, maths, business, and ethics (Emmert-Streib, 2021, p. 294).

Regarding the link between AI and knowledge, we think it is worth mentioning that this technology can be a support in the Knowledge Management process, for example, as a technology capable of extracting knowledge that is encoded in unstructured documents, detecting anomalies, optimising solutions to complex problems, among others (Sigov et al., 2022, p. 3).

Organisations themselves need to be prepared for technological change and may need more or less time to adapt, depending on the maturity in terms of technological knowledge and exposure of individuals and work teams. More traditional organisations may need more time to adapt their work processes, equipment and, of course, to sensitise individuals and teams to the purposes of technological change, and may need to take more cautious and attentive steps in terms of their scale and complexity (Malik et al., 2021, p. 14).

Thus, some barriers to implementing Industry 4.0 in organisations have been identified, such as the challenges posed by the need for continuous development of technology, problems with data management and the quality of the data obtained, poor coordination on the part of leadership and mismatched skills on the part of individuals (Hossain et al., 2023, p. 4).

3.4 Data Presentation and Analysis of the Exploratory Interviews and Case study

The organisation that is the subject of the study, according to the data provided during the study carried out in 2021, is located in the north of the country, belongs to a multinational group and is a textile industry with more than 70 years of history dedicated to the development and production of textile solutions for various industries, being a partner in the group to which it belongs, in R&D activities (Research and Development) for new solutions (processes and products) and innovation and technology, being certified by the ISO9001 and ISO14001 standards. In 2021, it had sales of around 107.3 million euros, and its turnover was 41.6% in Portugal, 17.3% in Germany and 15% in Slovakia and the rest in the Czech Republic, France, China, Poland and the USA. In 2021, the company had 268 employees, 20% were in technical positions and 80% in operational positions directly related to the shop floor. The average age was around 38 and 23% were female. In terms of HR indicators, that around 68.8% of employees had a secondary or higher education qualifications, 18.8% had a university degree and the staff turnover rate was around 3.1%.

In order to answer the research questions of 'How do the processes associated with Knowledge Management take place in organisations in the era of Industry 4.0?' and 'How does Knowledge Retention and Transfer occur in the Knowledge Management process in the era of Industry 4.0? we assessed the participants' perceptions of Knowledge Management in Industry 4.0, from the strategic organisational level (macro), through the strategic operational level (meso) with intermediate functions supporting the strategy (leaders and non-leaders) to the operational execution level on the shop floor.

The data is presented and analysed in a specific way according to the technique by which it was obtained (e.g. when analysing the interview, the data obtained through the interview is analysed).

Thus, in an initial approach, the exploratory interview was applied with the aim of exploring the perception of the phenomenon in a global way, by sector of activity and without any intentionality regarding the data to be obtained from the participants interviewed, which was the only technique used, collecting testimonies from managing directors, department heads and senior specialists.

As far as the case study phase is concerned, the data collection techniques were associated with the different organisational levels, *i.e.*, the interview was used at the strategic organisational level, the focus group considered for the strategic operational level, and the questionnaire survey applied to the operational execution level, the shop floor. Our decisions are illustrated (Table 3.3).

3.4.1 Data Processing and Sample Characterization

The data analysis takes into account the answers to the questions related Knowledge Management to Industry 4.0 and/or adaptation to technological innovation in the context of new technological challenges associated to upgrading in organisations.

The data from the interviews and focus groups were processed using the NVivo software and the data from the questionnaire survey were managed using the SPSS software. In all situations (interviews, focus group and questionnaire survey), the data obtained on Knowledge Management in Industry 4.0 will be analyzed in a descriptive way. Our analysis will accompany the data obtained during the interviews, focus

Table 3.3 Allocation of the data being analyzed

Tecnhical	Project phase	Sector	Instrument code I field	Data in analysis
Interview	Initial	Industry	IW1_IND_DIR I Humans relations IW2_IND_HDS I Humans relations	Answers to question 4 and answers to questions 6, 7 and 8 (interviews in Portuguese)
		Services	IW3_IND_CDA I Engineering	
		Reseach	IW4_IND_CDA I Biology	
Interview	Case study	Industry	IW5_IND_DIR I Human relations IW6_IND_DIR I Product Industrialization IW7_IND_DIR I Production IW8_IND_ADM IAdministration IW9_IND_DIR I Operations and logistic IW10_IND_DIR I Industrial engineering IW11_IND_DIR I Maintenance engineering IW12_IND_DIR I Quality IW13_IND_DIR I Controlling IW14_IND_ DIR I Production (foreign participants) IW15_IND_ DIR I Logistic (foreign participants)	Answers to questions 6, 7 and 8
Focus Group	Case study	Industry	FG1_IND; FG2_IND; FG3_IND I Controlling; logistics; product industrialization; maintenance engineering; industrial engineering; purchasing; production; quality, human relations	Answers to question 4
Questionnaire survey			QS1 I Production-impregnating	Answers to questions 3 and 4

groups and questionnaire survey with different approaches and reflections in order to enrich the reading.

We therefore decided that the most appropriate strategy for analysing the data would be Thematic Analysis, 'Template Analysis' (King et al., 2012), which allows the data to be structured into themes and sub-themes closely related to the concepts under study. To this end, the data was 'coded', based on its relationship to the themes/

Table 3.4 Items considered in data analysis

Interview questions Portuguese	Interview questions English
6. Do you think there is any relationship between knowledge management and organisational strategy? If so, which one(s)? In what way? 7. How important do you think knowledge management is in the new and different ways of managing work in the reality of revolution 4.0? 8. What relationship do you see between innovation and knowledge management?	4. How big of an impact do you think knowledge management will have in new and different forms of labor management in the revolution 4.0 reality?
Focus group	**Questionnaire survey**
4. What does it mean to you to be in the era of Industry 4.0? And what impact is this new era having on your function?	3. What is Industry 4.0? 4. Can everyone adapt to technological innovation?

concepts under study, and organised in a meaningful way, i.e. what it reveals about the participants' perceptions and/or experiences.

It should be noted that, in this chapter, the treatment of the data allows us to obtain a different analysis depending on the technique, i.e., in the case of the interviews, the focus was on the participants ' strategic perception of the relationship between knowledge management and the reality of Industry 4.0; in the case of the focus groups, the data reflected the opinion of the participants at a level close to the function, i.e. more operational, due to the influence of Industry 4.0 on their function, and in the case of the questionnaire survey, the orientation is clearly operational, where we can obtain frequencies of the participants' understanding of the concept of Industry 4.0 and the adaptation to the technological innovation resulting from the I.4.0 era.

Below we can see the questions that triggered the data obtained by the different data collection techniques, i.e. in the interviews, focus groups and questionnaire surveys, and which were considered in the analysis (Table 3.4).

The sample takes into account the characteristics of the target population of the entire research project, from the initial approach to the phenomenon of Knowledge Management, to the case study. The sample included the 15 interview participants (national and foreign participants), the 17 participants in the 3 focus groups held and the 20 participants in the questionnaire survey. Below we present some of the characteristics of the total sample, namely function and educational qualifications (Table 3.5).

3.4.2 *Data Analysis*

By analysing the data obtained throughout the research project, in its initial exploratory, and case study phases, we were able to see that the participants recognized that Knowledge Management in Industry 4.0 is a very important, controversial, and equally challenging topic for organisations. After deciding on the data collection

Table 3.5 Sample summary information

Sectors	Initial approach participants	Case study participants	Function	Academic qualification
Industry	–	1	Administrator	College = 100%
	3	8	Managing directors	College = 80% High School = 20%
	1	17	Heads of department/area Senior specialists	College = 76% High school = 24%
		20	Specialised operators and technicians	High school = 35% High school = 56% College = 9%
Services	1	–	Heads of department/area Senior specialists	College = 100%
Research	1	–	Heads of department/area Senior specialists	College = 100%

techniques to be used, the respective instruments were designed to help operationalise the collection process itself. The techniques to be used at each stage were meticulously defined, taking into account the purpose of the research, the target population, and the objectives specific to each moment. Thus, and aware of the importance of the diversity that different techniques can bring to data collection, we considered the interview to be the most appropriate for collecting data at the macro level, from managers and directors, insofar as it would allow us to capture the strategic understanding of the phenomenon at the highest hierarchical level. The focus group, due to its interactive nature, would make it easier to capture the theme in the population at the meso level, the middle managers, and the questionnaire survey, which due to its practicality on the shop floor, would collect evidence at the operational level. The data from the fully transcribed interviews were processed using NVivo software and the average planned duration of each interview, which was the actual duration, was 40 min.

3.4.2.1 Interviews

Knowledge Management in I4.0

Organisations are waking up to this new reality of Knowledge Management, and the participants pointed out that the use of Industry 4.0 technologies enables successful

Knowledge Management by allowing information to be easily accessed and effectively transferred. Also, this was possible when the contingency period resulting from COVID 19 was experienced.

> Industry 4.0 is very much in vogue right now, and it's a fashionable topic, so to speak. (...) Today, since we can't get together, we use teams, we create a channel and post it on the channel and then there's a person who will process this information and then pass it on to actions... Each person gives their input and good ideas emerge and are applied. Industry 4.0 is also at work here. It uses the internet in a way that adapts to the times we're going through. Now this knowledge is recorded more officially, and everyone can access it (...) the advantages I see in Industry 4.0 is also transparency in information, which is something I really like (...) in IW10_DIR_IND.

Technological Innovations

Industry 4.0 is seen as a process of natural industrial revolution, linked to social development that cannot be ignored. It could be a new way of learning and transforming knowledge and represents a major challenge for organisations, particularly in terms of innovation. On the other hand, technological innovation, due to the possibility of making a large amount of data and information available in a short period of time, increases awareness in organisations of the difficulty in managing this potential knowledge, both because of the quantity and the limited, finite time available to do so.

> (...) That's why 4.0 is something we have to keep our eyes on... it's an opportunity and a new way of learning and transforming knowledge. in IW1__IND_ DIR

> It's a big challenge, but we also embrace it, so we have an area more linked to innovation and development, research (...). In IW_3_SERV_HDS

> We have an initial difficulty, which is knowing what's important. We have access to so many different things and because of our time... it's finite... knowing what we can collect for ourselves. And then we have to have a filter to know what we're going to work on... it's very challenging! I think that knowledge drives innovation when we manage to manage what we know and what others know is almost an engine for us to have more information. in IW6_IND_DIR.

Artificial Intelligence

Knowledge Management using Industry 4.0 technology, such as Artificial Intelligence, can make it possible to add other correlations between data that are much more complex to execute, interconnections that will make it possible to segregate large amounts of data, control details, define priorities and ensure the efficiency and effectiveness ofactivities.

> (...) and e.g. the XXXX system is Ind. 4.0 for me, one of the advantages is that the information is available on site next to the machine, using the tablet to access a manual or an electrical diagram (...) in IW10_IND_DIR.

> (...) learning every day and now we're looking for special technology in the end that we'll create lot of transparency (...) which help us to face the upcoming challenges and sharing is this case is also a way to develop your knowledge because the others will help you to identify gaps...because you may not to see all the gaps yourself...it's a way how you can

> further improve...and collaborating with other systems and processes models...and collecting data from everywhere (...) in IW15_IND_DIR.
>
> (...) What happens is that the laboratory notebooks don't do calculations, they don't integrate data, they don't do anything (...) at the end of the day all we process and our research and our knowledge involve transferring data from the notebook to the computer and putting all the statistics in there... in IW4_ INV_HDS.

Collaborative Technologies

Our data also mentioned the importance of machine-human collaborative technologies (e.g. robots, CoBots), to ensure optimization and adaptation to organisational needs and changes.

> All the themes that are being addressed in Industry 4.0 they usually tend to eliminate that category of people who only know how to assemble and disassemble, load and unload and who don't want to do anything else... those are the operations that are going to be automated...Here in this field we're trying much more to encourage collaborative work between machine and person, trying to eliminate burdens and smooth out operations that are difficult, and we're trying to develop robots, CoBots and non-robots so that people's time is freed up and they can start taking on tasks that are more linked to the act of deciding...and not just the act of executing...and one of the most important things about Industry 4.0 is for our company and our production to adapt to the conditions of the moment. in I_W8_IND_DIR

Development of Individuals

Another topic is related to the opportunity for individuals to expand the scope of their roles, as the technologies of the Industry 4.0 era able to perform routine tasks,freeing and stimulating individuals to develop other activities and acquire new skills.

> (...) an example of an operator who, in a typical traditional role, needed to handle raw materials and equipment and needed to work harder, to make a more physical effort for the moment we're going through, since it's transforming these operations into machines, in the fact that it's machines doing these operations and the operator himself has to have other digital control, process and IT skills (...), the ability to think, to perceive the whole, to be more of a manager of his process than just a micro area.(…) the ability to think, the ability to see beyond what is their more micro area, to have a perception of the whole, to be more of a manager of their process than just an operator who handles a set of materials and equipment. in IW2_ IND_ HDS
>
> (...) we have operators from the shopfloor who were doing some of the test of department management did 20 years ago, they are analysing their attendances procedures, they are analysing the XXX rate , what their productivity was yesterday...and kind of reporting that up through management...that's a big change and a technological announcement of changing the thinking and empowering the people that they can do the biggest difference ...I don't run any machines...I can improve quality...but they can... so why don't we equip them with the information and the tools they need ... in IW14_DIR_IND

Obstacles

It is also interesting to note that the interviewees have a clear perception that there is a road ahead with human and material obstacles to overcome. For Industry 4.0 to become a reality in organisations, there are necessary behaviours to adopt, such as openness to change and motivation to learn, as well as the preparation of organisational processes, such as productive ones related to the simplification of activities and operations.

There is an awareness that this is a process of behavioral change in the making, but it is essential that automation processes are understood and accepted, ensuring a balance between machine and human action.

> I think that this path to Industry 4.0 is still a process in the making, although if I wanted to 'sell' this idea I could, because we have very advanced systems, but behaviour has yet to keep up with this change. in IW5_IND_DIR

> First of all, I think that the mindset of people has to be open, they should not be afraid of changes and that is something I've really seen for all my career (...) people have to see the change rather as an opportunity rather than as a threat (...) you are dealing with fear with psychology (...) and you have to be able as a leader to leading change and to take the fears away from the people (...). in I_14_DIR_IND

On the other hand, to be in the realty of I4.0 means to have simplified processes and operations.

> Look, I'm a bit critical of this, not because of the concept itself, but more because I don't know if people really understand what digital 4.0 means. For me, I may be thinking wrong, but that's how I think... and that's how I conceive things... I can't think of 4.0, or 5 or 10 or 15... If my processes, my operations, my activities aren't well consolidated and simplified... in IW7_IND_DIR

3.4.2.2 Focus Group

Impact of I4.0 Technologies

When analysing the data obtained in the focus groups, we can see that the participants generally consider that the new technological era is seen as something that will cause some unrest and instability in individuals and teams, mainly related with fears and lack of knowledge about I4.0.

The participants believe that the new technologies will make it possible to change the organisational strategy, influencing management to make decisions to reduce the number of employees for activities that can be provided by the new technologies. On the other hand, they realize that these could be new opportunities to develop new functions and create new "jobs".

> (...) The idea I have is that many people think that with industry 4.0 a lot of jobs will be lost because it will automate or digitize a lot of things, but they forget that a lot of other jobs will be created that don't exist today...for data processing, analysis and obtaining information, which at the moment is breaking down the barrier of what people don't know and think it will be and what it will be...having that balance of trying to meet industry 4.0. in FG2_IND_HDS

Knowledge Management and Obstacles

It is mentioned that sometimes, as a reaction and protection mechanism, the individual can influence the Knowledge Management process by adopting a position of not sharing knowledge.

> (...) people are a bit afraid of feeling that they are being replaced, often in this case by machines and automated processes, in Industry 4.0. Maybe they're even afraid to pass on some knowledge to other people (...). in FG1_ HDS

> However, it is worth emphasizing that the participants emphasize the uniqueness of knowledge in individuals and teams, and that it is complex to attribute human characteristics to machines, such as sensitivity to obstacles and creativity.

> (...) I think there has to be a balance, I agree, we talk about experience, and we can't give machines the sensitivity that people have... in FG_3_ HDS

Data and Artificial Intelligence

If, on the one hand, the large amount of data made available by the introduction of new technologies can be an obstacle to Knowledge Management in this technological context, especially if the organisation does not yet have the necessary conditions for managing this potential Knowledge, on the other hand, the importance of new technologies in supporting knowledge management is undeniable, due to the automation processes, AI, the ability to separate, transform and make accessible the available data for processing.

> (...) the first big revolution is going to be in data, because with Industry 4.0 the amount of data accessible to everyone is going to be so large that if you don't have the capacity to process it, it won't mean anything...a concrete example, we have XXX, everything is there, but if you don't know it's there...what we foresee in the future is that you'll have everything available everywhere, in other words, you can have everything there, but if you don't know how to get it there...you'll carry on doing what you've done up until now... in FG3_ IND_ HDS

> (...) in some things, yes, we've already felt it here in the company ...speaking when I was more in the technical area, many of the things we did, or we were dependent on the knowledge of a person for it to work well...we passed that on to the machine...we automated the machine in a certain way so that it always works well...so there we have an improvement...Machine learning... in FG3_IND_HDS

> (...) Power BI is one of the most talked about tools of the moment, it symbolizes Industry 4.0 and for me it also symbolizes some knowledge transfer. So, from the moment we have the raw data, whatever it is, and we start to analyse it, and from then on we transform the data and manage to interpret it (...) in FG1_IND_ HDS

The Role of Leadership

According to the technical experts who participated in the focus groups, the changes associated with Industry 4.0 must be accompanied by leaders who can play a role in unblocking fears and facilitating the definition of new paths and opportunities for the future of individuals and work teams.

> (...) I think the manager also plays a very important role within the team... (in preparing for change) ... in FG2_IND_HDS

What I feel is that people, especially older people, have a harder time with Ind.4.0 and it's going to be extremely overwhelming in psychological terms... in FG2_IND_HDS

For an overview of the data described during the interviews and focus group, below is a summary of the information.

3.4.2.3 Questionnaire Survey

The survey by questionnaire made it possible to collect of data on the understanding of the phenomenon of Knowledge Management in the era of Industry 4.0 by the population at the operational level, on the shop floor, complementing the analysis carried out through interviews and focus groups. For this analysis, the data obtained from the responses listed below were considered (Table 3.6).

In the table below, we present the data with the percentage of responses broken down according to the structure previously outlined. The data from the questionnaire survey can be consulted (on request to the research team) in its various forms of presentation, namely in pie charts and histogram (Table 3.7).

According to the results presented, in termos of workers' understanding of the topic Industry 4.0, there is a strong desire to be informed about what will happen in the future, with 35% of responses on this item. In addition, Industry 4.0 is understood as a revolution that will automate all processes and connect all data, with 30% of responses. The perception that these changes are already happening was not significantly evident, with only 5% of responses. Regarding the need to adapt to Technological Innovation, it is noteworthy that 40% of respondents believe that everyone can adapt, as long as they are trained to meet the demands of this Technological Innovation, placing the responsibility on the organization. Twenty five per cent of these participants are willing to learn about the most innovative equipment, with technology already being part of their daily routine.

Table 3.6 Specific items from the questionnaire survey analysis

Questionnaire survey
3. What is Industry 4.0?
(A) It's the 4th industrial revolution that will automate all processes and connect all data
(B) It's what's already happening in our company with the intelligent equipment that can replace us
(C) I don't know what it is. I'd like to know more, and have it explained to me what's going to happen in the future
4. Can everyone adapt to technological innovation?
(A) Yes, everyone can adapt, as long as they are trained to meet the demands of this technological innovation
(B) No, not everyone will adapt. The younger ones will, because they were born in the computer age
(C) I'm willing to get to know the most innovative equipment. Technology is already part of my day-to-day life

Table 3.7 Analysis of questionnaire survey responses

Questionnaire survey	% of responses
3. *What is Industry 4.0?*	
(A) It's the 4th industrial revolution that will automate all processes and connect all data	30,0
(B) It's what's already happening in our company with intelligent equipment that can replace us	5,0
(C) I'd like to know more, and have it explained to me what's going to happen in the future	35,0
(D) I don't know what it is	30,0
4. *Can everyone adapt to technological innovation?*	
(A) Yes, everyone can adapt, as long as they are trained to meet the demands of this technological innovation	40,0
(A) Yes, everyone can adapt, as long as they receive training to meet the demands of this technological innovation; (C) I'm willing to learn about the most innovative equipment. Technology is already part of my daily life	25,0
(B) No, not everyone will adapt. The younger ones will, because they were born in the computer age	5,0
(B) No, not everyone will adapt. The younger ones will, because they were born in the computer age; (C) I'm willing to get to know the most innovative equipment. Technology is already part of my daily life	5,0
(C) I'm willing to learn about the most innovative equipment. Technology is already part of my daily life	25,0
(D) I don't know if I'll be able to adapt	0,0

Thus, we can reflect that a significant percentage of the shop-floor population (from the sample) feels motivated to receive the technological changes characteristic of Industry 4.0 and is willing to experience the associated change processes, considering that the role of the organization will be essential to ensure successful conditions, such as the development of training to acquire the necessary skills.

Concluding note and directions for the future

In analyzing the data gathered in the research project, which focused on the concepts of Knowledge Management and Industry 4.0, we were able to assess that organizations in different sectors are aware of the urgent need for change in the way Knowledge is managed and in understanding its importance in responding to the challenges of this new reality, particularly due to the technological advances it brings, which make it unpredictably.

In fact, there is a common thread running through the testimonies gathered, which highlights that the future lies in considering the industrial revolution as a reality that is an integral part of our work routines and, as such, is imminent in the interactions between individuals and teams, between custmores and suppliers, and across different national and international markets. It is undeniable that this unprecedented change is already affecting the current state of organizations and will shape future planning, leading to new forms of Knowledge Management. These new forms could directly involve the use of emerging technologies, such as Artificial Intelligence, which encompasses concepts such as neural networks, machine learning, or deep learning, and are currently associated with terms like digital marketing, decision-making, Industry 4.0, and digital business (Ruiz-Real, 2021, p. 14).

In general, Industry 4.0 is seen by participants as a revolution that has changed the relationship between humans and technology, but it should be approached as a learning process with innovation potential through continuous research and development of new solutions for people's lives.

Regarding the phenomenon of Knowledge Management in Industry 4.0, it is often mentioned that the process of Knowledge Transfer could become more transparent, agile, efficient, and effective, as data, information, and knowledge become visible, measurable, and accessible in repositories and databases that can store vast amounts of information. The Knowledge Management process could benefit from this promised, and in some cases already proven, efficiency and effectiveness in resource management. Therefore, given the vast amount of data and information available, it is expected that fewer resources, including time, cost, raw materials, and people, will be required, thus increasing the likelihood of achieving the desired outcomes. At this point, it is important to highlight that technology brings with it concerns about the reduction of human resources, and human resource managers and organizational psychologists will have an significant role to play in guiding this new path, which, while natural, should be balanced and offer integrative solutions for human development. This could involve, for example, broadening areas of competence and enabling the performance of more stimulating activities to be carried out, increasing employee involvement in decision-making. Knowledge Management is somewhat presented as a method (Mohammed et al., 2018a, 2018b) that actively highlights expertise and knowledge, creating real value for organizations (Chen & Huang, 2009). This reality has been observed not only through the data collected from the participants but also through the literature review, which reinforces the importance of integrating knowledge management into various fields of knowledge, from social sciences to engineering and computer science.

The technological power and strength of the new industrial era in today's society is undeniable. Technological changes and the constant need to adapt immediately to new realities and/or multifaceted, highly automated contexts have greatly contributed to Industry 4.0 gaining its status, "promoting substantial gains in productivity and flexibility and transforming the nature of industrial work" (Junior & Saltorato, 2018, p. 743). It is worth mentioning the importance of having organizational preparation for the the use of data, the processing of information, and its transformation into potential Knowledge. Indeed, it is essential to create the conditions that serve as a source of motivation for embracing technological changes, as organizations will undoubtedly experience differentiated situations that require support of managers (Malik et al., 2021). This is confirmed by the data, which highlight the importance of the role of leadership as a guide in organizational change.

In organizational terms, systems can become more agile, increasing the ability to respond to the constant demands of the market, enhancing competitiveness, as in the case of the automation of certain tasks. Industry 4.0 technology is mentioned by as an advantage by the participants in the research project because it enables the correlation of complex data, which is essential for defining activities and actions to tbe aken at the moment, as well as supporting sustainable decision-making for the future.

In addition, we observed that the data shows that the technological processes associated with Industry 4.0 could provide individuals and teams with a time gain, as humans could be freed from more routine tasks and thus devote themselves to other activities, such as continuous improvement and quality control, being better prepared for new challenges. There is an obvious concern about maintaining a balance between the contribution of technology and human action, as it is often not possible to replace the human element.

The data also revealed that the role of leadership within organizations is crucial in guiding and facilitating openness to the necessary changes, aligning strategy with operations, and ensuring transparency in communicating of goals to ensure alignment with organizational culture. We believe that this topic is a potential avenue to explore and develop in future research on the phenomenon under study.

From the point of view of preserving knowledge within the organization, and given the new demands of the present and the future, we understand that it is essential to adapt and/or reinvent the process of Knowledge Management, which may involve the development of a specific technological tool for this organizational process, perhaps with the use of Artificial Intelligence.

Conclusion and Final Remarks

Finally, we wanted to reflect on the contributions and limitations of the research project. The aim was to analyse the phenomenon of Knowledge Management in Industry 4.0 from different perspectives and by different organisational actors. We wanted to find answers to the research questions of 'How do the processes associated with Knowledge Management take place in organisations in the era of Industry 4.0?' and 'How does Knowledge Retention and Transfer occur in the Knowledge Management process in the era of Industry 4.0?', processes that are fundamental to understanding the phenomenon and designing new organisational realities. And also to find clues or data that could confirm or disprove the hypotheses raised: (1) 'The Individual is a generator of knowledge, due to their experience as a professional and person in knowledge management within the organisation, in the era of Industry 4.0', (2) 'The Knowledge Retention and Transfer Processes adopted by the organisation influence the positioning and actions of Individuals in terms of the ways in which they use Knowledge', and (3) 'Knowledge Management, on the part of the organisation, is perceived as having an influence on organisational success and/or failure.'

In an initial approach, a chapter was dedicated to the topic 'Transferring and Sharing Knowledge in Organisations', with the aim of broadenbing the field of research in a first contact with the topic of Knowledge Management by exploring the perceptions of some players from organisations in the Industrial, Service and Research sectors.

The data collected confirmed the researchers interest and motivation to study a real case that was presented as 'Knowledge Management: A Case in Industry' where it was possible to find out, in a specific industrial context, how Knowledge Management takes place in an industry in the 4.0 technological era.

In the end, we wanted to understand the phenomenon of Knowledge Management in direct relation to the reality in Industry 4.0, and therefore developed an analysis based on the evidence obtained from the entire sample of the research project, whichthat is reflected in the current chapter 'Industry 4.0: Knowledge Management with Artificial Intelligence?'

The phenomenon of Knowledge Management is transversal to the organization, in which the individual plays a crucial role, as a leveraging element of Knowledge and Knowledge Management in the era of Industry 4.0.

In this journey, we believe we have made contributions to the literature, to organisational psychology and human resources, and to the target organisation of the case study.

Contributions to the Literature

Throughout the research project, literature reviews were carried out which, while on the one hand, provided an essential theoretical framework for understanding the phenomenon under study and, on the other hand, showed that there is still the possibility of substantially opening up the field of research with regard to the position and contribution of individuals in the knowledge management process in the era of Industry 4.0, and whether they add value to knowledge management within the organisation in the era of Industry 4.0.

Through our study, we were able to confirm and explain, based on the evidence obtained, that the participants were unanimous in their view that the individual is a knowledge-generating element, due to their experience as a professional and as a person, in knowledge management as a knowledge manager within the organisation in the era of Industry 4.0. In this context, we cannot fail to highlight the strategic importance of constant interaction in the Knowledge Management process, specifically in the acts of Knowledge Transfer and Sharing, and that this reality implies unequivocal participation and motivation by all the parties involved, i.e. the organisational and the individual/group (Liyanage et al., 2009; Paulin & Suneson, 2012).

In the context of the technological revolution, it was possible to conclude that the individual is a fundamental actor in organisations, and that technology cannot replace human nature in terms of critical analysis, creative and innovative thinking, the ability to deal with the unexpected, emotional management and the resolution of various problems of a professional and personal nature.

It was found that the Knowledge Retention and Transfer processes adopted by the organisation have a positive influence on the position and actions of individuals in terms of how they use knowledge, if the organisation manages to create the conditions for knowledge retention to take place by providing ways of learning content and recording and applying lessons learned, as we explained in the case study.

On the other hand, the way in which Knowledge Management is carried out by the organisation is perceived to have an influence on organisational success and/or failure. Thus, it was possible to ascertain from the data obtained, i.e. the participants' perceptions, that when knowledge is retained and transferred between individuals and work teams, organisational success increases to the extent that it boosts the success of the team's operation in terms of the process of sharing knowledge between individuals, with gains in learning time, problem-solving and proposals for improvement. On the other hand, 'lame' knowledge management in terms of structure and strategy leads to demotivation and disorientation in the functioning of teams and a high probability of misalignment with organisational objectives.

Contributions to Organisational Psychology and Human Resources

The literature review has shown that the phenomenon under study, Knowledge Management in Industry 4.0, is a topic of interest to various areas of knowledge, such as Economics, Systems Engineering, Computer Science, Industrial Engineering, Manufacturing Engineering, Management, Physics, Human Resources, Psychology, among others. If this is not the case, let's look at examples of this ilink in the studies carried out in the field of Manufacturing Engineering by designing intelligent technological systems and equipment, using Artificial Intelligence, and, intrinsically, Knowledge and Knowledge Management in organisations. As a result, we have, for example, Cobots-collaborative robots designed to work alongside humans and to carry out certain tasks of high precision and/or technical complexity, such as moving loads. The phenomenon is thus perceived and captured in different dimensions and is applied in each of these areas of knowledge, revealing its nature of eclectic social applicability.

We can't fail to mention that in this process of delving deeper into the literature, we found that Organisational Psychology is one of the least present fields in terms of analysis and reflection on the phenomenon, or at least our research did not return confirmation of research in this area by Psychology, which we believe is an interesting point for reflection and development in future research.

Capturing individuals' perceptions of the concepts of Knowledge, Knowledge Management, Knowledge Retention and Transfer and Industry 4.0 through the use of different data collection instruments will make it possible to increase the scope of knowledge about the phenomenon of Knowledge Management in the current technological era.

In this way, we believe that our study will add value by presenting the evidence gathered, which will allow us to analyse reality, particularly with regard to the importance of organisational communication in Knowledge Management, as well as the role of leadership in guidance and action, and also the influence of technology on work design, hoping to contribute another drop of knowledge to be highlighted by Organisational Psychology.

Throughout our project, we were able to conclude that organisations must necessarily wake up to the importance of Knowledge and Knowledge Management in this new era of Industry 4.0, which requires a more agile and attentive approach, given the volatility of situations and organisational changes.

We must also point out that, although this is not a concept analyzed in our project, we have seen that organisations are still adapting to the challenges of Industry 4.0, Japan (in 2016) has extrapolated the concept of Industry 4.0 to a new model of society, Society 5.0, which comes from the digital transformation of the I4.0 era. Society 5.0 is seen as a beneficiary of the I4.0 model and is considered as a social revolution that will make it possible to put technologies, intelligent systems, at the service of humans to improve their quality of life, leaving the focus on industries, unlike what happens in the Industry 4.0 revolution (Guimarães et al., 2019). The still scarce studies on the subject indicate that the Society 5.0 model will consider, among other things, the development of solutions for ageing, human longevity, disaster forecasting and solutions.

References

Abbate, S., Centobelli, P., Cerchione, R., Oropallo, E., & Riccio, E. (2023). Investigating healthcare 4.0 transition through a knowledge management perspective. *IEEE Transactions on Engineering Management, 70*(5), 3297–3310. https://doi.org/10.1109/TEM.2022.3200889

Adhiatma, A., Hakim, A., Fachrunnisa, O., & Hussain, F. K. (2023). The role of social media business and organizational resources for successful digital transformation. *Journal of Media Business Studies, 21*(1), 23–50. https://doi.org/10.1080/16522354.2023.2203641

Ahmad, A., Bosua, R., & Scheepers, R. (2014). Protecting organizational competitive advantage: A knowledge leakage perspective. *Computers & Security, 42*, 27–39. https://doi.org/10.1016/j.cose.2014.01.001

Alashwal, A. (2018). Developing a knowledge integration conceptual model for sustainable construction projects: literature review. In *Proceedings of the ASEAN Post Graduate Conference (APGC).*

Alavi, M., & Leidner, D. E. (2001). Review: Knowledge management and knowledge management systems: Conceptual foundations and research issues. *In Source: MIS Quarterly, 25*(1).

Amjad, M. S., Rafique, M. Z., Khan, M. A., Khan, A., & Bokhari, S. F. (2024). Blue ocean 4.0 for sustainability–harnessing blue ocean strategy through industry 4.0. *Technology Analysis & Strategic Management, 36*(4), 797–812. https://doi.org/10.1080/09537325.2022.2060072

Anderson, J. R. (1995). *Cognitive psychology and its implications* (4th ed.). W. H. Freeman and Company.

Antal, A. B. (2000). Types of knowledge gained by expatriate managers. *Journal of General Management, 26*(2), 32–51. https://doi.org/10.1177/030630700002600203

Aquilani, B., Piccarozzi, M., Abbate, T., & Codini, A. (2020). The role of open innovation and value co-creation in the challenging transition from industry 4.0 to society 5.0: Toward a theoretical framework. *Sustainability* (Switzerland), *12*(21), 1–21. https://doi.org/10.3390/su12218943

Arena, S., Manca, G., Murru, S., Orrù, P. F., Perna, R., & Reforgiato Recupero, D. (2022). Data science application for failure data management and failure prediction in the oil and gas industry: A case study. *Applied Sciences* (Switzerland), *12*(20). https://doi.org/10.3390/app122010617

Argote, L., & Ingram, P. (2000). Knowledge transfer: A basis for competitive advantage in firms. *Organizational Behavior and Human Decision Processes, 82*(1), 150–169.

Arias-Pérez, J., & Vélez-Jaramillo, J. (2022). Ignoring the three-way interaction of digital orientation, not-invented-here syndrome and employee's artificial intelligence awareness in digital innovation performance: A recipe for failure. *Technological Forecasting and Social Change, 174*, 121305. https://doi.org/10.1016/J.TECHFORE.2021.121305

Arshad, R., & Ismail, I. R. (2018). Workplace incivility and knowledge hiding behavior: does personality matter? *Journal of Organizational Effectiveness: People and Performance, 5*, 278e288. https://doi.org/10.1108/JOEPP-06-2018-0041

Bag, S., Pretorius, J. H. C., Gupta, S., & Dwivedi, Y. K. (2021). Role of institutional pressures and resources in the adoption of big data analytics powered artificial intelligence, sustainable manufacturing practices and circular economy capabilities. *Technological Forecasting and Social Change, 163*, 120420. https://doi.org/10.1016/J.TECHFORE.2020.120420

BAĞIŞ, M., Karagüzel, E. S., Kryeziu, L., & ARDIÇ, K. (2019). A longitudinal analysis on intellectual structure of human resources management: Theoretical foundations and research trends. *Mehmet Akif Ersoy Üniversitesi İktisadi ve İdari Bilimler Fakültesi Dergisi, 6*(2), 796–814. https://doi.org/10.30798/makuiibf.569513

Bajic, B., Suzic, N., Moraca, S., Stefanović, M., Jovicic, M., & Rikalovic, A. (2023). Edge computing data optimization for smart quality management: Industry 5.0 perspective. *Sustainability (Switzerland), 15*(7). https://doi.org/10.3390/su15076032

Balog, K. (2020). The concept and competitiveness of agile organization in the fourth industrial revolution's drift. *Strategic Management, 25*(3), 14–27. https://doi.org/10.5937/straman2003014b

Battaglia, D., Galati, F., Molinaro, M., & Pessot, E. (2023). Full, hybrid and platform complementarity: Exploring the industry 4.0 technology-performance link. *International Journal of Production Economics, 263*, 108949. https://doi.org/10.1016/J.IJPE.2023.108949

Battistella, C., de Toni, A. F., & Pillon, R. (2016). Inter-organisational technology/knowledge transfer: A framework from critical literature review. *Journal of Technology Transfer, 41*(5), 1195–1234. https://doi.org/10.1007/s10961-015-9418-7

Bennett, N., & Lemoine, J. (2014). What VUCA really means for you. *Harvard Business Review, 92*(1/2).

Biclesanu, I., Savastano, M., Chinie, C., & Anagnoste, S. (2023). The role of business students' entrepreneurial intention and technology preparedness in the digital age. *Administrative Sciences, 13*(8). https://doi.org/10.3390/admsci13080177

Bock, G.-W., & Kim, Y.-G. (2002). Breaking the myths of rewards: An exploratory study of attitudes about knowledge sharing. *Information Resources Management Journal, 15*(2), 14–21.

Bock, G.-W., Zmud, R. W., Kim, Y.-G., & Lee, J.-N. (2005). Behavioral intention formation in knowledge sharing: Examining the roles of extrinsic motivators, social-psychological forces, and organizational climate. *MIS Quarterly, 29*(1), 87–111.

Braun, V., & Clarke, V. (2006). Using thematic analysis in psychology. *Qualitative Research Psychology, 3*(2), 77–101. https://doi.org/10.1191/1478088706qp063oa

Cabrera, E. F., & Cabrera, A. (2005). Fostering knowledge sharing through people management practices. *The International Journal of Human Resource Management, 16*(5), 720–735.

Cabrera, A., Collins, W. C., & Salgado, J. F. (2006). Determinants of individual engagement in knowledge sharing. *The International Journal of Human Resource Management, 17*(2), 245–264.

Carayannis, E. G., Christodoulou, K., Christodoulou, P., Chatzichristofis, S. A., & Zinonos, Z. (2022). Known unknowns in an era of technological and viral disruptions—implications for theory, policy, and practice. *Journal of the Knowledge Economy, 13*(1), 587–610. https://doi.org/10.1007/s13132-020-00719-0

Cardoso, L., et al. (2007). Para uma revisão da teoria do conhecimento de Michael Polanyi. *Revista Portuguesa De Pedagogia, Ano, 41–1*, 41–54.

Cardoso, L., et al. (2011). Gestão do conhecimento em equipas: Desenvolvimento de um instrumento de medida multidimensional. *Psychologica, 55*, 79–83.

Cardoso, L., Gomes, A. D., & Rebelo, T. (2003). Gestão do conhecimento: Dos dados à informação e ao conhecimento. *9*.

Cardoso, L (2007a). Gestão do Conhecimento e competitividade organizacional: Um modelo estrutural. *Comportamento Organizacional e Gestão, 13*(2), 191–211.

Caruso, L. (2018). Digital innovation and the fourth industrial revolution: epochal social changes?. *Ai & Society, 33*(3), 379–392. https://doi.org/10.1007/s00146-017-0736-1

Casa-Nova, M. J. (2009). Etnografia e produção de Conhecimento. Lisboa: ACIDI.

Černe, M., Nerstad, C. G. L., Dysvik, A., & Skerlavaj, M. (2014). What goes around comes around: Knowledge hiding, perceived motivational climate, and creativity. *Academy of Management Journal, 57*(1), 172–192.

Chen, C.-J., & Huang, J.-W. (2009). Strategic human resource practices and innovation performance the mediating role of knowledge management capacity. *Journal of Business Research, 62*(1), 104–114. https://doi.org/10.1016/j.jbusres.2007.11.016

Chen, J., Lim, C. P., Tan, K. H., Govindan, K., & Kumar, A. (2021). Artificial intelligence-based human-centric decision support framework: An application to predictive maintenance in asset management under pandemic environments. *Annals of Operations Research.* https://doi.org/10.1007/s10479-021-04373-w

Chen, K. & Chiu, Y. (2018). Correlation between innovation strategy and operational performance in tourism based on competitive advantage. *Revista De Cercetare Si Interventie Sociala, 62.*

Chen, M. H., Wang, H. Y., & Wang, M. C. (2018). Knowledge sharing, social capital, and financial performance: The perspectives of innovation strategy in technological clusters. *Knowledge Management Research and Practice, 16*(1), 89–104. Palgrave Macmillan Ltd. https://doi.org/10.1080/14778238.2017.1415119

Chiarini, A., & Cherrafi, A. (2023). Integrating ISO 9001 and industry 4.0. An implementation guideline and PDCA model for manufacturing sector. *Total Quality Management & Business Excellence, 34*(13–14), 1629–1654. https://doi.org/10.1080/14783363.2023.2192916

Chinying Lang, J. (2001). Managerial concerns in knowledge management. *Journal of Knowledge Management, 5*(1), 43–59. https://doi.org/10.1108/13673270110384392. COG, *13*(2), 191–211 (n.d.).

Cohen, W. M., & Levinthal, D. A. (1990). Absorptive capacity: A new perspective on learning and innovation. *Administrative Science Quarterly, 35*(1).

Collins, C. J., & Smith, K. G. (2006). Knowledge exchange and combination: The role of human resource practices in the performance of high-technology firms. *Academy of Management Journal, 49*(3), 544–560.

Connelly, C. E., & Kelloway, E. K. (2003). Predictors of employees' perceptions of knowledge sharing cultures. *Leadership & Organization Development Journal, 24*(5), 294–301.

Constant, D., Kiesler, S., & Sproull, L. (1994). What's mine is ours, or is it? a study of attitudes about information sharing. *Information Systems Research, 5*(4), 400–421.

De Azambuja, A. J. G., Plesker, C., Schützer, K., Anderl, R., Schleich, B., & Almeida, V. R. (2023). Artificial intelligence-based cyber security in the context of industry 4.0—a survey. *Electronics (Switzerland), 12*(8). https://doi.org/10.3390/electronics12081920

De Luca, P., & Cano Rubio, M. (2019). The curve of knowledge transfer: A theoretical model. *Business Process Management Journal, 25*(1), 10–26. https://doi.org/10.1108/BPMJ-06-2017-0161

DeLong, D. W., & Fahey, L. (2000). Diagnosing cultural barriers to knowledge management. *Academy of Management Perspectives, 14*(4), 113–127.

Dey, P. K., Chowdhury, S., Abadie, A., Vann Yaroson, E., & Sarkar, S. (2023). Artificial intelligence-driven supply chain resilience in Vietnamese manufacturing small-and medium-sized enterprises. *International Journal of Production Research, 1*, 40. https://doi.org/10.1080/00207543.2023.2179859

Dias, C. C, & Figueiredo B. (2019). Sleep-wake behaviour during the first 12 months of life and associated factors: a systematic review. *Early Child. Development and Care.* https://doi.org/10.1080/03004430.2019.1582034

Doronin, D., Lei, S., & Shah, S. H. H. (2020). Reconsidering concept of knowledge sharing: Search for quality dimensions. *Kybernetes.* https://doi.org/10.1108/K-11-2019-0767

Durst, S., & Zieba, M. (2020). Knowledge risks inherent in business sustainability. *Journal of Cleaner Production*, 251. https://doi.org/10.1016/j.jclepro.2019.119670

Durst, S., & Wilhelm, S. (2011). Knowledge management in practice: Insights into a medium-sized enterprise's exposure to knowledge loss. *Prometheus, 29*(1), 23–38.
Durst, S., & Zieba, M. (2019). Mapping knowledge risks: Towards a better understanding of knowledge management. *Knowledge Management Research and Practice, 17*(1), 1–13. https://doi.org/10.1080/14778238.2018.1538603
Duvivier, F., Peeters, C., & Harzing, A. W. (2019). Not all international assignments are created equal: HQ-subsidiary knowledge transfer patterns across types of assignments and internal types of knowledge. *Journal of World Business, 54*(3), 181–190. https://doi.org/10.1016/j.jwb.2019.02.003
Emmert-Streib, F. (2021). From the digital data revolution toward a digital society: Pervasiveness of artificial intelligence. *Machine Learning and Knowledge Extraction, 3*(1), 284–298. https://doi.org/10.3390/make3010014
Eslami, Y., Ashouri, S., & Lezoche, M. (2021). Semantic interoperability and sustainability an industry 4.0 product life cycle issue. *IFIP International Conference on Advances in Production Management Systems, 633*, 418–426. https://doi.org/10.1007/978-3-030-85910-7_44
Farbiz, F., Habibullah, M. S., Hamadicharef, B., Maszczyk, T., & Aggarwal, S. (2023). Knowledge-embedded machine learning and its applications in smart manufacturing. *Journal of Intelligent Manufacturing, 34*(7), 2889–2906. https://doi.org/10.1007/s10845-022-01973-6
Fauzi, M. A., Nya-Ling, C. T., Thursamy, R., & Ojo, A. O. (2019). Knowledge sharing: Role of academics towards research productivity in higher learning institution. *VINE Journal of Information and Knowledge Management Systems, 49*(1), 136–159. https://doi.org/10.1108/VJIKMS-09-2018-0074
Feinberg, S. E., & Gupta, A. K. (2004). Knowledge spillovers and the assignment of R&D responsibilities to foreign subsidiaries. *Strategic Management Journal, 25*, 823e845. https://doi.org/10.1002/smj.396
Field, A. (2020). *Descobrindo a estatística usando o SPSS-5*. Penso Editora.
Fischer, C. (2018). Motivated to share your knowledge? development of a scale to measure knowledge sharing motives of public employees. *Permanent Study Group III Public Personnel Policies.*
Ford, J. K., Baldwin, T. P., & Prasad, J. (2017). Annual review of organizational psychology and organizational behavior transfer of training: The known and the unknown. https://doi.org/10.1146/annurev-orgpsych
Gagné, M. (2009). A model of knowledge-sharing motivation. *Human Resource Management, 48*(4), 571–589.
Galaskiewicz, J., & Wasserman, S. (1989). Mimetic processes within an interorganizational field: An empirical test. *Administrative Science Quarterly, 28*, 22–39.
Garcia-Morales, V. J., Jimenez-Barrionuevo, M. M., & Gutierrez-Gutierrez, L. (2012). Transformational leadership influence on organizational performance through organizational learning and innovation. *Journal of Business Research, 65*(7), 1040–1050.
García-Sánchez, P., Díaz-Díaz, N. L., & de Saá-Pérez, P. (2019). Social capital and knowledge sharing in academic research teams. *International Review of Administrative Sciences, 85*(1), 191–207. https://doi.org/10.1177/0020852316689140
Ghobakhloo, M., Iranmanesh, M., Tseng, M. L., Grybauskas, A., Stefanini, A., & Amran, A. (2023). Behind the definition of Industry 5.0: a systematic review of technologies, principles, components, and values. *Journal of Industrial and Production Engineering, 40*(6), 432–447. https://doi.org/10.1080/21681015.2023.2216701
Ghoshal, S., & Bartlett, C. (1988). Creation, adoption and diffusion of innovations by subsidiaries of multinational corporations. *Journal of International Business Studies, 19*(3), 365–388.
Girard, J. P. (2006). Where is the knowledge we have lost in managers? *Journal of Knowledge Management, 10*(6), 22–38. https://doi.org/10.1108/13673270610709198
Girard, J. P., & McIntyre, S. (2010). Knowledge management modeling in public sector organizations: A case study. *International Journal of Public Sector Management, 23*(1), 71–77. https://doi.org/10.1108/09513551011012330

Goswami, A. K., & Agrawal, R. K. (2018). A reflection on knowledge sharing research: Patterns and trends. *VINE Journal of Information and Knowledge Management Systems, 48*(3), 352–372. https://doi.org/10.1108/VJIKMS-08-2017-0049

Guimarães, D. C., Cruz, C. D., Miranda, D. D., & Russo, S. L. (2019). Produção científica sobre a Sociedade 5.0. In *10th International Symposium on Technological Innovation. 10th International Symposium on Technological Innovation.* https://doi.org/10.7198/S2318-3403201900010918

Gupta, S., Justy, T., Kamboj, S., Kumar, A., & Kristoffersen, E. (2021). Big data and firm marketing performance: Findings from knowledge-based view. *Technological Forecasting and Social Change, 171*, 120986. https://doi.org/10.1016/J.TECHFORE.2021.120986

Hislop, D. (2003). Linking human resource management and knowledge management via commitment: A review and research agenda. *Employee Relations, 25*(2), 182–202.

Holten, A.-L., Robert Hancock, G., Persson, R., Marie Hansen, Å., & Høgh, A. (2016). Knowledge hoarding: antecedent or consequent of negative acts? The medi-ating role of trust and justice. *Journal of Knowledge Management, 20*, 215e229. https://doi.org/10.1108/JKM-06-2015-0222

Holton E. F., Bates, R. A., Seyler, D. L., & Carvalho, M. B. (1997). Toward construct validation of a transfer climate instrument. *Human Resources Development Quarterly, 8*(2), 95−113.

Hossain, S., Hassan, S., & Karim, R. (2023). Assessment of critical barriers to industry 4.0 adoption in manufacturing industries of Bangladesh: An ISM-based study. *Brazilian Journal of Operations and Production Management, 20*(3). https://doi.org/10.14488/BJOPM.1797.2023

Huang, P. C., Yao, C. L., & Chen, S. (2019). Development of the organizational resources towards innovation strategy and innovation value: Empirical study. *Revista de Cercetare Si Interventie Sociala, 64*, 108–119. https://doi.org/10.33788/rcis.64.9

Ismail, M., Sobri, S. S., Zulkifly, N. A., Hamzah, S. R., & Yamato, E. (2016). Knowledge transfer between expatriates and host country nationals: Contribution of individual and social capital factors. *Organizations and Markets in Emerging Economies, 7*(2), 65–87. https://doi.org/10.15388/omee.2016.7.2.14208

Jasinska-Biliczak, A., Štverková H., & Winiarczyk, I. (2017). Foreign investment of hi-tech enterprises as the source of knowledge transfer to the region—opolskie voivodship. *Case Study, Business.*

Jeong, H. (2001). The MNC knowledge transfer, subsidiary absorptive capacity and HRM; Jocumsen, G. (n.d.). *How small businesses make their strategic marketing decisions.*

Joe, C., Yoong, P., & Patel, K. (2013). Knowledge loss when older experts leave knowledge-intensive organisations. *Journal of Knowledge Management, 17*, 913e927. https://doi.org/10.1108/JKM-04-2013-0137

Joia, L. A., & Lemos, B. (2010). Relevant factors for tacit knowledge transfer within organisations. *Journal of Knowledge Management, 14*(3), 410–427.

Junior, G. T., & Saltorato, P. (2018). Impactos da indústria 4.0 na organização do trabalho: uma revisão sistemática da literatura impacts of the industry 4.0 on work organization: a systematic review of the literature. *18*(2), 743–769.

Junior, G., & e Saltorato, P. (2017). Impactos da Indústria 4.0 na Organização do Trabalho: Uma revisão sistemática da literatura. *Revista Cientifica Electrônica de Engenharia de Produção*, 767. produção on line, ISSN 1676–1901.

King, N. (2012). Doing template analysis. *Qualitative Organizational Research: Core Methods and Current Challenges, 26*, 426.

Kleinknecht, A. (2020). The (negative) impact of supply-side labour market reforms on productivity: An overview of the evidence. *Cambridge Journal of Economics, 44*(2), 445–464. https://doi.org/10.1093/cje/bez068

Krzywdzinski, M., & Butollo, F. (2022). Combining experiential knowledge and artificial intelligence the digital transformation of a traditional machine-building company. *Management Revue, 34*(2), 161–184. https://doi.org/10.5771/0935-9915-2022-2-161

Lee, C. S., & Tajudeen, F. P. (2020). Usage and impact of artificial intelligence on accounting: Evidence from Malaysian organisations. *Asian Journal of Business and Accounting, 13*(1), 213–239. https://doi.org/10.22452/ajba.vol13no1.8

Lee, C. Y., & Chien, C. F. (2022). Pitfalls and protocols of data science in manufacturing practice. *Journal of Intelligent Manufacturing, 33*(5), 1189–1207. https://doi.org/10.1007/s10845-020-01711-w

Leite, M. S. R., & Gasparotto, A. M. S. (2018). Análise SWOT e as suas funcionalidade: O autoconhecimento da empresa e sua importância. *Revista Interface Tecnológica, 15*(2), 184–195.

Lenart-Gansiniec, R. (2019). Organizational learning in industry 4.0. *Problemy Zarządzania, 17*(2 (82)), 96–108.

Lerch, C. M., Heimberger, H., Jäger, A., Horvat, D., & Schultmann, F. (2022). AI-readiness and production resilience: Empirical evidence from German manufacturing in times of the Covid-19 pandemic. *International Journal of Production Research, 1*, 22. https://doi.org/10.1080/00207543.2022.2141906

Li, P., Bastone, A., Mohamad, T. A., & Schiavone, F. (2023). How does artificial intelligence impact human resources performance. evidence from a healthcare institution in the United Arab Emirates. *Journal of Innovation & Knowledge, 8*(2), 100340. https://doi.org/10.1016/J.JIK.2023.100340

Liao, J., Welsch, H., & Stoica, M. (2003). Organizational absorptive capacity and responsiveness: An empirical investigation of growth-oriented SMEs. *Entrepreneurship Theory and Practice, 28*(1), 63–86.

Lin, C.-P. (2007). To share or not to share: Modeling tacit knowledge sharing, its mediators and antecedents. *Journal of Business Ethics, 70*(4), 411–428.

Lin, H.-F., & Lee, G. (2004). Perceptions of senior managers toward knowledge-sharing behaviour. *Management Decision, 42*(1), 108–125.

Lin, L. H., & Wang, K. J. (2022). Talent retention of new generations for sustainable employment relationships in work 4.0 era—assessment by fuzzy delphi method. *Sustainability (Switzerland), 14*(18). https://doi.org/10.3390/su141811535

Liyanage, C., Elhag, T., Ballal, T., & Li, Q. (2009). Knowledge communication and translation-a knowledge transfer model. *Journal of Knowledge Management, 13*(3), 118–131. https://doi.org/10.1108/13673270910962914

Lundvall, B. A., & Johnson, B. (1994). The learning economy. *The Learning Economy and the Economics of Hope, 107*.

Madreiter, T., Kohl, L., & Ansari, F. (2021). A text understandability approach for improving reliability-centered maintenance in manufacturing enterprises. In *IFIP International Conference on Advances in Production Management Systems* (pp. 161–170). Springer International Publishing. https://doi.org/10.1007/978-3-030-85874-2_17

Malik, N., Tripathi, S. N., Kar, A. K., & Gupta, S. (2021). Impact of artificial intelligence on employees working in industry 4.0 led organizations. *International Journal of Manpower, 43*(2), 334–354. https://doi.org/10.1108/IJM-03-2021-0173

Manual NVivo Pro for Windows, Version 11.4, Copyright © 1999–2017 QSR International Pty Ltd. ABN 47 006 357 213.

Matzler, K., Renzl, B., Mooradian, T., Von Krogh, G., & Mueller, J. (2011). Personality traits, affective commitment, documentation of knowledge, and knowledge sharing. *The International Journal of Human Resource Management, 22*(2), 296–310.

McAfee, A., & Brynjolfsson, E. (2017). *Machine, platform, crowd: Harnessing our digital future*. WW Norton & Company.

McKinnon, J. L., Harrison, G. L., Chow, C. W., & Wu, A. (2003). Organizational culture: Association with commitment, job satisfaction, propensity to remain, and information sharing in taiwan. *Journal of International Business Studies, 11*(1), 25–44.

Mohammed, A. A., Hafeez-Baig, A., & Gururajan, R. (2019). Designing a knowledge management measurement for educational institutions: A qualitative research. *International Journal of Information Systems and Change Management, 11*(2), 157–187. https://doi.org/10.1504/IJISCM.2019.104626

Mohammed, A. A., Gururajan, R., & Hafeez-Baig, A. (2018). An exploratory qualitative research to address processes that are utilised for managing knowledge: a case study in a Queensland

Regional University. *International Journal of Higher Education and Sustainability, 2*(3). https://www.researchgate.net/publication/335566279

Mohammed, A. A., Hafeez-Baig, A., Gururajan, R., & Mohammed, A. A. (2018). Exploring processes that are used for managing knowledge in the higher education environment: A case study in a queensland regional university. 10.24088

Moher, D., Liberati, A., Tetzlaff, J., Altman, D. G., & The PRISMA Group. (2009). Reprint—preferred reporting items for systematic reviews and meta-analyses: The PRISMA statement. *Physical Therapy, 89*(9), 873–880. https://doi.org/10.1093/ptj/89.9.873

Monavvarian, A., & Kasaei, M. (2007). A KM model for public administration: The case of labour ministry. *Vine, 37*(3), 348–367. https://doi.org/10.1108/03055720710825654

Muskat, B., & Mair, J. (2020). Knowledge sharing and power in the event workforce. *Event Management, 24*(5), 597–609. https://doi.org/10.3727/152599519X15506259856228

Narwane, V. S., Raut, R. D., Yadav, V. S., Cheikhrouhou, N., Narkhede, B. E., & Priyadarshinee, P. (2021). The role of big data for supply chain 4.0 in manufacturing organisations of developing countries. *Journal of Enterprise Information Management, 34*(5), 1452–1480. https://doi.org/10.1108/JEIM-11-2020-0463

Nonaka, I. (1994). A dynamic theory of organisational knowledge creation. *Organisation Science, 5*, 14–37.

Nonaka, I., & Takeuchi, H. (1996). The knowledge-creating company: How japanese companies create the dynamics of innovation. *Long Range Planning, 29*(4), 592–610.

Noreen, U., Shafique, A., Ahmed, Z., & Ashfaq, M. (2023). Banking 4.0: Artificial intelligence (AI) in banking industry & consumer's perspective. *Sustainability (Switzerland), 15*(4). https://doi.org/10.3390/su15043682

Nourmohammadi, A., Fathi, M., & Ng, A. H. C. (2022). Balancing and scheduling assembly lines with human-robot collaboration tasks. *Computers & Operations Research, 140*, 105674. https://doi.org/10.1016/J.COR.2021.105674

Ouédraogo, M., & Rinfret, N. (2019). The sharing and management of managers' knowledge: A structured literature review. *Knowledge and Process Management, 26*(3), 244–261. https://doi.org/10.1002/kpm.1606

Pan, M., Sikorski, J., Kastner, C. A., Akroyd, J., Mosbach, S., Lau, R., & Kraft, M. (2015). Applying industry 4.0 to the Jurong Island eco-industrial park. *Energy Procedia, 75*, 1536–1541. https://doi.org/10.1016/J.EGYPRO.2015.07.313

Parker, H., (2012). Knowledge acquisition and leakage in inter-firm relationships involving new technology-based firms. *Management Decision, 50*, 1618e163. https://doi.org/10.1108/00251741211266714

Paulin, D., & Suneson, K. (2012). Knowledge transfer, knowledge sharing and knowledge barriers–three blurry terms in KM. *The Electronic Journal of Knowledge Management, 10*(1), 81–91.

Presti, C., de Santis, F., & Bernini, F. (2023). Value co-creation via machine learning from a configuration theory perspective. *European Journal of Innovation Management, 26*(7), 449–477. https://doi.org/10.1108/EJIM-01-2023-0104

Putro, B. E., & Ilmaniati, A. (2018). Developing conceptual model for the effectiveness of knowledge sharing in business innovation for Indonesian MSMEs performance. In *3rd International Conference on Digital Arts, Media and Technology, ICDAMT 2018* (pp. 1–6). https://doi.org/10.1109/ICDAMT.2018.8376485

Quadri et al. (2019). Perceived effects of ICT on knowledge sharing among librarian in South-West Nigeria: A UTAUT theoretical approach. *Journal of Balkan Libraries Union, 6*(1), 38–46.

Queiroz, J., Leitão, P., Pontes, J., Chaves, A., Parra, J., & Perez-Pons, M. E. (2020). A quality innovation strategy for an inter-regional digital innovation hub. *ADCAIJ: Advances in Distributed Computing and Artificial Intelligence Journal, 9*(4), 31–45.

Radiancies, P., De Roure, D., Nicolescu, R., Huth, M., & Santos, O. (2022). Digital twins: Artificial intelligence and the IoT cyber-physical systems in Industry 4.0. *International Journal of Intelligent Robotics and Applications, 6*(1), 171–185.

Rahman, S., Islam, M. Z., & Abdullah, A. D. A. (2017). Understanding factors affecting knowledge sharing: A proposed framework for Bangladesh's business organizations. *Journal Internal of Science and Technology Policy Management, 8*(3), 275–298. https://doi.org/10.1108/JSTPM-02-2017-0004

Rana, J., & Daultani, Y. (2023). Mapping the role and impact of artificial intelligence and machine learning applications in supply chain digital transformation: A bibliometric analysis. *Operations Management Research, 16*(4), 1641–1666. https://doi.org/10.1007/s12063-022-00335-y

Rinfret, N., Bernier, L., Houlfort, N., Lemay, L., & Mercier, D. (2010). Défis et enjeux des connaissances: La réalité des cadres de la fonction publique québécoise. *Telescope, 16*(1), 208–224.

Rizzi, F., Pellegrini, C., & Todaro, N. (2017). Knowledge dynamics and resource efficiency in international business relations. In *Global Opportunities for Entrepreneurial Growth: Coopetition and Knowledge Dynamics within and across Firms* (pp. 199–227). Emerald Group Publishing Ltd. https://doi.org/10.1108/978-1-78714-501-620171014

Rodrigues, E., Veloso, A., Gonçalves, S., Silva, I., Santos, J., & e Brandão, C. (2023). A Gestão do Conhecimento no contexto da pandemia da Covid-19. *Tendências no Trabalho e na Gestão de Pessoas, Edição 1*, 271–292. Lisboa, Editora RH, Lda.

Ruggero, S. M., dos Santos, N. A., Estender, A. C., & da Silva, M. T. (2021). Industry 4.0: Expectations, impediments and facilitators. In *Advances in Production Management Systems. Artificial Intelligence for Sustainable and Resilient Production Systems: IFIP WG 5.7 International Conference, APMS 2021, Nantes, France, September 5–9, 2021, Proceedings, Part IV* (pp. 673–680). Springer International Publishing. https://doi.org/10.1007/978-3-030-85910-7_71

Ruiz-Real, J. L., Uribe-Toril, J., Torres, J. A., & Pablo, J. D. E. (2021). Artificial intelligence in business and economics research: Trends and future. *Journal of Business Economics and Management, 22*(1), 98–117. https://doi.org/10.3846/jbem.2020.13641

Rulke, D. L., Zaheer, S., & Anderson, M. H. (2000). Sources of manager's knowledge of organizational capabilities. *Organizational Behaviour and Human Decision Processes, 82*(1), 134–149.

Sackmann, S. A. (1992). Culture and subcultures: An analysis of organizational knowledge. *Administrative Science, Quarterly, 3*, 140–161.

Saßmannshausen, T., Burggräf, P., Wagner, J., Hassenzahl, M., Heupel, T., & Steinberg, F. (2021). Trust in artificial intelligence within production management–an exploration of antecedents. *Ergonomics, 64*(10), 1333–1350.

Schramm, W. (1971). Notes on case studies of instructional media projects. In *Working paper the Academy for Educational Development*, Washington, DC.

Schwaer, C., Biemann, T., & Voelpel, S. (2012). Antecedents of employee's preference for knowledge-sharing tools. *The International Journal of Human Resource Management, 23*(17), 3613–3635.

Shahbaznezhad, H., Rashidirad, M., & Vaghefi, I. (2019). A systematic review of the antecedents of knowledge transfer: An actant-object view. *European Business Review, 31*(6), 970–995. https://doi.org/10.1108/EBR-07-2018-0133

Sigov, A., Ratkin, L., Ivanov, L. A., & Xu, L. D. (2022). Emerging enabling technologies for industry 4.0 and beyond. *Information Systems Frontiers*, 1–11. https://doi.org/10.1007/s10796-021-10213-w

Da Silva, C. W. B., Nascimento, J. S., Ferraz, M. A., Roberto, J. C. A., & Soares, M. C. (2023). Qualidade 4.0: tecnologias emergentes e suas aplicações. *Revista de Gestão e Secretariado (Management and Administrative Professional Review), 14*(7), 12116–12132. https://doi.org/10.7769/gesec.v14i7.2525

Silva, S. L. (2002). Informação e competitividade: a contextualização da gestão do conhecimento nos processos organizacionais. *Ci Inf. Brasília, 31*(2), 142–151, maio/ago.

Silva, E. L. e Menezes, E. M. (2005). *Metodologia da pesquisa e elaboração de dissertação* (4th ed.). UFSC.

Sordi, V. F., Nakayama, M. K., Cunha, C. J. C. de A., & Binotto, E. (2017). Fatores determinantes e possíveis barreiras ao compartilhamento de conhecimento nas organizações. *Revista Eletrônica de Estratégia & Negócios, 10*(2), 225. https://doi.org/10.19177/reen.v10e22017225-246

Sordi, V. F., Nakayama, M. K., & Binotto, E. (2018). Compartilhamento de Conhecimento nas Organizações: Um Modelo Analítico sob a Ótica da Ação Cooperativa. *Perspectivas Em Gestão & Conhecimento, 8*(1), 44–66. https://doi.org/10.21714/2236-417x2018v8n1p44

Souza, L. (2019). Pesquisa com análise qualitativa de dados; conhecendo a Análise Temática. *Arquivos Brasileiros De Psicologia: Rio De Janeiro, 71*(2), 51–57.

Tangaraja, G., Mohd Rasdi, R., Abu Samah, B., & Ismail, M. (2016). Knowledge sharing is knowledge transfer: a misconception in the literature. *Journal of Knowledge Management, 20*(4), 653–670. Emerald Group Publishing Ltd. https://doi.org/10.1108/JKM-11-2015-0427

Toledo, L., & Shiaishi, G. (2009). Estudo de caso em pesquisas exploratórias qualitativas: Um ensaio para a proposta de protocolo do estudo de caso. *Revista FAE, Curitiba, 12*(1), 103–119.

Tseng, M. L., Tran, T. P. T., Ha, H. M., Bui, T. D., & Lim, M. K. (2021). Sustainable industrial and operation engineering trends and challenges toward industry 4.0: A data driven analysis. *Journal of Industrial and Production Engineering, 38*(8), 581–598. https://doi.org/10.1080/21681015.2021.1950227

Turner, C. J., Emmanouilidis, C., Tomiyama, T., Tiwari, A., & Roy, R. (2019). Intelligent decision support for maintenance: An overview and future trends. *International Journal of Computer Integrated Manufacturing, 32*(10), 936–959. https://doi.org/10.1080/0951192X.2019.1667033

Upadhyay, A., Balodi, K. C., Naz, F., di Nardo, M., & Jraisat, L. (2023). Implementing industry 4.0 in the manufacturing sector: Circular economy as a societal solution. *Computers & Industrial Engineering, 177*, 109072. https://doi.org/10.1016/J.CIE.2023.109072

Usoro, A., Sharratt, M. W., Tsui, E., & Shekhar, S. (2007). Trust as an antecedent to knowledge sharing in virtual communities of practice. *Knowledge Management Research & Practice, 5*(3), 199–212.

Vergara, S. C. (2004). *Projetos e relatórios de pesquisa em administração* (5th ed.). Atlas.

Verma, S., & Singh, V. (2022). The employees intention to work in artificial intelligence-based hybrid environments. *IEEE Transactions on Engineering Management, 71*, 3266–3277. https://doi.org/10.1109/TEM.2022.3193664

Vuković, M., & Thalmann, S. (2022). Causal discovery in manufacturing: A structured literature review. *Journal of Manufacturing and Materials Processing, 6*(1). MDPI. https://doi.org/10.3390/JMMP6010010

Wang, R., Milisavljevic-Syed, J., Guo, L., Huang, Y., & Wang, G. (2021). Knowledge-based design guidance system for cloud-based decision support in the design of complex engineered systems. *Journal of Mechanical Design, 143*(7), 072001. https://doi.org/10.1115/1.4050247

Webster, J., & Watson, R. (2002). Analysing the past to prepare for the future: Writing a literature review. *MIS Quarterly, 26*(2), 13−23.

Weiwei et al. (2020). A model for knowledge transfer in a multi-agent organization based on lattice kinetic model. *Journal of Systems Engineering and Electronics, 31*(1), 156–167.

Williams, C., & Durst, S. (2019). Exploring the transition phase in offshore outsourcing: decision making amidst knowledge at risk. *Journal of Business Research, 103*, 460e471. https://doi.org/10.1016/j.jbusres.2018.01.013

Wolniak, R. (2021). The concept of operation and production control. *Production Engineering Archives, 27*(2), 100–107. https://doi.org/10.30657/pea.2021.27.12

Yaqub, M. Z., & Alsabban, A. (2023). Industry-4.0-enabled digital transformation: Prospects, instruments, challenges, and implications for business strategies. *Sustainability (Switzerland), 15*(11). https://doi.org/10.3390/su15118553

Yin, R. K. (1981). The case study crisis: Some answers. *Administrative science quarterly*, 26(1), 58–65. http://www.jstor.org/about/terms.html.

Yin, R. K. (2005). *Estudo de Caso: Planejamento e Métodos*. Porto Alegre: Bookman-Artmed.

Zahidul Islam, M., Said, T. F., Sumardi, W. A., & Rahman, S. (2020). Factors affecting organizational effectiveness: A proposed framework for Brunei's public sector. *International Journal of Asian*

Business and Information Management, 11(2), 15–27. https://doi.org/10.4018/IJABIM.2020040102

Zeba, G., Dabić, M., Čičak, M., Daim, T., & Yalcin, H. (2021). Technology mining: Artificial intelligence in manufacturing. *Technological Forecasting and Social Change, 171*, 120971. https://doi.org/10.1016/J.TECHFORE.2021.120971

Zebre, M., Ismael, G. Y., Nakabandi, O. A., Saleh, S. S., & Aqel, M. (2020). Impact of innovation technology in enhancing organizational management. *Studies of Applied Economics, 38*(4).

Zhang, W., & Zhang, W. (2018). Knowledge creation through industry chain in resource-based industry: Case study on phosphorus chemical industry chain in western Guizhou of China. *Journal of Knowledge Management, 22*(5), 1037–1060. https://doi.org/10.1108/JKM-02-2017-0061

Chapter 4
Generative Artificial Intelligence in Candidate Selection

Joana Santareno

Abstract This chapter contains a study that explores the impact of Generative Artificial Intelligence on candidate selection, analyzing its potential to reshape recruitment in Human Resources Management. The methodology involved two interviews with a Human Resources professional and a thematic content analysis. Subsequently, prompts were developed and tested in *ChatGPT 4.0* to simulate key steps in hiring an Accounting Technician. Findings indicate that Generative Artificial Intelligence enhances automation, agility, accuracy, and objectivity. Furthermore, it enables detailed feedback, personalized communications, and customized selection instruments. These benefits enrich the candidate experience and bolster employer branding. Ethical and legal considerations, especially transparency and data privacy, remain crucial. A balanced mix of human expertise and Artificial Intelligence innovation can improve decision-making, boost productivity, and drive organizational competitiveness. This synergy underscores the ongoing need to adapt Human Resources practices in the digital era.

Keywords Generative artificial intelligence · Candidate selection · Human resource management

4.1 Introduction

With the integration of Artificial Intelligence, particularly of the generative type, into the workforce, disruptive changes are emerging in the configuration of professions, in how tasks are performed, and in the growing need for technological expertise (World Economic Forum, 2023). The development of Generative Artificial Intelligence likewise introduces new perspectives for Human Resources Management, encompassing planning, practices, processes, platforms, and productivity (Budhwar et al., 2023).

J. Santareno (✉)
Porto Business School, Porto, Portugal
e-mail: joanasantareno@gmail.com

C. Machado (ed.), *Human Resource Management in a Disrupted World*, Lecture Notes on Multidisciplinary Industrial Engineering,
https://doi.org/10.1007/978-3-031-89948-5_4

In the field of recruitment and selection specifically, a recent study by LinkedIn Corporation (2024)–surveying senior and executive-level professionals on recruitment and selection management–reports that a significant 62% express optimism regarding this technology's impact on recruitment processes. They anticipate that Generative Artificial Intelligence will enhance candidate sourcing, enable more personalized approaches, and improve matching between candidates and open positions. The same study also shows a 14% increase, compared to 2023, in Artificial Intelligence-related skills in their LinkedIn profiles.

The emergence of this reality has led the European Parliament and the Council of the European Union to establish the first law regulating the use of Artificial Intelligence, known as the *AI Act*, with implications for recruitment and selection activities (European Union, 2024). The *AI Act* aims to promote fairness and accountability in recruitment by regulating the use of Artificial Intelligence systems classified as high-risk under this regulation, because they influence people's access to employment opportunities. While it imposes compliance burdens on employers and Artificial Intelligence providers, it also enhances trust and ensures ethical hiring practices, ultimately benefiting job seekers and the labor market as a whole. The OECD (Organization for Economic Cooperation and Development) also advocates for responsible Artificial Intelligence usage (OECD, 2024). Accordingly, Human Resources departments of organizations may play a pivotal collaborative role in complying with this legal framework, thereby preventing contingencies associated with the use of Artificial Intelligence in the workplace (Leitão, 2024).

Combining these contextual factors with the motivation to deepen theoretical and empirical knowledge on the subject resulted in defining the aim of this study: to analyze the impact of Generative Artificial Intelligence on candidate selection. It is likewise hoped that this work will serve as a positive contribution to the organization that provided insights into its operational context.

This chapter reflects the development of the study and is divided into several sections. This introduction offers background on the topic and outlines the study's objective. The second section provides the theoretical framework, focusing on three key concepts. The third section outlines the methodology employed. The fourth section presents the results. The fifth section discusses the findings. Finally, the sixth section presents the conclusions derived from the complete study.

4.2 Theoretical Background

4.2.1 Generative Artificial Intelligence

Generative Artificial Intelligence is a technology that uses algorithms capable of creating entirely new content in a range of formats (e.g., text, audio, images, video, etc.) (Boston Consulting Group, 2024; McKinsey & Company, 2024b). This distinctive functionality sets it apart from other forms of Artificial Intelligence, which are focused primarily on data analysis and outcome prediction (McKinsey & Company, 2024b). Generative Artificial Intelligence systems can also accelerate manual or repetitive tasks and customized content and information (Boston Consulting Group, 2024).

More specifically, Generative Artificial Intelligence models are *Large Language Models (LLMs)*, which are capable of processing and generating natural language (McKinsey & Company, 2024a) in a manner akin to human communication. Their ability to produce entirely new content relies on algorithms that learn patterns from a vast volume of data, enabling the development of computational models that can produce results consistent with those patterns or even exceed them in novel ways (McKinsey & Company, 2024b). However, the effectiveness of these outcomes largely depends on the quality, scope, and diversity of the dataset used. On the other hand, this underscores the need to pay attention to protecting the privacy of training data, safeguarding intellectual property, reducing the risk of occurring biases, ensuring transparency regarding the original sources in generated outputs (Boston Consulting Group, 2024), and conducting real human checks of the outputs generated (McKinsey & Company, 2024b).

The core of Generative Artificial Intelligence is thus anchored in Deep Learning, an evolutionary form of Machine Learning, which itself is a branch of Artificial Intelligence (McKinsey & Company, 2024a). Models such as *Transformers*—including the widely recognized *Generative Pre-Trained Transformers (GPTs)*—are exposed to extensive amounts of data from which they extract patterns with contextual connections and statistical correlations (McKinsey & Company, 2024c). In doing so, they become capable of generating new examples aligned with what they have learned. Nonetheless, issues such as the called "hallucinations" by developers of this type of models—wherein models like this generate seemingly coherent yet factually incorrect content—and biases rooted in skewed training data (Boston Consulting Group, 2024) pose significant challenges to the use of Generative Artificial Intelligence tools.

Finally, it is important to note that, while this technology offers significant potential for innovation and optimization across diverse organizational and business contexts, including Human Resources, it must also be accompanied by ethical and regulatory considerations (McKinsey & Company, 2024b). Those considerations will help to harness Generative Artificial Intelligence's transformative capabilities without compromising fundamental values, such as privacy, equity, and trust.

4.2.2 Candidate Selection

Within the context of a Human Resources Recruitment and Selection process, *Selection* constitutes the stage in which individuals most suited to perform a particular role, in a specific organizational setting, are chosen from among the candidates identified in the recruitment phase. It is carried out through the application of selection methods and techniques, based on established criteria, and concludes with the chosen candidate's acceptance of the job offer (Ployhart, 2006).

Concretely, the selection process aims to assess the compatibility between candidates' profiles (both technical and behavioral, as well as any relevant prior experience), the requirements of the position to be filled, and the characteristics and goals of the hiring organization (Ployhart, 2006). At the same time, for each candidate, the act of selection should consider the projected merit, integration, and overall fit that will best serve the organization's functioning (Rego et al., 2020).

In this regard, organizations can draw on a variety of selection tools, which allow for the evaluation of candidates' attributes either individually or collectively. By way of illustration, though not exhaustively, given continuous developments in this area (as expressed by Rego et al., 2020):

- Personality inventories, cognitive ability tests, knowledge tests, curriculum analyses, work-sample tests, simulations, and recommendation letters and references are among the *specific selection methods.* These focus on one or more candidate attributes, thereby preventing a comprehensive assessment of candidates.
- Interviews and assessment centers are *holistic selection methods* providing a global evaluation of candidates, as they measure multiple elements simultaneously.

It is essential that choices in this domain be made based on a set of interrelated factors that ensure the efficacy of the selection process. These factors include: (a) the objectives underlying the intended selection; (b) the characteristics of the position to be filled; (c) the predictive capacity of the adopted selection mechanisms with respect to candidates' future overall performance, as well as each mechanism's specific purpose; (d) the organizational resources available; and (e) the implementation channels, which are increasingly shaped by advanced technological innovations (as will be noted in the following section).

Still, regarding factors (c) and (e) specifically, Ployhart (2006) highlights the importance of using validated selection methods to improve selection decisions, reduce potential execution bias, and enhance the legitimacy of the implementation process. Moreover, the evolution of technologies and predictive analytics, alongside their integration into selection processes, optimizes competency assessment and shortens the time required to identify the best talent fit.

Additionally, selection processes must consider issues such as:

- The promotion of diversity, equity, and inclusion as a way to build a diverse workforce and enable organizations to benefit from its advantages (subjects explored in Shore et al., 2018).
- And the perceptions formed by candidates (whether selected or rejected) about their experience in the selection process, in terms of both procedural justice and the treatment they receive (Rego et al., 2020).

Thus, as noted by Ployhart (2006) and Rego et al. (2020), the selection decision significantly influences an organization's ability, through its human capital, to achieve its objectives and establish a sustainable competitive advantage in its market, including the labor market. Moreover, since selection processes also impact on organizational culture, attractiveness, reputation and identification and retention of talented employees, they represent a pivotal dimension of strategic people management.

4.2.3 *Candidate Selection Using Generative Artificial Intelligence*

The methods typically employed in the candidate selection process, such as those mentioned in the preceding section, require substantial time and are susceptible to bias when conducted solely through human intervention. Generative Artificial Intelligence, by contrast, offers innovative technological solutions aimed at optimizing the selection process in terms of equity, impartiality, accuracy, efficiency, and time management. Despite the limited number of sources specifically addressing Generative Artificial Intelligence in selection processes, it is still possible to highlight some relevant applications, which include:

- *Résumé screening*: Through the extraction of relevant information from résumés, candidates can be ranked according to predefined criteria, with an exclusive focus on factors relevant to the position's responsibilities, and additional justifications for each classification. This approach ensures more objective evaluations while reducing both the workload and the time required to complete this type of selection task (Glazko et al., 2024).
- *Profile matching*: By analyzing multiple variables (e.g., experience, competencies, and qualifications), candidates' profiles can be matched to the job requirements and organizational compatibility. This approach increases the likelihood of identifying the best-fit candidates, as it considers alignment with organizational culture. Additionally, this method reduces the time and effort typically required for résumé screenings while ensuring precision in profile matching (Budhwar et al., 2023).
- *Simulations of interviews*: Asynchronous video interviews allow for the evaluation of candidates while simultaneously predicting their future job performance through voice analysis, chatbots, and facial expression recognition (Hemalatha et al., 2021).
- *Automation of routine tasks*: Streamline routine tasks through automation such as scheduling interviews, sending notifications, and managing candidate databases, enabling more efficient resource allocation within Human Resources departments (Hemalatha et al., 2021). Consequently, Human Resources professionals can dedicate greater attention to strategic undertakings (Budhwar et al., 2023; Hemalatha et al., 2021).

Beyond the benefits evident in these various applications of Generative Artificial Intelligence, it remains crucial to consider ethical issues, as noted by Hill and Narine (2023). In this regard, integrating ethical guidelines into organizational governance models is essential to ensure transparency in human decision-making processes assisted by this technology, as well as for protecting candidates' data privacy. Furthermore, measures must be taken to safeguard against the biases that may be embedded in the algorithms of Generative Artificial Intelligence models. This underscores the importance of carefully designing and testing such systems to mitigate these risks, particularly when decisions affect individuals' opportunities and influence an organization's human capital quality and capacity to achieve desired outcomes, such as those arising from hiring in a selection process.

4.3 Methodology

To address the objective established for this exploratory study—analyze the impact of Generative Artificial Intelligence on candidate selection—, an experimental methodological framework was developed, incorporating a qualitative approach organized into several phases.

The process began with two structured, open-ended interviews conducted via email with a Human Resources professional who oversees recruitment and selection for a multinational group of businesses that employs over one thousand people. The first interview was exploratory in nature, providing a broad overview of the organization's recruitment and selection process. Drawing on the data collected, a second, more in-depth interview was then conducted to gather detailed information on the company's specific recruitment and selection practices.

For confidentiality reasons, neither the identity of the Human Resources professional nor that of the organization can be disclosed. Likewise, the selection instruments mentioned in the results section must remain confidential.

For the analysis of the interview data, a Thematic Content Analysis (based on Bardin, 2016) was employed, implemented through a manual coding of the interview content using a defined category system. This category system comprises two categories and eleven subcategories, each of which is accompanied by operational definitions. The recorded units in each category highlight themes within the interviewee's written responses, thus representing information pertinent to the study's subject matter. Furthermore, both an inductive review of the literature and a deductive reading and analysis of the interview data informed the construction of this category system.

In a subsequent, separate phase, four prompts were designed for submission to version 4.0 of the software known as *ChatGPT* (*Chat Generative Pre-Trained Transformer*). This software functions as a chatbot that utilizes Generative Artificial Intelligence models, enabling it to generate content based on user-provided prompts and the dataset on which it was trained (Budhwar et al., 2023). Notably, in the case of ChatGPT, its training dataset encompasses a broad cross-section of internet-based text up to a specified cutoff date. The purpose of choosing this tool was to test aspects of the organization's selection process, particularly those geared toward hiring for the position of Accounting Technician.

The construction of these prompts adhered to guidelines issued by OpenAI, the company behind ChatGPT. These guidelines, which are selectively summarized below in accordance with the OpenAI Developer Platform (2024), concern prompt engineering strategies aimed at optimizing interaction between the model and the user, thereby achieving the best possible results.

Providing clear instructions, including detailed guidance, is essential for eliminating ambiguities and aligning the system's responses with user expectations. This principle increases the likelihood that the model will understand the context provided by the user and generate relevant, accurate outputs. One way to advance this principle is to apply three distinct roles in the formulation and organization of prompts: *system*, *user*, and *assistant*. Each role contributes to achieving the final objective:

1. *System Role*: In the system role, a set of rules and guidelines is provided to the model, such as specifying the persona, the tone to be used, and the desired format. This role establishes the overarching framework that shapes how the model is respond.
2. *User Role*: In the user role, the request or task is presented for the model to carry out. This role specifies the problem, context, or question that needs to be addressed, giving the model a clear directive.
3. *Assistant Role*: In the assistant role, it's possible to illustrate or enforce how the model should respond to the previous instructions. It's possible to provide an output, highlight preferred structures, or demonstrate the style and detail expected in the final answer. This role can also serve as an example or a guide for the model on how to present the content it generates.

Once this structure is in place, a response is produced that is shaped by both the system and user instructions, further refined (if necessary) by the assistant role.

Another tactic to consider when designing prompts is supplying reference texts, as a way to prevent inconsistent or implausible answers. It is also advisable to divide complex tasks into simpler subtasks in order to obtain more reliable responses. Furthermore, allowing the model to "think"—that is, requesting a logical reasoning process rather than an immediate answer—often yields more accurate and thorough outcomes.

Another strategy involves supplying outputs deemed more reliable and efficient, generated by other Generative Artificial Intelligence tools, as reference examples for the model. By doing so, multiple models are employed rather than relying exclusively on a single outcome.

The final strategy involves systematically testing modifications, which entails making prompt adjustments that result in satisfactory responses for the user.

Lastly, several factors support the chosen methodological options:

- *Accounting Technician position*: This position was selected because it is the role in which the organizational department, where the interviewee works, has the greatest influence on recruitment and selection process.

- *Curricula selection*: Résumés were sourced from publicly accessible LinkedIn profiles. Subsequently, all names appearing in the generated files were removed to protect privacy, and the résumés were then pseudonymized.
- *Interview format*: The interviews were conducted via email because, as this method was preferred by the respondent for answering the interview questions.
- *Choice of ChatGPT*: The decision to employ ChatGPT, a tool incorporating Generative Artificial Intelligence models, was guided by several considerations. These include: (a) its broad accessibility; (b) its capacity to generate content with cognitive performance comparable to that of a human (being); (c) its versatility, applicability, and ease of use; (d) its ability to swiftly draw on a large, continuously updated body of information; and (e) its aptitude for providing answers to complex questions effectively.

4.4 Results

Based on the analysis of the interview data (see the category system outlined in the Methodology section and presented in Table 4.1), and guided by the study's objective, the Company's selection process for external candidates is detailed. It focuses on the selection process of an Accounting Technician, highlighting the stages where the Company's People Management Department is directly involved (see Table 4.2). This process provided the foundation for constructing the *ChatGPT Prompts*, following a logical progression. These prompts, along with their respective outputs, are also described in this section of the paper. Additionally, illustrative excerpts from the interviewee's written testimony further enrich the results presentation.

The *first ChatGPT Prompt* (see Table 4.3) aimed to simulate and test the initial phase of the selection process under study—résumé screening—which, according to the interviewee, seeks "*(…) to assess candidates' résumés in terms of the criteria and requirements established during the profile creation (…)*." To achieve this, ten résumés were uploaded (the maximum number of documents that could be attached in ChatGPT at that time), carefully selected so that some of them, a priori, did not meet the requirements for the target position. The ChatGPT was then instructed to perform the screening according to the specified criteria. Those elements were derived from a job advertisement for an Accounting Technician position, provided by the Company. The prompt also requested the inclusion of the corresponding screening justifications.

ChatGPT's response to the first prompt (see Fig. 4.1) met the intended objectives by classifying and ranking the candidates based on how well their profiles aligned with the predefined criteria, while also providing justifications for its choices.

Although no interview information was obtained regarding feedback provided to applicants during the résumé screening phase, it is believed that including this element would have enhanced their overall candidate experience.

The *second ChatGPT Prompt* (see Table 4.4) combined the second and third stages of the selection process: conducting the interview and preparing the corresponding evaluation report. In the words of the interviewee, these stages aim "*(…) to*

Table 4.1 Thematic content analysis category system

Recruitment	This category covered the recorded units corresponding to the stages of the candidate recruitment process (the subcategories listed below), which may vary depending on the function to be filled and the type of recruitment (internal, external, or mixed). It may involve the participation of different stakeholders, including representatives from the functional areas related to people management, communication and marketing, business units, and administration	
	Recruitment need	This subcategory comprises the recorded units that address the procedure leading to the identification of the need to fill a position within the organization
	Profile to recruit	This subcategory involves the recorded units covering the elements (e.g., technical and behavioral competencies) that form the professional profile required for the role. Candidates must match these elements as closely as possible
	Job posting and communication	This subcategory encompasses the recorded units concerning the initial check of the internal database and the preparation of a job posting. This posting typically includes the core requirements, a concise job description, the work location, any contractual terms, and the projected number of positions to fill, to be communicated both internally and externally
	Position-specific recruitment details	This subcategory consists of the recorded units relating to the specifics of the recruitment process for the accounting technician position
Selection	This category encompasses the recorded units corresponding to each stage of the candidate selection process (the subcategories listed below). These stages may vary according to the position being hired for and may involve various stakeholders (representatives of people management, business units, and administration)	
	Curriculum screening	This subcategory encompasses the recorded units related to evaluating candidates' résumés in accordance with the professional profile defined during recruitment
	Interviews	This subcategory involves the recorded units concerning the conduction of interviews aimed at assessing and validating the unique aspects that candidates possess

(continued)

Table 4.1 (continued)

	Interview reports	This subcategory refers to the recorded units covering reports on the outcomes of interviews, which constitute decision-making tools for the final selection stage
	Evaluation tests	This subcategory covers the recorded units pertaining to psychometric assessments and theoretical or practical tests that evaluate competencies and capabilities applied by the candidates
	Decision-making	This subcategory encompasses the recorded units related to choosing the candidate(s) to be hired, including presentation of the job offer
	Communication	This subcategory incorporates all recorded units involving communications addressed to the candidates over the course of the selection process
	Position-specific selection details	This subcategory consists of the recorded units relating to the specifics of the selection process for the accounting technician position

Table 4.2 Selection process for the position of accounting technician with the direct involvement of the people management department

1	*Curriculum screening*
	Evaluation of external candidate's résumés against the criteria defined in the position profile
	Notification of candidates regarding their eligibility for the subsequent stage of the process
2	*Scheduling and conducting interviews*
	Scheduling interviews with candidates selected in the previous selection stage, during which a representative of the people management department serves as the interviewer
	Notification of candidates regarding their eligibility for the subsequent stage of the process
3	*Preparation of interview evaluation reports*
	Compilation of information for evaluation reports reflecting the outcomes of the interviews conducted, using a predefined template
	Notification of candidates regarding their eligibility for the subsequent stage of the process
4	*Presentation of the employment offer*
	The final selected candidate is notified of their approval and presented with an employment offer, which is subject to the candidate's acceptance
	Upon the candidate's acceptance of the employment offer, the remaining candidates are informed of their exclusion from the selection process

Table 4.3 First ChatGPT prompt

System	You are a specialist in using generative artificial intelligence for candidate selection. You have 20 years of experience and work in the human resources department of [insert the company's name]. Consult information about the company through the URL [insert the URL] The company seeks to hire an accounting technician to join its accounting department. The selected candidate must meet all the following mandatory requirements: 1. A university degree in management, accounting, or economics (excluding vocational degrees) 2. Solid knowledge of Microsoft office from a user perspective 3. At least 12 months of professional experience in the accounting field 4. Residence in [insert the location] Candidate résumés are attached
User	Screen the candidates' résumés, classifying each candidate as "qualified" or "not qualified" for the accounting technician position. Provide a justification for each decision
Assistant	Create a fully visible on-screen table with the results of your screening. These results must be ordered by candidates classified as qualified first, and then by those classified as not qualified Generate a list of candidates classified as qualified, ordered from most qualified to least qualified, including the justification for each candidate's classification

validate the technical and behavioral skills and experiences the candidate mentions on the résumé (...)" and "*(...) to create a space where the candidate can gain a better understanding of the organization.*" This second prompt also fulfills the requirement to produce competency assessment reports, as the interview stated: "*After the interviews have taken place, competency assessment reports are then presented* (...)." Accordingly, a request was made for an interview guide that would facilitate the completion of the interview evaluation grid provided by the Company, which was made available in ChatGPT for reference.

The result (see Fig. 4.2) was as expected, presenting questions appropriate for an interview, enabling a comprehensive evaluation of the candidate, and following the items included in the attached evaluation grid.

In this case, it was noted that it was not possible to determine whether the interview was guided by a generalized or a customized guide. Nevertheless, a solid and customized interview guide can be an invaluable tool in conducting interviews.

The *third ChatGPT Prompt* (see Table 4.5) pertains to the stage of the selection process following the interview: "*To proceed with the selection process, the People Management Department sends a selection notification to the top-ranked candidate, together with a hiring proposal/offer.*" To address this requirement, an email was requested, addressed to the selected candidate, communicating the final selection outcome and including the pertinent job offer. The organization's final selection communication template and proposal presentation served as references.

The email generated by ChatGPT (see Fig. 4.3) met expectations and introduced additional elements, such as the work location, schedule, and start date for the position.

Candidate Screening for the Accounting Technician Position at ______

The classification was carried out based on the established requirements.

Table of Results

Candidate	Classification	Justification
2nd CV.pdf	Qualified	Bachelor's degree in Business Administration, relevant professional experience in accounting, solid Microsoft Office skills, resident _____.
1st CV.pdf	Qualified	Bachelor's degree in Economics, more than 3 years of experience in accounting, resident _____.
3rd CV.pdf	Qualified	Bachelor's degree in Management, experience in accounting, resident _____.
4th CV.pdf	Qualified	Bachelor's degree in Economics, 1 year and 6 months of experience in accounting, resident _____.
5th CV - Lic.pdf	Not Qualified	Completed a professional training course in accounting, does not hold a university degree.
6th CV - Lis.pdf	Not Qualified	Lives in _____, does not meet the _____ residency requirement.
7th CV - Lis.pdf	Not Qualified	Lives in _____, does not meet the _____ residency requirement.
8th CV - Intern.pdf	Not Qualified	Adequate academic background, but has less than 12 months of accounting experience.
9th CV - HR.pdf	Not Qualified	Main experience in payroll processing rather than accounting.
10th CV - HR.pdf	Not Qualified	Main experience in payroll processing, does not have the necessary accounting experience.

Fig. 4.1 Output of the first ChatGPT prompt

Finally, the *fourth ChatGPT Prompt* (see Table 4.6) addresses the communication component, which is integral to every stage of the organization's selection process. According to the interviewee, "*At each stage of the recruitment process, candidates are notified whether they have been selected or not for the next phase.*" In this instance, the organization's general rejection notice model was used as a reference. The objective was to reformulate the usual communication, providing additional feedback to candidates participating in the selection process, following résumé review and the interview.

ChatGPT responded accordingly (see Fig. 4.4), offering illustrative justifications for communicating decisions to candidates.

Consistent with the feedback considerations mentioned for the résumé screening phase, incorporating this element into communications informing candidates of their

List of Qualified Candidates (Ordered)

1. **2nd CV.pdf**
   - **Education:** Bachelor's degree in Business Administration.
   - **Experience:** 3 years of accounting experience at ______ and other companies.
   - **Location:** ______.
   - **Justification:** Meets all requirements, with substantial experience and an appropriate academic background.
2. **1st CV.pdf**
   - **Education:** Bachelor's degree in Economics.
   - **Experience:** Over 3 years as an Accounting Technician at ______.
   - **Location:** ______.
   - **Justification:** Strong academic and professional background in accounting, plus residency in ______.
3. **3rd CV.pdf**
   - **Education:** Bachelor's degree in Management.
   - **Experience:** Several years of experience in various accounting roles.
   - **Location:** ______.
   - **Justification:** Fulfills all necessary requirements with accounting experience and a university degree.
4. **4th CV.pdf**
   - **Education:** Bachelor's degree in Economics.
   - **Experience:** 1 year and 6 months as an Accounting Technician.
   - **Location:** ______.
   - **Justification:** Satisfies all requirements, with both the training and experience needed in the field of accounting.

Fig. 4.1 (continued)

exclusion from the process significantly contributes to fostering a positive perception of the selection experience.

Table 4.4 Second ChatGPT prompt

System	The next stage of the candidate selection process consists of an interview designed to validate the information the candidate shares in their résumé, evaluate their personality and potential for the role, and assess their capacity to adapt to the organizational culture and the team they will join. Another key objective is to create an opportunity for the candidate to learn more about the organization
User	Prepare an interview guide for me to use during the selection interviews with candidates deemed qualified in my capacity as a representative of the human resources department. The questions must allow all fields in the attached interview evaluation grid to be completed, without exception
Assistant	Present the guide in a table format, taking into account every field in the attached grid and following the order of those fields

4.5 Discussion

Analyzing the selection process with and without Generative Artificial Intelligence, the findings suggest that integrating this technology can optimize the overall candidate selection process in multiple ways:

- *Automation of the selection process*: The already structured selection process, which follows prescribed practices at each phase and for each situation, can be further automated through predefined prompts, including any necessary customization. Such automation can yield time, cost, and resource efficiencies in filling vacancies, while also promoting greater agility by facilitating swift adaptation to the organization's evolving needs.
- *Speed in handling significant volumes of data*: The stages of data analysis, processing, and communication are expedited, allowing for more efficient management of large datasets.
- *Enhanced accuracy and objectivity*: By promoting greater precision and impartiality in selection-related actions, the likelihood of minimizing biases that might otherwise affect those conducting the selection process is increased.
- *Personalized messages and feedback*: Customizing communication and providing tailored feedback to candidates enrich the overall candidate experience and foster more meaningful interactions between the organization and applicants.
- *Customization and generation of selection instruments*: This technology also enables the creation and refinement of tools used in various selection methods and techniques, such as interview scripts and interview evaluation reports, as demonstrated in the present study.

In this way, organizations—particularly Human Resources professionals—can gain insights into the potential for integrating Generative Artificial Intelligence into the Candidate Selection process. It could also provide an interesting basis for deeper discussion regarding the applicability, challenges, and future perspectives of Generative Artificial Intelligence in Recruitment and Selection.

Interview Guide for Accounting Technician

Evaluation Field	Question	Notes
1. Overall Candidate Impact	**Clarity of Communication** Could you tell us a bit about yourself and your professional experience?	
2. Oral Expression	**Pitch – Tell us about yourself** How would you describe yourself professionally? What are your main strengths?	
3. Professional Conduct and Personal Presentation	**Relationship-Building Skills** How would you describe your teamwork style? Could you give an example of a situation in which you collaborated effectively with colleagues? **Proactivity and Dynamism** Could you describe a situation in which you had to be proactive in solving a work-related problem? **Attention to Details in Interview Preparation** How did you prepare for this interview? What information about our company did you find most relevant?	
4. Motivation	**Motivation for the Role** What motivated you to apply for this position at _____? **Alignment with the Company's Mission and Values** What do you know about the mission and values of _____? How do you see yourself aligned with them?	
5. Technical Competencies	**Knowledge and Experience in the Role** Could you detail your previous accounting experience? Which technical competencies do you consider most important for this role? **Proficiency in Microsoft Office** What is your level of proficiency with Microsoft Office? Could you give examples of how you have used these tools in your work?	
6. Adaptability	**Flexibility and Resilience** Could you share a moment when you had to adapt quickly to a change at work? **Stress Management** How do you handle high-pressure situations and tight deadlines? Could you give a concrete example?	

Fig. 4.2 Output of the second ChatGPT prompt

7. Company Knowledge	**Understanding of the Company** What attracted you to ____? Which aspects of our company do you find most interesting?	
8. Potential for Growth	**Ambition and Career Objectives** What are your medium- and long-term professional goals? How does this position fit into those objectives?	

This guide covers all **evaluation fields** from the interview grid and enables a **comprehensive assessment** of the candidate. Feel free to adjust the questions or add further notes according to your needs.

Fig. 4.2 (continued)

Table 4.5 Third ChatGPT prompt

System	The candidate selection process concludes with the choice of the candidate who will fill the vacancy. The type of employment contract to be signed with the chosen candidate is a permanent contract
User	Write a selection notification to be sent to the chosen candidate by the human resources department. Include a detailed hiring proposal The tone should be formal and persuasive Use as a reference the attached model for selection communication and the attached model for presenting the hiring proposal
Assitant	Email format

Dear [Candidate Name],

It is with great pleasure that we inform you that you have been selected for the position of **Accounting Technician** at ______, following a thorough selection process. Your professional experience, technical skills, and alignment with our company's values were the key factors in our decision.

We would like to invite you to join our team and provide below the details of our offer:

Offer Details

- **Position:** Accounting Technician
- **Work Location:** ______, Portugal
- **Contract Type:** Permanent Employment Contract
- **Start Date:** [Date to be Agreed Upon]
- **Annual Gross Salary:** [Annual Salary Amount] Euros

Benefits

- **Meal Allowance:** [Amount of Allowance]
- **Health Insurance**
- **Pension Plan**
- **Ongoing Training and Development**

Main Responsibilities

- Carry out general and analytical accounting tasks
- Prepare financial statements
- Prepare and submit tax filings
- Perform bank reconciliations and account analyses
- Assist in the preparation of the annual financial report

Next Steps

We kindly ask that you confirm your acceptance of this offer by **[Response Deadline]** via email to **[Contact Email]**. Please feel free to reach out to us with any questions you might have.

We believe your joining ______ will be extremely valuable, and we look forward to working with you to achieve our shared goals.

Thank you for your attention, and we await your confirmation.

Sincerely,

[Your Name]
[Your Position]
Human Resources Department

Fig. 4.3 Output of the third ChatGPT prompt

Table 4.6 Fourth ChatGPT prompt

System	After each stage of the selection process, a communication is sent to candidates deemed not qualified, informing them of their non-continuation in the process
User	Draft the communication described above in a personalized manner, including the reason for exclusion from the process following the résumé screening stage and the interview with the human resources department. Take into account the importance of communication in shaping the candidate's experience Use a formal and empathetic tone. Refer to the attached non-selection communication template
Assitant	Email format

Non-Continuation Notification in the Selection Process

Subject: Recruitment Process for Accounting Technician

Dear [Candidate Name],

[Company Name] thanks you for your interest in joining our organization and for taking part in the selection process for the Accounting Technician position.

After a thorough review of your résumé/interview with the Human Resources department, we regret to inform you that your application was not selected to move forward in the selection process for the [Business Unit/Central Structure].

Reason for Exclusion (Résumé): During the résumé screening, we noted [specific reason, such as insufficient minimum experience, lack of a required degree, etc.].

Reason for Exclusion (Interview): During the interview, we found [specific reason, such as technical competencies that do not align with the position's requirements, differing career expectations, etc.].

Following this decision, all your personal data and submitted documents will be deleted from our systems, in accordance with our privacy policy.

If you would like us to add your details to our candidate database for future job opportunities, please send us your résumé at [jobs..pt] **or via the link below:**

https://jobs..pt/

We thank you once again for your interest and wish you every success in your endeavors.

With our best regards,

Human Resources Department
[Name]
[Contact Phone Number]
[Contact Email]

Fig. 4.4 Output of the fourth ChatGPT prompt

4.6 Conclusion

The use of Generative Artificial Intelligence is redefining the traditional modus operandi of candidate selection processes within organizations. Although it entails benefits, risks, implementation challenges, and the need to fulfill legal requirements, it exerts a transformative impact given its potential for enhanced effectiveness. This technology yields significant gains in activities that constitute the selection stage of the entire recruitment process, ranging from résumé screening to the final choice of candidates.

Observations indicate that the synergy of human and technological forces promotes higher profitability by boosting both organizational productivity and performance. It also shapes the positioning organizations seek to attain in competitive markets (both commercial and labor). Keeping pace with future-of-work trends requires anticipation and adaptability, ensuring that organizational objectives are met in a context of continual improvement and evolution.

Accordingly, understanding the environment in which Generative Artificial Intelligence can be applied to candidate selection contributes to deeper insights into its implications, while fostering ongoing updates, agility, and adaptability among Human Resources professionals tasked with identifying and securing valuable talent. The effectiveness of candidate selection exerts a broad impact on organizational trajectory and performance. Moreover, examining the selection process and the candidate experience facilitates constant adjustment and refinement. About this task, Generative Artificial Intelligence can play an active, significant role through its capacity to generate innovative and tailored content. This, in turn, influences both the candidate experience and the employer brand image of the organizations.

From a broader perspective, focusing on skills related to Generative Artificial Intelligence is critical to ensuring that Human Resources professionals are prepared to curate information and help design Generative Artificial Intelligence-driven systems. Such preparation promotes continuous calibration, adaptation, and monitoring, thereby supporting organizations in achieving optimal performance. In parallel, concerns such as diversity, equity, inclusion, and possible algorithmic biases must be addressed as a matter of urgency.

4.6.1 Limitations and Future Directions

In light of the study's development process and the results presented, this work concludes by identifying the following research limitations and suggesting directions for future research in (evidence-based) Human Resources Management:

- A key limitation lies in the fact that not all questions from the interview scripts received comprehensive responses. It is believed that an alternative method, rather than electronic correspondence, could overcome this issue by leveraging the benefits of real-time interaction between interviewer and interviewee.
- Related to the limitation described above, the study was unable to fully replicate the entire company's selection process (and, ideally, the recruitment process as well) due to the unavailability of the partnering company's data.
- Potential future directions include:

  (a) Evaluating the impact of Generative Artificial Intelligence involvement in candidate recruitment and selection through dedicated metrics.

(b) Replicating the recruitment and selection process both with and without Generative Artificial Intelligence, using equivalent data, and conducting a comparative analysis based on each final process's final outcomes.
(c) And addressing issues related to the preferential use of ChatGPT in research activities, as indicated by Budhwar et al. (2023).

By incorporating these additional lines of inquiry, researchers and practitioners alike can continue to refine the integration of Generative Artificial Intelligence into recruitment and selection processes, balancing innovation with robust ethical, legal and methodological standards.

Acknowledgements This chapter originates from a group project submitted as the final assignment in the Postgraduate Program in People Management at Porto Business School in 2024. The present version incorporates revisions, updates, and improvements derived from subsequent reviews.

References

Bardin, L. (2016). *Análise de conteúdo*. Edições 70.
Boston Consulting Group. (2024). Generative AI. Retrieved January 1, 2025, from https://www.bcg.com/capabilities/artificial-intelligence/generative-ai
Budhwar, P., Chowdhury, S., Wood, G., Arguinis, H., Bamber, G. J., Beltran, J. R., et al. (2023). Human resource management in the age of generative artificial intelligence: Perspectives and research directions on ChatGPT. *Human Resource Management Journal*. https://doi.org/10.1111/1748-8583.12524
European Union. (2024). Regulation (EU) 2024/1689 of the European parliament and of the council of 13 June 2024 laying down harmonised rules on artificial intelligence and amending regulations (EC) No 300/2008, (EU) No 167/2013, (EU) No 168/2013, (EU) 2018/858, (EU) 2018/1139 and (EU) 2019/2144 and directives 2014/90/EU, (EU) 2016/797 and (EU) 2020/1828 (artificial intelligence act). *Official Journal of the European Union*. Retrieved January 1, 2025, from http://data.europa.eu/eli/reg/2024/1689/oj
Glazko, K., Mohammed, Y., Kosa, B., Potluri, V., & Mankoof, J. (2024). Identifying and improving disability bias in GPT-based resume screening. In *Proceedings of the 2024 ACM Conference on Fairness, Accountability, and Transparency (FAccT '24), Rio de Janeiro, Brazil* (pp. 1–21). ACM. https://doi.org/10.1145/3630106.3658933
Hemalatha, A., Kumari, P. B., Nawaz, N., & Gajenderam, V. (2021). Impact of artificial intelligence on recruitment and selection of information technology companies. In *Proceedings of the International Conference on Artificial Intelligence and Smart Systems (ICAIS-2021), Coimbatore, India* (pp. 60–66). IEEE. https://doi.org/10.1109/ICAIS50930.2021.9396036.
Hill, P. A., & Narine, L. K. (2023). Ensuring responsible and transparent use of generative AI in extension. *The Journal of Extension*. https://doi.org/10.34068/joe.61.02.13
Leitão, P. (2024). Artigo Técnico: O AI Act e o Futuro do Mundo do Trabalho. RHmagazine. Accessed 23 June 2024.
LinkedIn Corporation. (2024). The future of recruiting 2024. Retrieved June 23, 2024, from https://business.linkedin.com/talent-solutions/resources/future-of-recruiting
McKinsey & Company (2024a). What is deep learning?. Retrieved January 1, 2025, from https://www.mckinsey.com/featured-insights/mckinsey-explainers/what-is-deep-learning.
McKinsey & Company. (2024b). What is generative AI?. Retrieved June 23, 2024, from https://www.mckinsey.com/featured-insights/mckinsey-explainers/what-is-generative-ai

McKinsey & Company (2024c). What is machine learning?. Retrieved January 1, 2025, from https://www.mckinsey.com/featured-insights/mckinsey-explainers/what-is-machine-learning

OECD. (2024). Recommendation of the council on artificial intelligence. Retrieved June 23, 2024, from https://legalinstruments.oecd.org/en/instruments/OECD-LEGAL-0449

OpenAI Developer Plataform. (2024). Prompt engineering. Retrieved November 19, 2024, from https://platform.openai.com/docs/guides/prompt-engineering

Ployhart, R. E. (2006). Staffing in the 21st century: New challenges and strategic opportunities. *Journal of Management*. https://doi.org/10.1177/0149206306293625

Rego, A., Cunha, M. P., Gomes, J. F. S., Cunha, R. C., Cabral-Cardoso, C., & Marques, C. A. (2020). *Manual de Gestão de Pessoas e do Capital Humano*. Edições Sílabo.

Shore, L. M., Cleveland, J. N., & Sanchez, D. (2018). Inclusive workplaces: A review and model. *Human Resource Management Review*. https://doi.org/10.1016/j.hrmr.2017.07.003

World Economic Forum. (2023). The future of jobs report 2023. Retrieved June 23, 2024, from https://www.weforum.org/publications/the-future-of-jobs-report-2023/digest/

Part III
Thematic Area: Evolution of Work and the Workforce

Chapter 5
A Performance Evaluation System in a Non-technological Portuguese SME: How Far Do We Need to Run?

Bárbara Balbina Macedo Marques and Ana Paula Ferreira

Abstract This study examines the performance evaluation of a small to medium-sized enterprise (SME) in Portugal, focusing on its performance evaluation system (PES) and the corresponding impacts based on employee perceptions. Given that PESs can serve multiple purposes and influence various aspects of a company, it is essential to utilize them effectively to maximize their benefits. However, the existing PES of the company in question primarily functions to differentiate employees based on their remuneration. The objective was to (1) assess the current Performance Evaluation System (PES) of the company, (2) pinpoint its advantages and disadvantages, (3) examine the implications of the PES on the company's other HRM practices and overall job satisfaction, and (4) propose recommendations and improvement actions that could facilitate organizational change and environmental adaptation through the performance evaluation system.

Keywords Performance evaluation · Human resource management · Implications · Change

5.1 Introduction

The Human Resource Management (HRM) practices in Portuguese SMEs, as discussed by Carolina Machado and Pedro Novo (Machado & Novo, 2014; Novo & Machado, 2013), often reflect a limited and underdeveloped approach. This is particularly evident in performance evaluation systems, which are frequently informal, inconsistent, or absent. Such practices can hinder the ability of these organizations to adapt effectively to disruption and change, which are becoming increasingly prevalent in today's volatile business environment.

B. B. M. Marques · A. P. Ferreira (✉)
School of Economics, Management and Political Science, University of Minho, Braga, Portugal
e-mail: aferreira@eeg.uminho.pt

C. Machado (ed.), *Human Resource Management in a Disrupted World*, Lecture Notes on Multidisciplinary Industrial Engineering,
https://doi.org/10.1007/978-3-031-89948-5_5

Novo and Machado (2013) emphasize that small and medium enterprises (SMEs) lack structured HRM systems partly due to resource constraints and managerial focus on immediate operational priorities. However, this absence limits SMEs' capacity to align individual and organizational goals, identify skill gaps, and foster a culture of continuous improvement. Paauwe and Boselie (2005)highlight that robust HRM practices, including performance evaluation, are critical enablers of organizational agility and adaptability, particularly during periods of disruption and transformation.

In the context of Portuguese SMEs, the informal nature of HRM practices can exacerbate resistance to change and impede innovation (Machado & Novo, 2014; Novo & Machado, 2013). Implementing structured HRM systems, even in resource-constrained contexts, could, therefore, serve as a lever for agility and resilience. Since SMEs are uniquely positioned to respond flexibly to market shifts, addressing these HRM shortcomings could significantly enhance their competitiveness in the face of disruptive forces.

Performance evaluation (PE) is critical in organizations (Assis, 2021). Given the daily realities experienced by organizations, they are tasked with developing a performance evaluation system (PES) that ensures strong performance while addressing and improving less favorable aspects. This approach focuses on continuous evolution and success.

This work focuses on the PES of a Portuguese SME called Verdy (a fictitious name). It appears that Verdy is not fully capitalizing on the potential benefits of its performance evaluation. The company conducts annual performance evaluations based on various indicators; however, these evaluations are currently used solely to differentiate payments related to the variable component of employee remuneration, serving as the basis for awarding annual bonuses.

According to Sousa et al. (2006) and Chiavenato (2014), the contributions of a performance evaluation system should be diverse. They should Assist in promotions and salary adjustments, provide employee performance feedback, offer professional guidance, and design development strategies. Unfortunately, Verdy's current performance evaluation system does not leverage its full potential.

The current work aims to analyze the existing PES at Verdy and explore how PES data can be utilized for other organizational practices, particularly HRM. This aligns with the previously expressed idea that establishing structured HRM systems, even in resource-limited environments, can serve as a catalyst for agility and resilience. By addressing these HRM challenges, SMEs—given their inherent flexibility in adapting to market changes—can significantly enhance their competitiveness in the face of disruptive challenges.

5.2 Theoretical Framework

5.2.1 The Importance of Performance Evaluation

Performance evaluation is a cornerstone of HRM, playing a critical role in aligning individual goals with organizational objectives, enhancing employee satisfaction, and fostering adaptability in times of disruption and change.

PE is fundamental for organizational success as it provides a systematic approach to assessing employee contributions and identifying areas for improvement. As highlighted by DeNisi and Smith (2014), PE serves as a feedback mechanism, enabling employees to understand their strengths and weaknesses while aligning their efforts with organizational goals. It also offers managers valuable insights to make informed decisions regarding promotions, compensation, and development needs.

Organizations benefit from well-designed PE systems as they promote accountability, enhance productivity, and support strategic planning (Aguinis, 2013). For managers, PE is a vital tool for monitoring team performance, identifying training gaps, and fostering a culture of continuous improvement. Organizations risk inefficiencies, reduced employee morale, and missed growth opportunities without structured evaluations.

This practice is intricately connected to other HRM practices, such as recruitment, training and development, and compensation management. These practices create a cohesive framework that drives organizational effectiveness and employee engagement. For instance, PE provides essential data to inform training programs, ensuring they address specific skill gaps identified during evaluations (Noe et al., 2017).

Moreover, PE is crucial in aligning recruitment efforts with organizational needs. By identifying high-performing employees and understanding their characteristics, HR teams can refine their recruitment strategies to attract similar talent (Brewster et al., 2016). Additionally, PE influences compensation systems by providing a basis for merit-based pay and rewards, thus reinforcing a performance-driven culture.

In the context of talent management, PE is indispensable. It ensures that succession planning efforts are grounded in accurate employee potential and performance assessments. As emphasized by Cascio and Boudreau (2016), an integrated HRM approach that includes robust PE systems is key to achieving long-term organizational sustainability.

The relationship between performance evaluation and job satisfaction is well-documented in HRM literature. Employees who receive regular, constructive feedback are more likely to feel valued and motivated, contributing to higher levels of job satisfaction (Judge & Kammeyer-Mueller, 2012). PE fosters open communication between employees and managers, helping to build trust and clarify expectations.

However, the design and implementation of PE systems significantly influence their impact on job satisfaction. Transparent and fair evaluation processes enhance employee perceptions of organizational justice, a critical determinant of satisfaction (Colquitt et al., 2001). Conversely, poorly executed evaluations can lead to dissatisfaction, reduced morale, and turnover.

Managers play a pivotal role in ensuring that PE systems are perceived positively. Providing timely and actionable feedback, focusing on development rather than criticism, and involving employees in the goal-setting process are strategies that enhance the effectiveness of PE and its contribution to job satisfaction (Bakker & Demerouti, 2007).

In today's dynamic business environment, disruption and change are inevitable. PESs can be instrumental in helping organizations navigate these challenges. By identifying high-performing and adaptable employees, PE supports the development of a resilient workforce capable of responding to changing demands (Ulrich & Dulebohn, 2015).

The integration of PE with strategic change management initiatives is particularly important during periods of disruption. For instance, during the COVID-19 pandemic, organizations with flexible PE systems were better equipped to realign employee objectives with rapidly shifting priorities (Gigauri, 2020). Similarly, in the context of digital transformation, PE enables organizations to assess and develop the digital competencies required for success in a technology-driven world.

Disruption also presents opportunities for innovation, and PE can facilitate this by fostering a culture of continuous improvement. As Pulakos et al. (2000) and Ybarra (2023) highlighted, organizations that use PE to identify and reward innovative behaviors are better positioned to thrive in uncertain environments. Additionally, the feedback provided through PE systems helps employees adapt to new roles and responsibilities, enhancing their ability to contribute to organizational goals during times of change.

5.2.2 The Performance Evaluation System

Performance Evaluation Systems are effective tools that can significantly enhance both the quality of work and the professional lives of employees (Chiavenato, 2014). Reifschneider (2008) highlights the importance of carefully planning the development of these systems. This involves identifying the areas to be assessed, selecting appropriate data collection instruments, determining the evaluation timeframe, and establishing key performance indicators.

Considering the possible and different purposes of an evaluation, a unique criterion should not be used to calculate performance with the same purpose, for example, to ascend to a remuneration level or achieve a bonus. This tool should improve organizational and professional performance (Schedler & Schneider, 2015). So, to build an evaluation system, there is a set of important premises to guarantee the procedure's success.

Fairness and accuracy ensure that all participants perceive the system as just and precise (Almeida, 1996). Relevance is crucial for ensuring that the results of the PE can be utilized in decision-making, preventing the evaluation from being seen as irrelevant or unnecessary by those being assessed. The PES must encompass specific and credible measures to maintain reliability (Enes, 2019). It must be consistent, fulfilling a designated function. Evaluators should apply the same evaluation criteria to all participants to avoid favoritism (Choon & Embi, 2012). Additionally, all pertinent aspects of the function must be included to create a valid PES (Enes, 2019).

Simplicity is also an important consideration when constructing this type of system (Almeida, 1996). It is vital to minimize subjectivity (Enes, 2019), meaning that the criteria should accurately reflect the actual performance of those being evaluated (Choon & Embi, 2012), ensuring the process's accuracy.

The system should enable transparent communication of both the evaluation criteria and the results (Enes, 2019). The outcomes of the evaluations must be disseminated to the relevant individuals, and the information should be directed appropriately (Almeida, 1996).

Creating a flexible system allows evaluators to choose their preferred techniques during the evaluation process. Ultimately, the PES should be used to identify and reward successes rather than suppress failures, fostering skill development (Almeida, 1996).

The incorrect construction of the evaluation system can lead to errors that are difficult to resolve, highlighting a wide range of perceptual errors (Aguinis, 2013; Antonioni, 1994; London & Smither, 1995). In this construction process, it is important to consider the sources and evaluation methods, data collection instruments, periodicity, etc.

Performance evaluation sources refer to the origins of the information used to assess an employee's performance. These sources vary depending on the goals and approach of the evaluation system.

Self-evaluation means that employees assess their performance, reflecting on their achievements, competencies, and areas for improvement. This encourages self-reflection and responsibility and enhances engagement in the process (London & Smither, 1995), but may be subjective or biased, depending on the employee's self-perception.

Supervisors often have a comprehensive view of employees' tasks and behaviors (Aguinis, 2013), so direct supervisors evaluate their subordinates' performance based on observation and set goals–*Evaluation by Immediate Supervisor*. However, this source may be biased or inconsistent if supervisors are not adequately trained.

Peer Evaluation consists of having colleagues at the same hierarchical level assessing each other's performance. It is helpful for roles requiring teamwork and provides multiple perspectives (Brutus et al., 1998), but it can be influenced by favoritism or interpersonal conflicts.

Subordinate Evaluation implies that employees provide feedback to their managers, particularly regarding leadership and team management. It encourages improvement in leadership practices and offers valuable insights into management (Antonioni, 1994). Subordinates may feel uncomfortable providing honest feedback, especially in hierarchical environments.

Feedback from Clients or Stakeholders can also be used as a source. Colleagues from other departments (internal) or clients (external) evaluate employees' performance in terms of service delivery. This offers insights into service quality and customer satisfaction (Parasuraman et al., 1988), although feedback may be influenced by individual experiences or factors outside the employee's control.

Feedback can be gathered from multiple sources: supervisors, peers, subordinates, and even clients–*360-degree Feedback*. This can provide a more comprehensive and balanced view of performance (Edwards & Ewen, 1996) but is time- and resource-intensive and challenging to implement consistently.

Direct Observation, or Objective Data, involves using performance indicators (KPIs), productivity metrics, or direct observation of employee activities. It provides objective and measurable data (Aguinis, 2013) but may overlook qualitative aspects of performance, such as teamwork or attitude.

The ideal source of performance evaluation depends on several factors, such as (a) performance evaluation objectives (development, compensation, or strategic decision-making); (b) nature of the work (some sources are better suited for collaborative roles, while others work well for individual tasks); (c) organizational culture (open cultures may benefit more from approaches like 360-degree feedback).

By selecting the right combination of sources, organizations can ensure a fair, accurate, and effective evaluation process that aligns with their strategic goals.

Common approaches to **performance evaluation methods** include judgmental, absolute standards, and results-oriented methods (Lunenburg, 2012). Specific techniques range from traditional methods like ranking and graphic rating scales to modern approaches such as 360-degree appraisal and Management by Objectives (MBO). These systems aim to improve work performance, communication, and employee potential (Aggarwal & Thakur, 2013).

Traditional performance evaluation methods, such as ranking and graphic rating scales, offer simplicity but can be subjective and biased (Ghutke, 2014). Modern approaches like 360-degree feedback provide a more comprehensive view by incorporating multiple perspectives (Calle-Quezada et al., 2024; Kanaslan & Iyem, 2016). While traditional methods focus on personality traits, modern techniques emphasize work results and non-financial indicators (Drogomyretska, 2014; Narkunienė & Ulbinaitė, 2018). The balanced scorecard and performance prism are modern methods combining financial and non-financial metrics (Narkunienė & Ulbinaitė, 2018). However, traditional approaches can be static and outdated, failing to provide insights into why results occur (Rosová & Balog, 2012). Artificial intelligence is emerging as a tool for more objective and efficient performance evaluation (Calle-Quezada et al., 2024). Ultimately, the choice of method should consider the organization's specific needs and goals, as there is no one-size-fits-all solution (Hadžimehmedagić & Akbarov, 2013; Narkunienė & Ulbinaitė, 2018).

The **data collection instruments** used typically involve questionnaires and interviews. The former allows for the collection of information about the roles, responsibilities, skills, and abilities of those evaluated in a concise, clear manner, with preferably well-structured questions divided into categories (Almeida, 1996; Armbrust, 2023; Lima, 2009).

Interviews stimulate communication and encourage positive performance reinforcement and alignment of perspectives (Sousa et al., 2006). They can enhance competence and relationships, leading to a better understanding of the work already carried out and future tasks to be developed (Kubiak, 2022).

Aguinis (2013) emphasizes tailoring evaluation frequency to job requirements and organizational goals, with more frequent reviews for dynamic roles and annual evaluations for stable positions. London and Smither (1995) highlight that feedback quality and its developmental impact matter more than strict regularity, advocating for adaptable evaluation schedules. Cascio and Boudreau (2016) broadly support aligning performance evaluation practices, including frequency, with organizational objectives and job characteristics.

5.3 Verdy and Its Performance Evaluation System

5.3.1 The Company

Verdy was founded in early 2000 and operates in fruit production, preservation, importation/trading, and exportation. It is headquartered in the northern region of Portugal and is currently one of the largest fruit producers in the Iberian Peninsula.

Currently, it employs 60 full-time workers, 11 of whom perform administrative roles in the office, including the CEO, while 49 work in the warehouse. Of the 60 full-time employees, 43 are women, with 28 having completed basic education, 11 having completed secondary education, 1 holding secondary education and a technical course, and 3 holding higher education degrees. The remaining 17 employees are men, with 9 having basic education, 2 with secondary education, 1 with secondary education and a technical course, and 5 with higher education degrees.

The administrative roles performed by the 11 employees are divided among eight departments. Generally, the 11 workers engage in activities across multiple departments.

As for the roles performed by the warehouse employees, these are not divided into specific categories, as their tasks exclusively involve fruit selection and packaging, labeling, and machine handling. All 49 warehouse workers are, therefore, capable of performing any of these tasks.

The company does not yet have a formalized department in terms of the human resources area, although this is one of the organization's future goals. Five employees—two women and three men—perform administrative tasks and other duties related to this area, including the CEO.

5.3.2 The Company's Performance Evaluation System

Verdy's PES, used for 59 company employees, merely distinguishes employees remuneratively by allocating an annual bonus.

This PES includes a set of 13 criteria that are evaluated for all employees, regardless of their role. These criteria are leadership, attendance, punctuality, flexibility, availability, organization, teamwork, productivity, quality of work produced, integration, ethics, technical aptitude, and presentation. Each criterion has a description and behavioral indicators to aid the evaluation process, although they lack precision. For instance, in the indicator "arrives late (above average)," the average is not specified.

Each criterion is scored on a scale of 1 to 5, with 1 representing performance below any expectation and 5 representing exemplary behavior. This scoring is based solely on the evaluators' observations (without using any supporting grid or framework). Based on these criteria and the corresponding scores, the company awards a monetary bonus to the employee, using a scale from 0 to 100% of a predetermined value.

The only evaluators are the CEO and the warehouse team leader. The 10 office workers, including the warehouse team leader, are evaluated solely by the CEO. These two individuals conduct the evaluation because they perform supervisory roles, making this a supervisor-subordinate evaluation system.

The evaluation results are not publicly posted. However, when workers feel unfairly evaluated—often after conversations with colleagues—they approach the CEO, who is willing to clarify the assigned rating.

5.4 Methodology

To analyze the existing PES at Verdy and explore how PES data can be utilized for other HRM practices to enhance Verdy's competitiveness in the face of disruptive challenges, the present study is based on a case study, relying primarily on qualitative research.

For data collection, semi-structured interviews were conducted with employees and the CEO. Additionally, company documents were utilized as sources of data.

The study population comprises all 59 employees. Regarding the sample, given the theme of this investigation, it consisted of 20 Verdy employees subject to the PES, selected based on their tenure at the company and their professional roles. It is important to note that, regarding job functions, the distinction lies only between warehouse and office roles since, as previously mentioned, all warehouse employees are trained to perform the same tasks. Gender, educational background, and age were not considered for sample selection. This is because the workforce is predominantly female, and the educational background primarily corresponds to basic education, leaving little room for significant differentiation. Age information was obtained only during the interviews, so it was not a factor in sample selection.

A total of 20 interviews were conducted, with 10 involving employees hired on or before 2017 and 10 with employees hired after 2017. The year 2017 was chosen as it represents the average admission year. Among the 20 interviewees, 6 were office workers subject to the PES, out of the 10 total office employees.

The interviews took place in January 2024, with an average duration of approximately 20 min. The most extended interview lasted 42 min, while the shortest lasted 12 min. Of the 20 interviews, 16 corresponded to women and 4 to men, with ages ranging from 23 to 50 years. Regarding tenure at the company, employees had been with Verdy for periods ranging from 1 to 19 years, with 4 indicating that they were in their first job.

Concerning professional roles, 6 interviewees performed administrative or office roles, while the remaining 14 handled all warehouse functions.

Informed consent is one of the most critical tools for ensuring and demonstrating respect for study participants (Mack et al., 2005). Accordingly, Verdy interviewees were informed about the research problem and its objectives. The voluntariness and anonymity of the interviews were guaranteed. All participants gave informed consent and agreed to the audio recording their statements during the interviews.

5.5 Data Analysis and Discussion

The interviews were analyzed using content analysis. To carry out this process, the interviews were reviewed in alignment with the study's objectives, and the dimensions and subdimensions deemed relevant for analysis were built (See Appendix).

5.5.1 Perception About the Company's PES and Its Process

5.5.1.1 Existence of the Evaluation

Of the 20 interviewees, half initially believed they were not being evaluated, while the other half believed they were. Among the 10 participants who considered themselves to be evaluated, 7 gave a confident "yes," in contrast to 3 interviewees who were unsure in their responses. However, during the interviews, 6 of the 10 who initially said they were not evaluated contradicted their initial response, ultimately associating the bonus with performance evaluation.

Thus, the number of interviewees affirming the existence of the evaluation was 16. These 16 respondents became the total considered for the subsequent questions regarding the PES.

5.5.1.2 PES Functioning and Evaluation Criteria

Regarding the *purpose and intent* of the PE, 15 respondents indicated that it is merely used for awarding bonuses, excluding the presence of training, progression, or even improvement opportunities as outcomes of the evaluation. The only exception was one participant who held an office position and was also responsible for supervising warehouse employees.

Armbrust (2023) emphasizes the need to understand the purpose of evaluations, which can serve to plan training, recommend behaviors aimed at improving the evaluated individuals, or provide rewards. In the present case, the evaluation is perceived solely as a mechanism for awarding bonuses.

Performance evaluations should not be used merely to measure human performance for a single purpose, such as salary progression (Schedler & Schneider, 2015). Evaluations can serve many other objectives, such as ensuring the organization meets its strategic goals, improving average performance, and enhancing internal communication (Marques, 1992; Sousa et al., 2006). Furthermore, evaluations can positively impact HR practices, including recruitment and selection, career progression, training needs assessments, talent retention, and the development of new leaders. Combined with an adequate evaluation process, these practices contribute to better organizational performance (Serrano, 2010).

Regarding *who conducts the evaluations*, 13 respondents stated that it is carried out by the warehouse manager, who also supervises the warehouse employees, among many other responsibilities, and the CEO. This constitutes a supervisor-subordinate evaluation model.

Of these 13, 11 work in the warehouse, while the other two hold office positions, and for them, evaluations are perceived as a new process and a learning phase, so participants considered it logical for their direct supervisors—the two individuals mentioned above—to conduct the evaluations.

The remaining three respondents stated that they are evaluated solely by the CEO, as they hold office roles. The evaluations are conducted only through discussions between the evaluators.

A PE can (and should) be conducted by different *sources*, including supervisors and colleagues, and are upward, downward, or horizontal evaluations (Pereira, 2021). In Verdy's case, based on the responses, the production manager evaluates the warehouse employees through supervision, assisting the CEO in this role. The CEO, in turn, is also responsible for evaluating office employees. This setup reflects a hierarchical evaluation model. However, according to Pereira (2021), this type of evaluation can lead to negative outcomes if the evaluator lacks a good relationship with the subordinate.

Verdy does not incorporate any other source of information in PE. As a result, employees feel excluded from the process, as it is entirely decided and carried out by the evaluators. Using another source in Verdy's PE, as self-assessment, can allow the evaluator to gather more insights into the opinions and behaviors of the workers (Ferreira et al., 2023). Additionally, as Chiavenato (2014) points out, self-assessment enables surpassing goals, achieving results, and creating improvement opportunities, which are vital in an organization.

Regarding the *timing* of PE, all 16 respondents reported that evaluations are conducted annually. One participant noted that, despite being an annual evaluation, only the month preceding the bonus distribution seems relevant.

Sousa et al. (2006) state annual formal evaluations are typically conducted. However, the frequency could be quarterly or monthly for highly operational roles. Since Verdy's warehouse employees perform operational tasks, more frequent evaluations would be appropriate.

Regarding *how* PE is made, most respondents described similar processes, explaining that the bonuses awarded through the PE are not uniform but vary depending on the evaluators' perception of each individual's performance.

All participants stated that neither the evaluation criteria nor the results were published. Thus, the only way for employees to understand the criteria or their evaluation scores is to inquire directly with the CEO. This usually happens when employees, after discussing among themselves, feel that the results are unfair.

Only two respondents revealed that, although the criteria are unpublished, they had access to them because they perform HR-related duties in addition to their regular responsibilities. Of the remaining 14, only nine could describe some criteria because they had consulted the CEO or discussed the matter with colleagues. The rest could not describe any criteria.

Employees can only infer whether their evaluation was good or poor by comparing their bonuses with those of their colleagues. All respondents unanimously expressed dissatisfaction with the lack of feedback from evaluators. Criteria and evaluation results are only disclosed when employees actively seek them out.

It became evident that evaluation criteria are not communicated, and, except for those who inquire about their evaluation outcomes, most employees remain unaware of the process.

According to Marques (1992), an evaluation process must consistently demonstrate transparency and clear communication, assuring workers that the organization is attentive to their performance. Additionally, all evaluations must be perceived as accurate, with criteria and results always communicated transparently (Almeida, 1996; Enes, 2019). Choon and Embi (2012) stress the importance of ensuring that evaluation results are disclosed to the right people, which is not happening in this case.

Respondents also noted the absence of constructive feedback from evaluators. Employees are left to gauge their performance solely by comparing bonuses with their colleagues, a practice that Kubiak (2022) deems inappropriate. Constructive feedback from evaluators is essential for improvement.

At Verdy, the PE does not meet many principles underlying designing and implementing a functional and effective performance evaluation system.

5.5.2 *Impact and Recommendations Regarding PES*

5.5.2.1 *Positive and Negative Aspects*

The only positive aspect of the PE is the monetary bonus received. Beyond this, only one respondent noted that individuals' perception of their evaluation can impact them positively or negatively and thus considers it an important factor.

Regarding the weaknesses of the evaluation, the lack of transparency and the consequent ignorance of and inability to understand the evaluation criteria were the most highlighted issues.

Five respondents mentioned feeling that injustices occur during the evaluation process. They expressed the opinion that the CEO does not have accurate knowledge of their performance, relying solely on the supervision of the production manager, who, in their view, acts without impartiality and exhibits favoritism.

Other issues mentioned included the absence of feedback, the competitiveness stemming from comparing bonuses among colleagues, and the sense that the criterion of availability to work overtime plays a significant role in the decision to award the highest bonus.

A transparent and participatory evaluation process, where the evaluation criteria are disclosed and feedback is guaranteed, leads to the successful execution and application of a PE. These respondents emphasized that the lack of knowledge about the criteria and the absence of feedback need improvement.

Choon and Embi (2012) argue that evaluations should consistently measure each individual's actual performance, leaving no room for favoritism. At the same time, Enes (2019) advocates that evaluations should minimize criteria that allow for subjectivity, as it is important for the evaluation to reflect each individual's performance genuinely.

5.5.2.2 Premises

Regarding the understanding of essential premises for a PES, such as fairness and relevance, most participants clearly do not consider the evaluation fair, with only three individuals refraining from giving a concrete answer, as they associated their response with the amount of the bonus they receive, which typically corresponds to the maximum. Additionally, one participant stated they could not answer the question because they were unaware of the criteria or how they were classified against them.

Concerning relevance, most participants do not find this premise present in the evaluation process—the six who consider the evaluation relevant justified their perspective based on the monetary bonus they receive.

As a result, 10 of the interviewees mentioned they disagreed with the evaluation process conducted at Verdy. Furthermore, two participants did not express an opinion on the matter, citing their newness to the organization. The remaining four who agreed with the evaluation also noted that, despite their agreement, there is still significant room for improvement.

As Almeida (1996) noted, the evaluation system must be perceived as fair, and, according to Alves (2016), evaluations should be conducted justly, with professionalism and integrity, to avoid adverse consequences. Additionally, as Enes (2019) stated, evaluations must be relevant, with their results being utilized for a meaningful purpose. However, the participants do not perceive the PE as fair or relevant.

5.5.2.3 Implications and Recommendations for Improvement

Regarding the influence of the PE on the interviewees' performance, six stated that the PE does not impact their work.

Considering the weaknesses identified throughout their responses, the interviewees suggested that the PE process should undergo several changes and offered recommendations for improvement. Specifically, twelve participants emphasized the need to **publish the evaluation criteria**, ensuring that everyone understands what they are being assessed on. They also advocated for the implementation of **self-assessment**, the provision of **feedback**, greater **involvement of the CEO** in evaluating their performance, the **inclusion of more supervisors** in the evaluation process, and **better utilization of the PE** to facilitate salary progression and training opportunities. Additionally, one participant mentioned the lack of a dedicated Human Resources (HR) manager to assist with the evaluation process.

According to Schedler and Schneider (2015) and Alves (2016), PE is an organizational process that, when executed properly, contributes to employee performance improvement. However, as most interviewees indicated, this specific PE does not significantly influence their performance. This highlights the importance of using the evaluation to help individuals develop their competencies, recognize obstacles, and achieve set objectives (Almeida, 1996). Moreover, dissatisfaction with the PE in some cases has led to reduced effort in their roles, emphasizing the need for an enhanced performance evaluation system (PES).

Another significant issue is that employees are unaware of how they are evaluated, which prevents them from understanding areas for improvement—a critical shortfall in the current evaluation process.

Participants suggested several measures to address perceived injustices in the evaluation process, mainly **incorporating more evaluation sources**, such as self-assessment and **enhancing feedback mechanisms** to provide employees with constructive guidance. Incorporating self-evaluation could be valuable, allowing employees to express their opinions about their performance and provide evaluators with a broader perspective. Feedback would clarify for employees what they should maintain and improve in their performance while addressing misunderstandings and reducing unnecessary conflicts—issues mentioned frequently during interviews.

These suggestions aim to make the evaluation system more transparent and effective. For PE results to be meaningful for evaluators, employees, and the organization, they should identify specific areas for improvement or skill acquisition (Otlvelre-Cestro et al., 1996). Currently, this systematic approach appears absent.

By adopting such measures, the PE could function as a tool for pinpointing improvement opportunities and supporting training programs to foster employee development (Almeida, 1996).

Additionally, recruiting more supervisors to oversee employee performance and appointing a dedicated HR manager were recommended. Since, as noted by Assis (2021), HR management is responsible for ensuring performance across various dimensions, having a dedicated HR professional could bring greater attention to executing an effective PE process.

5.5.3 PES Impacts on Work-Related Satisfaction

5.5.3.1 *Perceptions of PE Satisfaction*

When asked about their satisfaction with the results of PE and, consequently, with their own performance, the majority confirmed feeling satisfied with the PE solely because they received the maximum bonus amount or due to the monetary reward itself, as money impacts their quality of life. Only five participants stated that they were not satisfied with their evaluation.

All participants reported feeling satisfied with their performance.

It is known that satisfaction can result from a performance evaluation process, with diverse rewards being provided in the work context (Lobo, 2020). Accordingly, most interviewees stated they felt satisfied with the reward received for their performance (bonus), and therefore, they expressed satisfaction with the evaluation conducted. According to Cunha et al. (2016), the perception of poor evaluation results tends to contribute to dissatisfaction, hindering organizational development. However, this was not evident in this case, as the participants reported being content with the bonus awarded.

5.5.3.2 Factors Influencing Job Satisfaction

When asked about the factors that most positively and negatively impact their job satisfaction, the majority cited enjoying their job role (12 respondents) as the most positive factor. This was followed by the work environment and relationships with colleagues (3 respondents) and, in equal measure, work conditions and the bonus (4 respondents, 2 for each). Additionally, one participant mentioned the company's location, which is close to their home, as the most positive factor.

Regarding the factors that most negatively impact job satisfaction, the most frequently mentioned were the work environment and a lack of organization

(13 responses). Seven participants highlighted the feeling of underappreciation, including insufficient salary recognition and rivalries among colleagues (3 responses for lack of appreciation, 4 for conflicts). Poor communication, excessive overtime, and deadline pressure were the least mentioned factors (5 responses).

Silva et al. (2021) note that several factors influence work-related satisfaction, affecting individuals differently. What brings satisfaction to one individual might cause dissatisfaction for another. The interview findings support this notion. For instance, while the majority identified the work environment as a negative factor, a minority considered it a significant source of satisfaction.

Some of these aspects of job satisfaction appear to be closely related to the perception that the PES (Performance Evaluation System) does not seem fair, as favoritism is perceived. The lack of knowledge about the criteria, corresponding scores, and subsequent feedback can explain the components of lower satisfaction.

5.5.3.3 *Recommendations to Increase Work-Related Satisfaction*

In 13 responses, suggestions for organizational improvements were highlighted, with some proposing establishing a daily/weekly plan to enable employees to know the tasks they need to perform at the start of each workday. This suggestion stemmed from many participants expressing that they often do not know what to do upon arriving at work each day. Additionally, the majority of interviewees mentioned that improved communication would increase satisfaction. Better communication would help avoid conflicts and misunderstandings and enhance organization.

Another proposal mentioned in 6 responses was the implementation of "salary differentiation," as participants disagreed with the largely uniform salaries despite differing performance levels. This sample revealed confusion regarding the remuneration components (fixed pay—salary—and variable pay—bonuses/incentives) (Nascimento et al., 2017).

To further increase satisfaction, 3 participants suggested having a more assertive warehouse manager, with one specifically emphasizing the importance of impartiality in such a role. Another 3 proposed stricter discipline, including the introduction of penalties. Lastly, 1 respondent suggested that implementing shifts in the company would boost satisfaction, while another proposed offering employee benefits, such as exempting workers from working on their birthdays.

According to Anwar and Abdullah (2021), employees work more efficiently when rewarded, whether through bonuses or other incentives. Suggestions such as granting a day off on one's birthday or providing variable bonus payments align with the types of rewards advocated by these authors. Similarly, the previously main emphasized proposals by interviewees–improving organization and communication, also promote employee satisfaction. As Cunha et al. (2016) noted, performance that is rewarded in various ways leads to job satisfaction, which drives greater commitment to the organization and its goals.

5.6 Conclusions and Recommendations

Based on the information collected from the sample in this study, it can be concluded that most participants did not have a detailed and concrete understanding of the Performance Evaluation System (PES). Thus, in response to the primary objective of this project, the following conclusions are drawn:

- The evaluation is not universally understood, with half of the participants initially stating they were not being evaluated.
- Regarding the fundamental principles of the evaluation system, most participants indicated that the PE is conducted primarily to award bonuses and serves only as a tool for monetary distinction.
- Concerning the sources and frequency of the evaluation, participants reported similar answers, noting that the PE is conducted through a supervisor-subordinate model involving two evaluators and occurs annually. There are no additional evaluation sources, nor is there a system for recording the results of the criteria assessed, as it is solely based on supervision by the warehouse manager and the CEO.
- The evaluation criteria are unknown to most employees, revealing a lack of dissemination and an absence of knowledge regarding the scores they receive in each criterion.
- Feedback is absent, as it only occurs when employees inquire about their bonus amount, initiating a discussion with their superiors.

When framing the perceived advantages and disadvantages of the PES:

- The bonus award was mentioned as the most significant advantage and, for most participants, the sole positive aspect of the evaluation.
- The lack of knowledge and dissemination of criteria, and consequently the resulting scores, prevents employees from understanding the aspects on which they are evaluated.

Additional findings show that most participants do not consider the evaluation fair or relevant and disagree with its purpose. They argued that the system should have broader applications beyond monetary awards. The evaluation impacts performance, with implications for the organization both positively—encouraging satisfaction—and negatively—leading to reduced effort and engagement.

Based on participants' insights, suggested improvements to the current PES can be disseminating the evaluation criteria and better utilizing PE to identify training needs, support salary progression, and increase supervisory roles in warehouse management.

Findings on work-related satisfaction reveal that most participants expressed satisfaction with the PE, which seems contradictory but is explained primarily by the bonus they receive. Regarding satisfaction with performance, not a single participant reported being dissatisfied. The majority are unsatisfied with their workplace, but those satisfied point to the work environment and relationships with colleagues, recognition/appreciation, and enjoyment of their job. Factors contributing to dissatisfaction include the work environment, lack of organization, lack of recognition, and conflicts among colleagues.

To improve satisfaction, the recommendations included better organization through, for example, implementing a daily/weekly task plan, effective communication to avoid conflicts, and recognition through bonuses.

To increase satisfaction, participants explicitly argued that in addition to performance-based bonuses, there should be a salary differentiation reflecting different performance levels. This recommendation highlights a lack of understanding that salaries are fixed compensation, independent of performance, whereas bonuses are variable and depend on the company's financial ability to award them (Chiavenato, 2014).

In general conclusions, it can be pointed out:

- Responses were generally uniform, showing consistent views on the PES, improvement suggestions, and satisfaction issues.
- The PES and its management could positively impact employee satisfaction by being repurposed for other Human Resources Management (HRM) practices and improving management aspects such as organization methods and communication capacity.
- It is worth emphasizing that apart from the two employees who perform office and HR roles, none of the participants could confidently detail all the evaluation criteria used, undermining one of the key aspects of an evaluation. This finding underscores that the organization still has significant progress to make.

5.7 The Journey of Verdy–How Far Do We Need to Run?

In 2022, approximately 945,649 SMEs were operating in Portugal, with the vast majority of these (901,496) being micro-sized enterprises that employed between zero and nine people. Around 38,156 small businesses had between 10 and 49 employees and 5,997 medium-sized businesses (Espinosa, 2025).

Research indicates that many of these organizations have not fully established formal HRM practices within Portuguese SMEs. Recruitment and selection are among the more commonly utilized HRM practices; however, other areas, such as training and development, performance appraisal, and strategic HR planning, are often less formalized (Machado & Novo, 2014).

This suggests that while SMEs play a crucial role in Portugal's economy, there is room for growth in adopting comprehensive HRM practices to enhance organizational effectiveness and employee development. Verdy is no exception in this scenario.

So… How far do we need to run? When asking ChatGPT if there is a particular significance of the expression, this was the response:

The expression "How far do we need to run?" can carry multiple layers of significance depending on the context in which it is used. Here are some interpretations:

1. *Literal Interpretation: The question is about the physical distance required to run, often asked in contexts like sports, training, or an emergency situation.*
2. *Metaphorical Significance:*

   Effort and Perseverance: It may reflect a deeper question about the effort and endurance required to achieve a goal or overcome a challenge.

   Escape or Avoidance: It could symbolize a desire to escape from a problem, difficulty, or threat, raising existential or psychological questions.

   Journey or Progress: It may ask how much more effort or time is needed to reach a particular milestone in life or a metaphorical destination.
3. *Emotional or Psychological Context:*

   Anxiety or Uncertainty: It might reflect feelings of doubt or overwhelm, questioning whether the effort required is worth it or achievable.

   Hope or Determination: Conversely, it could signify a hopeful inquiry, seeking clarity on the path forward to achieve success.
4. *Philosophical or Societal Implications:*

   Broader Challenges: This question might resonate in a social or global context, symbolizing the collective effort needed to address a significant issue like climate change, inequality, or progress.

The ChatGPT interpretation of "How far do we need to run?" that best aligns with Verdy's situation is the *Journey or Progress perspective*. This interpretation asks how much more effort or time is needed to reach a particular milestone, which is highly relevant to Verdy's need to improve its Performance Evaluation System (PES) and overall Human Resource Management (HRM) practices.

Verdy's PES is currently underdeveloped and primarily functions to differentiate employees based on remuneration. The company lacks structured HRM systems, which limits its capacity to align individual and organizational goals, identify skill gaps, and foster a culture of continuous improvement.

To address these shortcomings, Verdy needs to undertake a journey of significant organizational change:

Expanding PES Purpose: Verdy must evolve its PES beyond just awarding bonuses. The company needs to leverage the evaluation system for broader HRM practices, including promotions, salary adjustments, providing employee feedback, offering professional guidance, and designing development strategies.

Enhancing Transparency and Communication: A crucial step in Verdy's journey is to improve transparency and communication regarding the PES. This involves disseminating evaluation criteria, ensuring employees understand how they are assessed, and providing regular feedback.

Aligning with Strategic Goals: Verdy's journey should aim to create a PES that supports the company's ability to adapt to change and disruption. This involves identifying high-performing and adaptable employees and fostering a culture of continuous improvement.

Implementing Structured HRM Systems: To enhance its competitiveness in the face of disruptive forces, Verdy needs to implement more structured HRM systems. This journey involves addressing current HRM shortcomings to improve the company's agility and resilience.

The journey ahead for Verdy is substantial. The company needs to transform its PES from a simple bonus allocation tool to a comprehensive system that drives organizational effectiveness and employee engagement. This transformation requires significant effort in redesigning processes, training evaluators, and changing organizational culture.

The "Journey or Progress" interpretation of "How far do we need to run?" aptly captures Verdy's situation. The company has a considerable distance to cover in developing a PES that genuinely supports organizational agility and adaptability in the face of ongoing business disruptions. This journey requires sustained effort, clear milestones, and a commitment to continuous improvement in HRM practices.

Appendix

Theme	Subthemes	Excerpts
Perception about the company's PES and its process	Existence of evaluation	"Yes, I think so." (E4, F, 26, R, FA)
		"No, I don't think so." (E8, F, 50, A, FA)
	PES functioning and evaluation criteria	"It's a company policy for assigning bonuses and for the boss to understand what the employee does throughout the year." (E13, M, 24, R, FE)
		"The evaluation happens because there is a bonus at the end of the year." (E17, F, 36, A, FA)
		"The boss gives a bonus, and in my view, he evaluates people like, 'this one deserves this and that one deserves that.'" (E11, F, 45, A, FA)
		"We have training because the company has to provide it, but otherwise, honestly, we don't really go anywhere here." (E7, F, 23, R, FA)
		"Employees only think about the money as well." (E11, F, 45, A, FA)
		"At the time, he said he didn't do it alone; he asked for help from the manager here." (E8, F, 50, A, FA)

(continued)

(continued)

Theme	Subthemes	Excerpts
		"It's more related to the department; in the case of the office, it's mainly the boss." (E13, M, 24, R, FE)
		"It's once a year." (E9, F, 29, R, FA)
		"I know everyone receives it; some get more, others less. The minimum is a hundred and something, and the maximum is 80% of the base salary." (E4, F, 26, R, FA)
		"We don't have access to the parameters being evaluated." (E9, F, 29, R, FA)
		"They have access if requested; we don't post anything publicly." (E19, F, 37, A, FE)
		"I have them all here, but I won't read them. It goes from leadership, the ability of an individual who, for example, is usually downstairs organizing work, to attendance, flexibility, sometimes working overtime, doing multiple tasks, availability, integration, teamwork, organization, ethics, productivity, quality, and presentation." (E13, M, 24, R, FE)
		"I know it includes being versatile, being punctual, listening, being available, but I don't remember much else." (E4, F, 26, R, FA)
		"I asked about it once, and it's based on knowing how to do everything, not missing work, not being on leave, and all that." (E11, F, 45, A, FA)
		"Yes, employees ask each other, some share, some don't. When we see differences in bonuses, people start questioning the boss." (E9, F, 29, R, FA)
		"They receive a percentage of the salary, so if they do the math, they know what the percentage is." (E19, F, 37, A, FE)
Impact and recommendations regarding PES	Positive and negative aspects	"The bonus shows appreciation for our work, and it's always good for us." (E4, F, 26, R, FA)
		"Those who work well throughout the year usually receive a better bonus and feel more motivated." (E13, M, 24, R, FE)
		"I don't even know what the criteria are." (E2, F, 31, R, FA)

(continued)

(continued)

Theme	Subthemes	Excerpts
		"Favoritism is common; evaluations are not consistent." (E9, F, 29, R, FA)
		"There's no one to evaluate properly." (E11, F, 45, A, FA)
		"Another downside is the friction caused by the bonus system." (E13, M, 24, R, FE)
		"If someone takes a month off but works hard the rest of the year, they might still receive less due to that one month, which doesn't reflect the year as a whole." (E9, F, 29, R, FA)
	Premises	"Regarding the bonus, yes, it's fair." (E4, F. 26, R, FA)
		"No, it's not fair." (E8, F, 50, A, FA)
		"The evaluation here is only done for the bonus." (E5, F, 36, R, FA)
		"No, it's not pertinent." (E12, F, 39, A, FA)
		"I agree, but I think it should be more in-depth." (E16, F, 45, A, FE)
		"Regarding the bonus, yes, it's fair." (E4, F, 26, R, FA)
	Impact and recommendations for improvement	"No, if it's going to work like this, it's not worth it." (E13, M, 24, R, FE)
		"At the moment, I don't know my evaluation or score, but if I receive a bonus that I consider fair and on par with my peers, of course, I feel good because I put in the effort. However, if I get a mediocre bonus, I feel undervalued." (E5, F, 36, R, FA)
		"Of course, it influences; if the bonus is higher, it motivates me. If not, it demotivates me […]" (E6, M, 49, A, FA)
		"Knowing the criteria… the reasons, why I was evaluated in this way and not just being evaluated by the managers, it should also be done by the boss himself… that's my opinion." (E6, M, 49, A, FA)
		"If they had spoken to me, if they had talked to me before doing the evaluation, it would have been fairer […] I think that's what's missing here, communication is lacking." (E7, F, 23, R, FA)
		"I think so, because then people here, you see… I'm doing something wrong, and they don't say anything […]" (E10, F, 34, A, FA)

(continued)

(continued)

Theme	Subthemes	Excerpts
		"So how do I get the maximum bonus and still have the same salary as someone who gets 20%? It doesn't really make sense […]" (E4, F, 26, R, FA)
		"[…] responsibility, people with better knowledge, […] attendance, versatility, those who know how to organize the work and those who know how to organize, organization." (E6, M, 49, A, FA)
		"Quality in the work being done and organization." (E7, F, 23, R, FA)
		"I think it's important for everyone." [leadership criterion] (E8, F, 50, A, FA)
		"Ah yes, but only for the leadership. I think it also makes sense, you see, to try to understand the leader themselves […]" (E5, F, 36, R, FA)
PES impacts on work related satisfaction	Perceptions of PES Satisfaction	"Well, it's a good environment, […] our facilities are also good, the conditions are important, and the management as well […]" (E4, F, 26, R, FA)
		"For me, Job satisfaction is, as I said earlier, having a good work environment, teamwork […]" (E6, M, 49, A, FA)
		"Job satisfaction. Essentially, going to work should not feel like a punishment […] feeling happy doing it, enjoying what you do, feeling valued, and respected. All of this leads to satisfaction […]" (E20, M, 47, A, FE)
		"Yes. You could say so. [Job satisfaction]" (E19, F, 37, A, FE)
		"Girl, after 17 years, I'm still here. Because I'm fine here, right? My personal life is not very easy […]" (E12, F, 39, A, FA)
		"Day by day, I feel weaker here, because I see things, and working here becomes harder." (E8, F, 50, A, FA)
		"Right now, I'm more dissatisfied." (E10, F, 34, A, FA)
	Factors influencing job satisfaction	Main Causes of Satisfaction: "It's that I enjoy the work I do […]" (E11, F, 45, A, FA) "The positive factor is, well, the work. I actually like being here; I won't deny it." (E18, M, 45, A, FA)

(continued)

(continued)

Theme	Subthemes	Excerpts
		Main causes of dissatisfaction: "It's mostly the organization and then the environment; as I said, everyone is out for themselves here." (E9, F, 29, R, FA) "The work environment is the most negative aspect." (E10, F, 34, A, FA) "Organization, lack of organization, is the worst, even though I also dislike the gossiping and 'he said, she said.'" (E3, F, 28, R, FA) "It's the salary […] not being valued, they don't value you; whether you work more or less, it's all the same." (E4, F, 26, R, FA) "I think they create many unnecessary conflicts, perhaps due to the organization itself. I think above all, it's the organization's fault […]" (E19, F, 37, A, FE) "[…] I know that one of the points that should improve here in general is precisely communication." (E16, F, 45, A, FE)
	Recommendations to increase work-related satisfaction	"At least a meeting to clarify… I think they don't really know the workload that will come based on orders […] organization, for example, we arrive in the morning and often don't know where to start because we don't have access to the orders; the manager is the one who handles it." (E1, F, 23, R, FA)
		"First, better organization. Many of the arguments and so on happen because of poor organization." (E7, F, 23, R, FA)
		"Communication, because there are things that should go through me first but don't, and then there are delays, because I know the information goes astray, and then, well, the order gets delayed." (E17, F, 36, A, FA)
		"I think if he wants to keep someone and likes their work […] he should give them a salary increase to keep them motivated." (E8, F, 50, A, FA)
		"Salary progression […] having categories for each employee, something that doesn't exist here." (E18, M, 45, A, FA)

(continued)

(continued)

Theme	Subthemes	Excerpts
		"If there were another person managing everything, it would be easier because the lack of organization comes from her not having time for everything, and sometimes we start work at 8:20/8:30 because she hasn't arrived yet, and we don't know what the orders are." (E9, F, 29, R, FA)
		"Discipline […] there should be discipline; there isn't. There's no discipline for good manners, no discipline for boundaries […]." (E12, F, 39, A, FA)

References

Aggarwal, A., & Thakur, G. S. M. (2013). Techniques of performance appraisal-a review. *International Journal of Engineering and Advanced Technology, 2*(3), 617–621. https://www.ijeat.org/wp-content/uploads/papers/v2i3/C1188022313.pdf

Aguinis, H. (2013). *Performance management*. Pearson Education.

Almeida, F. N. (1996). *Avaliação de desempenho para gestores*. McGraw-Hill.

Alves, M. P. G. L. (2016). *Avaliação de Desempenho* [Dissertação de pós-doutoramento, Universidade Fernando Pessoa]. Repositório Institucional da Universidade Fernando Pessoa. https://bdigital.ufp.pt/handle/10284/5797

Antonioni, D. (1994). The effects of feedback accountability on upward appraisal ratings. *Personnel Psychology, 47*(2), 349–356.

Anwar, G., & Abdullah, N. N. (2021). The impact of Human resource management practice on Organizational performance. *International Journal of Engineering, Business, and Management, 5*(1), 35–47. https://doi.org/10.22161/ijebm.5.1.4

Armbrust, G. (2023). O que é avaliação de desempenho? Objetivos, tipos e como fazer. *Gupy Blog*. https://www.gupy.io/blog/avaliacao-de-desempenho#t3

Assis, F. (2021). Gestão do Desempenho. In: V. C. Paradela, R. L. B. Lopes, & A. P. G. Doro (Eds.), *Gestão de pessoas em ONGs* (pp. 96–108). Editora UFJF [Versão eletrónica]. https://repositorio.ufjf.br/jspui/bitstream/ufjf/13928/1/GestaodepessoasemONGs.pdf

Bakker, A. B., & Demerouti, E. (2007). The job demands-resources model: State of the art. *Journal of Managerial Psychology, 22*(3), 309–328.

Brewster, C., Chung, C., & Sparrow, P. (2016). *Globalizing human resource management*. Routledge.

Brutus, S., Fleenor, J. W., & London, M. (1998). Does 360-degree feedback work in different industries? *Human Resource Management, 37*(3), 245–256.

Calle-Quezada, P. A., Mora, S., Barragán-Landy, M. F. (2024). Modelos de evaluación del desempeño laboral: Una revisión sistemática de literatura. *Revista Gestión De Las Personas Y Tecnología, 17*(50), 88–117. https://doi.org/10.35588/0g6zzr18

Cascio, W. F., & Boudreau, J. W. (2016). *Investing in people: Financial impact of human resource initiatives*. Pearson Education.

Chiavenato, I. (2014). *Gestão de Pessoas. O novo papel dos recursos humanos nas organizações*. Editora Manole.

Choon, L. K., & Embi, M. A. (2012). Subjectivity, organizational justice and performance appraisal: Understanding the concept of subjectivity in leading towards employees´ perception of fairness in the performance appraisal. *Procedia–Social and Behavioral Sciences, 62*, 189–193. https://doi.org/10.1016/j.sbspro.2012.09.030

Colquitt, J. A., Conlon, D. E., Wesson, M. J., Porter, C. O., & Ng, K. Y. (2001). Justice at the millennium: A meta-analytic review of 25 years of organizational justice research. *Journal of Applied Psychology, 86*(3), 425–445.

Cunha, M. P., Rego, A., Cunha, R. C., Cabral-Cardoso, C., & Neves, P. (2016). *Manual de Comportamento Organizacional e Gestão*. Editora RH.

DeNisi, A. S., & Smith, C. E. (2014). Performance appraisal, performance management, and firm-level performance: A review, a proposed model, and new directions for future research. *Academy of Management Annals, 8*(1), 127–179.

Drogomyretska, M. I. (2014). The most promising employee evaluation methods in modern organizations. *Network Inteligence Studies, 2*(3), 38–44.

Edwards, M. R., & Ewen, A. J. (1996). *360-degree feedback: The powerful new model for employee assessment & performance improvement*. AMACOM.

Enes, S. C. (2019). *Implementação de um sistema de Avaliação de Desempenho numa PME*. [Projeto de Mestrado, Universidade do Minho]. RepositóriUM. https://repositorium.sdum.uminho.pt/handle/1822/60547

Espinosa, C. (2025). Retrieved January 2025, from https://www.ceicdata.com/en/portugal/number-of-companies/no-of-companies-micro-small--medium-sme

Ferreira, L. M., Alves, V. L. O., Lachi, H. M., & Cabral, D. F. (2023). *Avaliação de Desempenho*. Centro Universitário Anhanguera de São Paulo—Unidade Itaquera. https://repositorio.pgsscogna.com.br/handle/123456789/36062?filter_field_1=author&filter_type_1=equals&filter_value_1=ALVES%2C%20Vit%C3%B3ria%20Larissa%20Oliveira

Ghutke, S. (2014). Study of traditional methods of performance appraisal. *Abhinav-International Monthly Refereed Journal of Research in Management & Technology, 3*, 22–28.

Gigauri, I. (2020). Influence of COVID-19 crisis on human resource management and companies' response: The expert study. *International Journal of Management Science and Business Administration, 6*(6), 15–24.

Hadžimehmedagić, M., & Akbarov, A. (2013). *Traditional vs modern teaching methods. Advantages and Disadvantages*. Paper presented at the 3rd International Conference on Foreign Language Teaching and Applied Linguistics.

Judge, T. A., & Kammeyer-Mueller, J. D. (2012). Job attitudes. *Annual Review of Psychology, 63*(1), 341–367.

Kanaslan, E. K., & Iyem, C. (2016). Is 360 degree feedback appraisal an effective way of performance evaluation. *The International Journal of Academic Research in Business and Social Sciences, 6*, 172–182.

Kubiak, E. (2022). Increasing perceived work meaningfulness by implementing psychological need-satisfying performance management practices. *Human Resource Management Review, 32*(3), 1–16. https://doi.org/10.1016/j.hrmr.2020.100792

Lima, G. M. R. (2009). *Criação e validação de um questionário de satisfação com a avaliação de desempenho*. [Dissertação de Mestrado, Instituto Superior de Ciências do Trabalho e da Empresa]. Repositório do ISCTE. https://repositorio.iscte-iul.pt/handle/10071/2085

Lobo, V. L. C. P. (2020). *A cultura organizacional como gatilho da motivação e satisfação do trabalho*. [Dissertação de Mestrado, Instituto Politécnico de Tomar]. Repositório Comum. https://comum.rcaap.pt/bitstream/10400.26/36485/1/DISSERTA%C3%87%C3%83O%2020out.pdf

London, M., & Smither, J. W. (1995). Can multi-source feedback change perceptions of goal accomplishment, self-evaluations, and performance-related outcomes? *Personnel Psychology, 48*(4), 803–839.

Lunenburg, F. (2012). Performance appraisal: Methods and rating errors. *International Journal of Scholarly Academic Intellectual Diversity, 14*(1), 1–9.

Machado, C., & Novo, P. (2014). Human resource management challenges and changes in Portuguese SMEs. *Journal of Small Business and Enterprise Development, 21*(1), 125–139.

Mack, N., Woodsong, C., MacQueen, K. M.M., & Namey, E. (2005). *Qualitative research methods: a data collector´s field guide.* Family Health International.

Marques, J. (1992). *Recursos Humanos técnicas e gestão.* Texto Editora.

Narkunienė, J., & Ulbinaitė, A. (2018). Comparative analysis of company performance evaluation methods. *Entrepreneurship and Sustainability Issues, 6*(1), 125–138. https://doi.org/10.9770/jesi.2018.6.1(10)

Nascimento, A. L., Mercier, E. L., & Oliveira, F. E. (2017). Remuneração variável como fator motivacional para funcionários. *Rede Doctum de Ensino*, 1–23. https://dspace.doctum.edu.br/bitstream/123456789/1208/1/ARTIGO%20ALTERADO%20ATUALIZADO%2013%2012%202017.pdf

Noe, R. A., Hollenbeck, J. R., Gerhart, B., & Wright, P. M. (2017). *Human resource management: Gaining a competitive advantage.* McGraw-Hill Education.

Novo, P. R., & Machado, C. F. (2013). HRM in SMEs in Portugal: An innovative proposal of characterization. In C. Machado & P. Davim (Eds.), *Management and engineering innovation* (pp. 217–242). Wiley.

Otlvelre-Cestro, G. A., Lima, G. B. C., & Veiga, M. R. M. (1996). Implantação de um sistema de avaliação de desempenho: métodos e estratégias. *Revista de Administração, 31*(3), 38–52. https://repositorio.unb.br/handle/10482/1299

Paauwe, J., & Boselie, P. (2005). HRM and performance: What next? *Human Resource Management Journal, 15*(4), 68–82.

Parasuraman, A., Zeithaml, V. A., & Berry, L. L. (1988). SERVQUAL: A multiple-item scale for measuring consumer perceptions of service quality. *Journal of Retailing, 64*(1), 12–40.

Pereira, M. V. A. (2021). *Impacto das perceções de justiça organizacional na avaliação de desempenho sobre a satisfação no trabalho: Desigualdade de género.* [Dissertação de Mestrado, Instituto Superior de Contabilidade e Administração de Lisboa]. Repositório Científico. https://repositorio.ipl.pt/handle/10400.21/13970

Pulakos, E. D., Arad, S., Donovan, M. A., & Plamondon, K. E. (2000). Adaptability in the workplace: Development of a taxonomy of adaptive performance. *Journal of Applied Psychology, 85*(4), 612–624. https://doi.org/10.1037/0021-9010.85.4.612

Reifschneider, M. B. (2008). Considerações sobre avaliação de desempenho. *Ensaio: Avaliação e Políticas Públicas em Educação, 16*(58), 47–58. https://doi.org/10.1590/S0104-40362008000100004

Rosová, A., & Balog, M. (2012, November). *Traditional and modern methods and approaches to the evaluation of company performance* [Paper presentation] Carpathian Logistics Congress, Jeseník, Czech Republic, EU. [Slezská Ostrava]: TANGER, 2012 P. 1–7. ISBN 978-80-87294-33-8.

Schedler, M. F., & Schneider, E. (2015). Administração de recursos humanos—Política de avaliação de desempenho. *Revista Destaques Académicos, 7*(1), 20–30. http://www.univates.br/revistas/index.php/destaques/article/view/446/438

Serrano, O. M. E. (2010). *A Avaliação de Desempenho dos Trabalhadores da Administração Pública. O caso do Instituto Politécnico de Portalegre (IPP).* [Dissertação de Mestrado, Instituto Politécnico de Portalegre]. Repositório de Administração Pública. http://repap.ina.pt/handle/10782/563

Silva, B., Guimarães, L., & Machado, M. (2021). Estilos de Liderança e Satisfação no Trabalho. *Revista Multidisciplinar Humanidades e Tecnologias, 28*(1), 323–354. http://revistas.icesp.br/index.php/FINOM_Humanidade_Tecnologia/article/view/1532/1121.

Sousa, M. J., Duarte, T., Sanches, P. G., & Gomes, J. (2006). *Gestão de Recursos Humanos Métodos e Práticas.* Lidel–Edições Técnicas.

Ulrich, D., & Dulebohn, J. H. (2015). Are we there yet? What's next for HR? *Human Resource Management Review, 25*(2), 188–204.

Ybarra, O. (2023). The skills that help employees adapt: Empirical validation of a four-category framework. *PLoS ONE, 18*(2), e0282074. https://doi.org/10.1371/journal.pone.0282074

Chapter 6
Work-Family Conflict, Stress, and Job Satisfaction: The Mediating and Moderating Roles of Stress, Influence and Meaningful Work

Alexandra Fontes, José Carlos Pinho, Gina Gaio Santos, and Carlos Menezes

Abstract This study examines the impact of work-family conflict (WFC) on stress and its subsequent effect on job satisfaction among certified public accountants (CPAs). It also explores how influence at work and meaningful work moderate the relationship between WFC and stress. Data were collected through a structured online questionnaire administered to a sample of 4,843 CPAs in Portugal. The results indicate that stress acts as a mediator in the relationship between WFC and job satisfaction, emphasizing the role of stress as a critical factor that reduces job satisfaction due to WFC. Additionally, the findings suggest that influencing work lessens the negative effects of WFC on stress while experiencing meaningful work intensifies this relationship. This study contributes to the literature by providing a nuanced framework integrating WFC, stress, and job satisfaction within Portuguese CPAs. Organizations can utilize these insights to develop targeted strategies that enhance workplace well-being and mitigate the effects of WFC. By focusing specifically on CPAs—a crucial yet underexplored professional group—this research offers valuable implications for organizations aiming to support employee well-being and performance.

Keywords Work-family conflict · Stress · Job satisfaction · Certified public accountants

---

A. Fontes · J. C. Pinho · G. G. Santos (✉) · C. Menezes
School of Economics, Management and Political Science, University of Minho, Braga, Portugal
e-mail: gaiogina@eeg.uminho.pt

A. Fontes · J. C. Pinho
Centre for Research in Business, Markets & Society (iBMS), University of Minho, Braga, Portugal

G. G. Santos
CICS NOVA Research Centre, polo University of Minho, Braga, Portugal

C. Machado (ed.), *Human Resource Management in a Disrupted World*, Lecture Notes on Multidisciplinary Industrial Engineering,
https://doi.org/10.1007/978-3-031-89948-5_6

6.1 Introduction

Work-family conflict (WFC) significantly affects employees' well-being and job satisfaction in today's fast-paced and demanding work environment. Defined as the inter-role conflict arising when the pressures from work and family domains are incompatible, WFC can manifest in two directions: work interfering with family life and vice versa (Greenhaus & Beutell, 1985). This conflict has become increasingly prevalent due to the rise of dual-career families, evolving employment conditions, and heightened expectations from both work and family roles.

Previous research has established a strong link between WFC and several adverse outcomes, including job stress, burnout, and reduced job satisfaction (e.g., Le Fevre et al., 2003; Lee & Kleinman, 2003). Stress is 'the product of an imbalance between appraisals of environmental demands and individual resources' (Kinman & Jones, 2005, p. 101). Employees often face numerous stressors due to their professional responsibilities and family obligations (Park & Jang, 2017). Various sources of stress have been identified, including factors such as work overload, role ambiguity and conflict, poor workplace relationships, organizational climate, job insecurity, and WFC (Mansour & Tremblay, 2018; Richardson et al., 2008).

Despite the growing body of literature on WFC, a notable gap exists in understanding how personal resources, such as influence at work and meaningful work, moderate the relationship between WFC and stress. While existing studies have predominantly focused on work role overload and its direct effects (e.g., Edwards et al., 2002), this research focuses on the nuanced interplay between WFC, stress, and job satisfaction, particularly within the context of certified public accountants (CPAs).

This study contributes to the theoretical framework of the Conservation of Resources (COR) theory (Hobfoll, 1989, 2001, 2002) by examining how personal resources can buffer the adverse effects of WFC. It investigates the roles of workplace influence and meaningful work as moderating factors, suggesting that individuals' responses to WFC may vary based on these elements. The study emphasizes the importance of understanding how personal resources affect WFC and its outcomes (Lin et al., 2015; Mostafa, 2022). According to COR theory, there is a direct negative correlation between personal resources and stress (Wayne et al., 2007). This research expands theory by exploring the interaction between WFC and personal resources, focusing on how workplace influence and the sense of meaningful work can impact perceived stress levels.

This research fills a critical gap in the literature by focusing on the accounting profession—an underexplored area—and provides valuable insights into how organizations can implement targeted strategies to support employees in managing WFC. By addressing the moderating roles of influence at work and meaningful work, this study aims to enrich our understanding of the dynamics between WFC, stress, and job satisfaction, ultimately guiding organizations in fostering a healthier work environment.

This study proceeds as follows: We first present the theoretical background and hypotheses. Then, we present the method, followed by the results. We end with a discussion and conclusion that makes both theoretical and practical contributions.

6.2 Theoretical Background and Hypotheses

This study is grounded in Hobfoll's COR theory (Hobfoll, 1989, 2001), which posits that individuals are motivated to acquire, maintain, and protect resources they value to achieve their goals. Resources include tangible items (like cars and houses), personal traits (such as self-esteem and optimism), energy resources (like time and money), and conditions (such as job stability) (Barnett et al., 2012; Wayne et al., 2007).

According to COR theory, stress arises when individuals perceive a loss or a threat of losing these valued resources. This framework is particularly relevant in WFC, where the competing demands of work and family roles can lead to resource depletion, resulting in heightened stress levels. According to Greenhaus and Beutell (1985), WFC is a form of interrole conflict where the pressures from work and family domains are incompatible, leading to adverse outcomes such as increased stress and decreased job satisfaction. Previous research has established a robust link between WFC and stress, indicating that individuals who experience higher levels of WFC are more likely to report elevated stress levels (e.g., Calisir et al., 2011; Dodanwala et al., 2023; Pal & Saksvik, 2008; Singh & Nayak, 2015; Vickovic & Morrow, 2020). However, the role of personal resources in moderating this relationship has been less explored. This study aims to fill this gap by examining how two specific personal resources—meaningful work and influence—interact with WFC to affect stress levels.

6.2.1 The Relationship Between Work-Family Conflict, Stress and Job Satisfaction

The COR theory posits that individuals are motivated to acquire and protect resources they value to achieve personal goals (Hobfoll, 1989; Halbesleben et al., 2014). According to this theory, stress arises when individuals perceive a loss or potential loss of these valued resources (Grandey & Cropanzano, 1999; Hobfoll, 1989). Hobfoll (2001) describes it 'as a spiral of loss of resources', indicating that those lacking resources are at risk of further losses. Conversely, individuals with ample resources tend to gain more, leading to a 'spiral of gain of resources'. Thus, people aim to retain and enhance their resources to reduce stress (Braunstein-Bercovitz et al., 2012; Mauno et al., 2006).

According to Hobfoll's (1989) COR theory, professional accountants working in challenging and stressful environments (Pasewark & Viator, 2006; Persellin et al.,

2019) often deplete essential resources such as energy, time, and emotional resilience. As a result of Hobfoll's principle known as the 'spiral of loss of resources', individuals experience a reduction in available resources. This limitation complicates their ability to balance professional and family responsibilities, leading to WFC (Mansour & Tremblay, 2018). Gorgievski and Hobfoll (2008) further elaborate on this spiral, noting that the depletion of resources can trigger negative emotions and lead to a decline in both mental and physical health. According to the COR theory, individuals with limited resources are more vulnerable to additional losses; an initial loss can trigger a cascade of subsequent losses (Hobfoll, 2001). Consequently, WFC can result in further resource depletion, contributing to increased stress (Karatepe et al., 2010). Kossek et al. (2011) argued that resource depletion in the workplace can create challenges in meeting family obligations, further escalating WFC.

Stress arising from WFC often leads to job dissatisfaction (Carlson & Kacmar, 2000; Ford et al., 2007; Frone et al., 1992; Park & Jang, 2017; Singh & Nayak, 2015). Frone et al. (1992) suggested that stressors originating from family and work responsibilities can significantly affect job satisfaction. In this context, stressors related to work roles may conflict with family responsibilities, ultimately reducing overall job satisfaction. Because stress resulting from WFC is frequently linked to constraints in the workplace, it is expected to impact job satisfaction negatively. Numerous studies have shown a negative relationship between stress and job satisfaction (e.g., Guinot et al., 2014; Park & Jung, 2017; Singh & Nayak, 2015).

Drawing from the existing literature and COR theory, the following mediator hypotheses emerge:

Hypothesis 1: Stress mediates the relationship between WFC and job satisfaction, such that higher levels of WFC lead to increased stress, reducing job satisfaction.

6.2.2 Meaningful Work and Influence at Work as Moderating Factors

This study explores WFC as a critical stressor affecting well-being. The rise of dual-career families over the past thirty years has increased WFC, contributing to modern work stress (Braunstein-Bercovitz et al., 2012). WFC can cause stress by draining resources as individuals balance work and family roles. However, personal differences can influence how individuals manage these conflicts (Grandey & Cropanzano, 1999; Mostafa, 2022). The study suggests that meaningful work and a sense of influence can enhance coping abilities and reduce WFC-related stress.

Influence at work is characterized by the extent to which individuals can affect various aspects of their job, such as planning work and determining the order of tasks (Burr et al., 2019). Influence at work plays a fundamental role in individuals' well-being, as greater autonomy at work allows employees to manage work-related stress more effectively (Gözükara & Çolakoğlu, 2016; Lapierre & Allen, 2012; Li et al., 2021). It is a crucial job resource that can help mitigate the degree to which work interferes with family obligations since it can protect individuals from the stress and strain caused by work demands (Li et al., 2021). Thus, increased control over work responsibilities and decisions allows individuals to structure their jobs better, thereby minimizing work interference with family life (DiRenzo et al., 2011). Grönlund (2007) state that job control can transform stressful demands into positive challenges. The findings of their study suggested that job control may not only reduce strain but also yield benefits, particularly for women under conditions of high demand.

Based on the principles of COR theory and the literature, we propose the following hypothesis:

Hypothesis 2: Influence at work moderates the relationship between WFC and stress, such that the positive association between WFC and stress is weaker for individuals with higher levels of influence at work.

Meaningful work is 'the global judgement that one's work accomplishes significant, valuable, or worthwhile goals congruent with one's existential values' (Allan et al., 2019, p. 502). This definition suggests that when individuals perceive their work as meaningful, they are more likely to experience a sense of fulfilment. It involves a strong alignment between the demands of a work role and an individual's beliefs and values, showcasing an intrinsic interest in the work itself (Meng et al., 2023; Mostafa, 2022; Sánchez-Cardona et al., 2021). Therefore, the meaningfulness of work is recognized as an essential psychological resource that helps individuals cope effectively with the inherent challenges of their job (Meng et al., 2023; Mostafa, 2022; Sánchez-Cardona et al., 2021). In the study by Sánchez-Cardona et al. (2021), task variety, skill variety, and task significance are essential job resources that enhance meaningful work. These elements improve employees' sense of meaningfulness, leading to better engagement and performance. Additionally, employees who perceive more resources in their roles tend to find their work more meaningful and are more likely to wish to remain in their positions.

A key principle of the COR theory is that individuals must actively invest their resources to gain more or protect against losses (Hobfoll, 2001). Work meaningfulness is an essential resource that helps individuals accumulate additional resources, manage social interactions, and cope with job demands, ultimately enhancing their well-being (Fletcher, 2016). Allan et al. (2019) found that meaningful work directly predicts work engagement, commitment, and job satisfaction, influencing self-rated performance and organizational behaviors. Their findings also show that meaningful work is positively correlated with life satisfaction and general health while being negatively correlated with negative feelings. The authors highlight how meaningful work leads to positive attitude changes that drive behavioral shifts in the workplace (Allan et al., 2019).

Moreover, the holistic stress model (Simmons & Nelson, 2007) establishes that meaningfulness at work is a crucial indicator of eustress. This concept illustrates how an individual's cognitive appraisal of a situation or event can benefit them and significantly enhance their well-being (Simmons & Nelson, 2007, p. 45). In their study, Meng et al. (2023) underscore that meaningful work significantly moderates the relationship between hindrance stressors and burnout, indicating that those who perceive their work as meaningful experience less burnout.

According to Greenhaus et al. (2006), resources gained from work enhance one's non-work life through instrumental and emotional pathways. In the same vein, Steger and Dik (2010) emphasize that meaningful work is a vital resource that enriches life outside of work. Meaningful work clarifies one's roles in organizations and society, which helps reduce the psychological strain of balancing work and family responsibilities (Bragger et al., 2021). It also instils a strong sense of purpose, allowing individuals to align their actions across different life domains and develop a coherent life narrative (Johnson & Jiang, 2017).

Research shows that those who perceive their work as meaningful experience benefits that extend beyond the workplace, aiding them in navigating challenges, nurturing personal relationships, and enriching life (Johnson & Jiang, 2017). Ultimately, meaningful work facilitates resource transfer between roles, significantly reducing WFC. This suggests that meaningful work is a crucial protective factor against WFC. Chalofsky (2003) argues that work meaningfulness involves integrating our values into our professional lives. When work is meaningful, individuals can better manage pressure, develop skills, and achieve a balance between family and work demands.

Despite the growing body of research highlighting the importance of perceived meaningfulness at work, its potential moderating role in the relationship between WFC and stress remains underexplored (Minkkinen et al., 2020). Departing from COR theory, we propose that influencing and experiencing meaningful work can act as valuable resources that may diminish the relationship between WFC and stress.

These considerations lead us to the following hypothesis regarding moderation effects:

Hypothesis 3: Meaningful work moderates the relationship between WFC and stress, such that the positive association between WFC and stress is reduced for individuals who experience higher levels of work meaningfulness (Fig. 6.1).

6.3 Method

The study utilized a quantitative approach and gathered data through an online self-administered questionnaire using a cross-sectional design. A protocol was established with the National Institute of Chartered Accountants to access the list of certified accountants, as the Portuguese state regulates the profession. A preliminary questionnaire pretest was conducted with 172 certified accountants to validate the questionnaire. Subsequently, the questionnaire was distributed via email using Google Forms to a random sample of 5949 certified accountants by the National

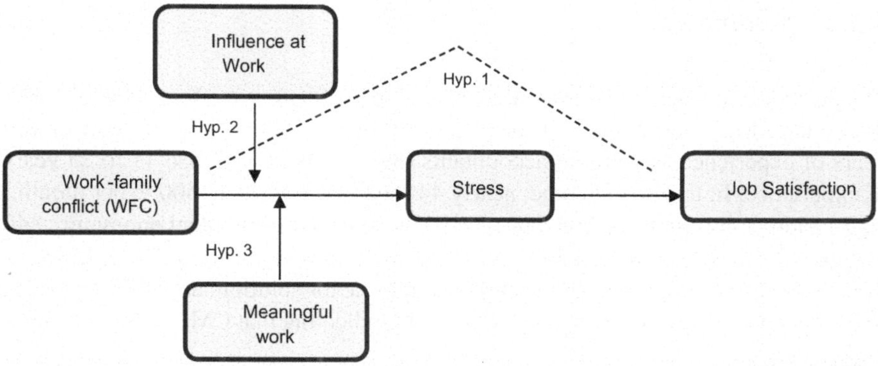

Fig. 6.1 (ALT-TEXT) shows the different variables under study and evidences their relationships. For example, Stress mediates the relationship between WFC and job satisfaction. Influence at Work and Meaningful Work moderate the relationship between WFC and Stress.

Institute of Chartered Accountants. The data collection process lasted two months, and two follow-up reminders were sent three and five weeks after the initial mailing to increase the response rate. After thorough scrutiny, 4833 valid questionnaires were considered for the research, representing nearly 80% of the initial sample.

6.3.1 *Measurement Scales*

We gathered most of the measurement scales used in this study from the relevant literature. Specifically, we considered various subdimensions of the Copenhagen Psychosocial Questionnaire, particularly those relating to Work-Family Conflict, Stress, Job Satisfaction, Influence at Work, and Meaning of Work.

Concerning WFC, it was measured by five items, two of which were taken from the work-like conflict sub-scale of the work-individual interface (Burr et al., 2019). An example of an item is: "Do the requirements of your work interfere with your private and family life?" We drew the remaining three items upon the questionnaire pretest, in which we interviewed six experts in the field and four academics. We used the DASS-21 measurement scale relative to the stress variable. The original scale was taken from Lovibond and Lovibond (1995). Pais-Ribeiro et al. (2004) adapted and validated it for the Portuguese context. An example of a stress item was: "I felt that I was using a lot of nervous energy." We have also included a subscale about "Influence at Work" of Burr et al. (2019), comprising four items. An example is: "Do you have a large degree of influence on the decisions concerning your work." Finally, from the same authors (Burr et al., 2019), we have included in the study a subscale of "Meaningful Work" which includes three items. An example of one item is: "Is your work meaningful?".

6.3.2 Sample Profile

The demographic analysis showed that 62% of respondents were female, while 38% were male. Most respondents were over forty-one (77%) with more than eleven years of experience as certified accountants (89%). About 42% had 11 to 25 years of experience. In terms of income, nearly 44% earned less than 1500 € per month.

To address common method bias (CMB), we ensured respondent anonymity and conducted a pretest, followed by a post hoc analysis using Harman's single-factor test (Podsakoff et al., 2003). We found no single-factor solution, and the VIF values of the dependent variable were all less than 3.0, indicating that CMB was not a threat to the study.

6.4 Results

Our analysis used the PLS-SEM approach with Smart PLS 4.0 software (Ringle et al., 2024) and conducted non-parametric bootstrapping with 5000 replications. This method provided robust confidence intervals using bias-corrected and accelerated bootstrap. The analysis followed two steps. Firstly, we evaluated the measurement model, focusing on the reliability and validity of reflective constructs. Secondly, we examined the structural model, considering R^2, path coefficients, and the standardized root mean square residual (SRMSR) as an indicator of model fit.

6.4.1 Assessment of the Measurement Model

The measurement model assesses the reliability and validity of various sub-dimensions. Validity encompasses two types: convergent and discriminant validity. Similarly, reliability comprises several significant coefficients: composite reliability (rho_c and rho_a) and average variance extracted (AVE) (Hair et al., 2017). (Table 6.1).

Discriminant validity was thoroughly examined using (1) The Fornell-Larcker criterion and (2) the Heterotrait-Monotrait (HTMT) criterion. In the first method, the AVE value was squared and rooted and compared against the inter-correlation of the constructs. All diagonal values were found to be higher than their non-diagonal counterparts. Regarding HTMT, previous research has suggested establishing discriminant validity thresholds of 0.85 and 0.90 (Henseler et al., 2015). Table 6.2 indicates no issues with discriminant validity in both cases.

Table 6.1 Results for the measurement model and reliability measures

Construct/dimension/item	Loading factors
Work-family conflict Cronbach's Alpha = 0.961; CR (Rho_A) = 0.961; AVE = 0.867	
There are moments that you need to work and manage your family time at the same time	0.831
Do you feel your work drains so much energy that it harms your private life?	0.961
Do you feel your work takes so much time that it harms your private life?	0.965
The requirements of your work interfere with your private and family life	0.959
Due to your work, you have to change your family plans	0.933
Stress Cronbach's alpha = 0.921; CR (Rho_A) = 0.923; AVE = 0.679	
I found it hard to calm down after something upset me	0.872
I tended to over-react to situations	0.868
I felt that I was using a lot of nervous energy	0.796
I found myself agitated	0.804
I found it difficult to relax	0.792
I was intolerant of anything that kept me from getting on with what I was doing	0.777
I felt that I was rather touchy	0.854
Job satisfaction Cronbach's alpha = 0.886; CR (Rho_A) = 0.914; AVE = 0.687	
Opportunities for promotion	0.906
Satisfaction with the job in general	0.913
Satisfaction with present pay	0.766
Satisfaction with working conditions	0.716
Satisfaction with one own capacity	0.825
Influence at workCronbach's alpha = 0.933; CR (Rho_A) = 0.934; AVE = 0.832	
Do you have a large degree of influence on the decisions concerning your work?	0.894
Can you influence the amount of work assigned to you?	0.892
Do you have any influence on what you do at work?	0.943
Do you have any influence on how you do your work?	0.921
Meaningful of work Cronbach's alpha = 0.927; CR (Rho_A) = 0.930; AVE = 0.872	
Is your work meaningful?	0.944
Do you feel that the work you do is important?	0.934
Do you feel engaged with your work?	0.923

6.4.2 Assessment of Structural Model and Hypotheses Testing

The findings suggest that the structural model effectively predicts WFC (dependent variable). The blindfolding procedure assessed the model by calculating the cross-validated redundancy index (Q^2). A Q^2 value greater than zero indicates the model's

Table 6.2 Correlations (descriptives, Fornell-Larcker and HTMT)

	Mean	S.D	1	2	3	4	**5**
1. Influence at work	3,57	1.20	**0.912**				
2. Job satisfaction	3.33	1.02	0.023 (0.026)	**0.829**			
3. Meaning of work	4.07	0.983	0.312 (0.335)	0.039 (0.043)	**0.934**		
4. Stress	1.03	0.70	−0.153 (0.165)	−0.255 (0.274)	−0.060 (0.065)	**0.824**	
5. Work-family conflict	3.11	1.25	−0.265 (0.280)	−0.070 (0.074)	−0.170 (0.182)	0.204 (0.216)	**0.931**

Note(s) Numbers in bold in the diagonal row are Square Roots of AVE, while numbers in the same column are values of correlations of reflective constructs. Numbers in brackets (2nd line) are constructs' **HTMT** values

predictive relevance. The structural model demonstrates satisfactory predictive relevance for all endogenous constructs. The coefficient of determination (R^2) shows that 60% of the total variance for Job Satisfaction is accounted for. Additionally, we computed the effect size (Cohen f^2) for each endogenous variable, and all values significantly exceeded the base level of 0.02. Furthermore, the model exhibits a good fit, with an SRMR of 0.66, well below the conventional cut-off of 0.08 (Hair et al., 2017), confirming the validity of our results.

Results indicate a non-significant relationship between WFC and Job Satisfaction ($\beta = -0.018$; $p > 0.05$). On the other hand, WFC's direct effect on Stress is positive and significant ($\beta = 0.178$; $p < 0.001$). Finally, the study confirmed the Stress's direct impact on Job satisfaction (($\beta = -0.251$; $p < 0.001$), which is significant although negative. (Table 6.3).

6.4.3 *Mediation Analysis*

Our study examined the pivotal role of stress in the relationship between WFC and Job Satisfaction. We followed the analytical approach outlined in Zhao et al.'s (2010) methodology, evaluating the significance of the indirect effects mediated by stress. We used bootstrapping to assess this mediating effect and tested for significance using a five- to ninety-percent confidence interval. This thorough approach underscores our confidence in our findings.

As shown in Table 6.3, Hypothesis H1, which examines the mediation effect of Stress between WFC and Job Satisfaction, is supported. The indirect effect is significant, with a positive path coefficient ($\beta = 0.045$; $p < 0.001$), suggesting that Stress is a mediating factor in reducing job satisfaction in the presence of WFC. The mediation is complete, as the direct effects of WFC on Job Satisfaction become

Table 6.3 Mediation and moderation effects

	Path coefficient	t-values (bootstrap)	Confidence interval	Decision
Mediation effect (indirect effects)				
H1: WFC → Stress → Job Satisfaction	0.045	10.308	[−0.053–(−) 0.036]	Supported
Moderation effects				
H2: control over influence at work X WFC → Stress	−0.017	1.486	[−0.038–0.005]	N_Supported
H3: control over meaning at work X WFC → stress	−0.040	3.328	[−0.064–(−) 0.016]	Supported
Direct effects				
WFC → job satisfaction	−0.018	1.265	[−0.048–0.009]	N_Supported
WFC → stress	0.178	12.820	[0.150–0.205]	Supported
Stress → job satisfaction	−0.251	17.005	[−0.280–(−) 0.222]	Supported

Based on t (4999); *$p < 0.05$; t (0.05; 4999) = 1.645; **$p < 0.01$; t(0.01; 4999) = 2.327; ***$p < 0.001$; t (0.001; 4999) = 3.092; † bias-corrected and accelerated (BCa) BCI; one-tailed

non-significant (Total effect: $\beta = -0.063$; $p < 0.001$ → Direct effect: $\beta = -0.018$; $p > 0.05$). In other words, there is full mediation.

6.4.4 Moderation Analysis

Our study explored the interaction effect of Influence at Work and the Meaningful Work on the WFC- Job Satisfaction relationship. The results do not support H2, which states that Influence at Work moderates the relationship between WFC and stress. In other words, results indicate no evidence that Influence at Work interacts with the WFC - Stress relationship ($\beta = -0.017$; $p > 0.05$). (See Fig. 6.2).

On the other hand, hypothesis H3 proposed that Meaningful Work experience moderates the relationship between WFC and stress. Specifically, it suggests that the positive association between WFC and Stress is lessened for individuals who perceive their work as Meaningful (see Fig. 6.3). This hypothesis is supported by showing a significant effect ($\beta = -0.040$; $p < 0.001$). This finding indicates that meaningful work can mitigate the negative effects of WFC on Stress levels.

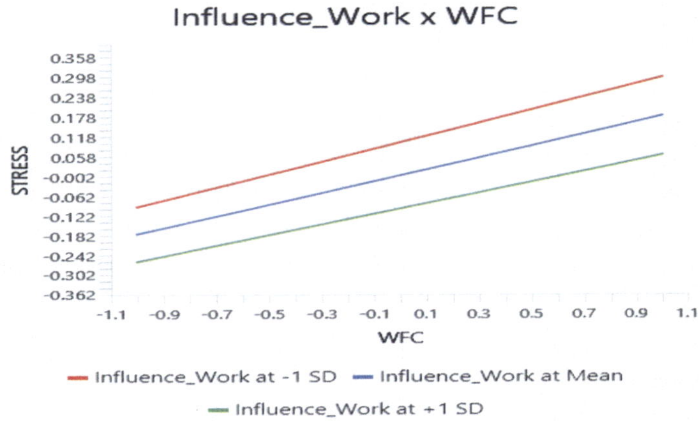

Fig. 6.2 (ALT-TEXT) This figure shows the interaction effect of influence at work on the WFC-Stress relationship. Results indicate that there is no evidence that Influence at Work interacts with the WFC–Stress relationship

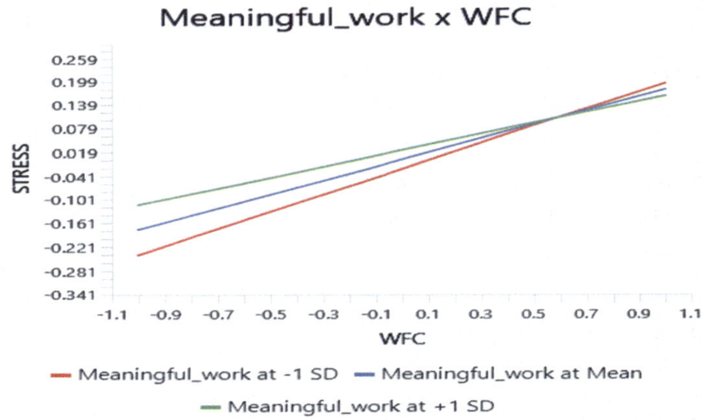

Fig. 6.3 *(ALT*-TEXT) This figure shows the interaction effect of meaningful work on the WFC-Stress relationship. Results suggest that the positive association between WFC and Stress is lessened for individuals who perceive their work as Meaningful

6.5 Discussion and Conclusions

This study explored the influence of WFC on the job satisfaction of CPAs, considering the mediating role of stress in this relationship. Additionally, it examined the impact of influence at work and meaningful work on the association between WFC and stress. In alignment with established research findings (Chen et al., 2014; Parker & Jang, 2017; Singh & Nayak, 2015), we assert that stress mediates the relationship between WFC and job satisfaction. This interference between work and family roles significantly impacts mental health, resulting in diminished job satisfaction.

In line with previous research (e.g., Bragger et al., 2021; Meng et al., 2023; Mostafa, 2022; Sánchez-Cardona et al., 2021), this study establishes the critical role of personal resources in shaping the relationship between WFC and stress. The findings demonstrate that the effects of WFC on stress levels differ significantly based on individual perceptions of meaningful work.

Furthermore, this study firmly supports the COR theory, which asserts that personal resources—such as a strong sense of purpose and control at work—enhance individual well-being and promote the development of additional resources, like stronger social relationships. This combination effectively safeguards against the adverse effects of work interference with family life and enhances overall well-being (Allan et al., 2019; Fletcher, 2016; Mostafa, 2022).

One significant finding is that influence at work, which refers to how employees perceive their ability to impact workplace decisions and outcomes, did not reduce stress levels. While theoretically having influence may provide individuals with greater autonomy and control, research indicates that the stress-buffering effects of autonomy can vary depending on specific contexts (Liao et al., 2019). In highly demanding situations like WFC, influence at work may not adequately address the emotional and cognitive strain caused by competing work and family demands. Additionally, the research by Lapierre and Allen (2012) showed that control at work was strongly related to reductions in work interference with family but only among employees who used more planning behavior.

Studies suggest (e.g., Liao et al., 2019; Marzocchi et al., 2024) that the buffering effects of workplace influence are often more effective in task-related stressors rather than stressors stemming from personal and role conflict, such as WFC. Influence alone may not directly alleviate the emotional toll of managing conflicting work and family roles but might require additional support mechanisms (e.g., organizational resources or social support).

Meaningful work gives employees a sense of purpose and intrinsic motivation, enhancing their psychological resilience (e.g., Bragger et al., 2021; Meng et al., 2023; Mostafa, 2022). The ability to derive meaning from work can help individuals reframe stressors as manageable or worthwhile, reducing their negative impact. This finding aligns with the COR model, which posits that resources like meaningful work mitigate the adverse effects of job demands. Organizations can leverage this by fostering a work environment that enhances employees' sense of purpose and alignment with organizational goals.

This study's results show that not all workplace resources are equally effective in mitigating stress. Influence at work may not directly address the personal and emotional strain caused by WFC. In contrast, meaningful work provides a deeper psychological buffer by enhancing resilience and reducing the perceived burden of stress. These findings underscore the importance of tailoring interventions to the specific nature of workplace stressors and leveraging resources that resonate with employees' intrinsic values.

6.5.1 Theoretical Contributions

This study makes a significant and necessary contribution to the literature on WFC, workplace stress, and job satisfaction by enhancing theoretical and practical understanding. It unequivocally establishes stress as a key mediator between WFC and job satisfaction, highlighting its critical role in addressing the harmful effects of imbalances between work and life. This finding strengthens theoretical models by demonstrating how WFC indirectly undermines job satisfaction. The direct relationships among WFC, stress, and job satisfaction provide compelling insights, particularly regarding the substantial impact of stress on job satisfaction. It highlights stress as a critical factor affecting employee well-being, emphasizing the urgent need for specific interventions.

Additionally, the significant moderation effect of control over meaning at work on the WFC-stress relationship highlights the importance of meaningful work in mitigating stress. This contribution enriches the literature on workplace resilience and aligns with positive organizational psychology perspectives.

6.5.2 Practical Implications

This study highlights vital implications for organizations. It suggests that they should implement strategies to reduce stress from WFC, which is shown to impact job satisfaction negatively. Effective interventions should focus on building coping skills to help individuals manage WFC-related stress and enhance their sense of control. Furthermore, improving planning and organizational skills can aid in better time and energy allocation between work and family responsibilities (Lapierre & Allen, 2012). At the organizational level, maintaining open communication with employers and family members about demands and conflicts can lead to collaborative solutions and improve work-life balance. It is also important to seek social support from colleagues, supervisors, or support networks when needed, as this can help mitigate the effects of WFC and improve job satisfaction (Li et al., 2021).

Furthermore, the findings indicate that enhancing meaningfulness at work can weaken the positive relationship between WFC and stress. Therefore, organizations should focus on strategies such as offering professional development opportunities, increasing job autonomy, improving leadership quality, communicating the significance of work, and updating career development programs to align with employee aspirations. By investing in employee meaning and autonomy, organizations can help reduce WFC (Bragger et al., 2021; Meng et al., 2023; Mostafa, 2022).

Conducting both formal and informal interviews, as well as periodic surveys, can help organizations assess levels of WFC, work meaningfulness, and influence at work. This approach allows organizations to identify employees who need family-friendly policies and determine the most appropriate measures. By targeting these specific needs, organizations can potentially reduce the costs associated with providing these benefits while also increasing their effectiveness in supporting employees (Mansour & Tremblay, 2018).

Management training should focus on fostering a supportive work environment. Managers must recognize WFC and stress, support employees in balancing responsibilities, and highlight the importance of each team member's role. Finally, if it is impossible to reduce WFC directly, managers can reduce stress by fostering a workplace climate of social support (Rahim, 1997). According to the COR Theory, social support is an essential environmental resource that can help reduce stress (Hobfoll, 1989). In addition, implementing work-life balance programs can further reduce stress levels (Park & Jung, 2017; Zheng et al., 2015).

6.5.3 Limitations and Future Research Directions

Our study has several limitations that future research should address. Since we surveyed only accountants, generalizing the results is challenging. Future studies should examine the applicability of these findings in other professions, like nursing and education, and consider conducting research in different countries to assess the impact of cultural values. Additionally, demographic factors such as gender, marital status, number of children, and spouse's employment should also be explored in future studies.

The non-significant findings regarding control over influence at work indicate that there are opportunities to investigate contextual factors or alternative moderating variables that could better clarify these dynamics. Additionally, the quantitative methods may not capture all individuals' opinions on specific issues. Future research employing qualitative data collection techniques, such as in-depth interviews and observations, could address this gap (Singh & Nayak, 2015).

One limitation of this study is that the findings come from data collected from the same respondents simultaneously, raising concerns about common method bias. Future research could address this by using data from multiple sources and periods. Additionally, this study focused only on work meaningfulness and influence as moderating factors; exploring other variables like professional efficacy, job complexity, or self-sacrificing disposition could be valuable.

References

Allan, B. A., Batz-Barbarich, C., Sterling, H. M., & Tay, L. (2019). Outcomes of meaningful work: A meta-analysis. *Journal of Management Studies, 56*(3), 500–528. https://doi.org/10.1111/joms12406

Barnett, R. C., Brennan, R. T., Gareis, K. C., Ertel, K. A., Berkman, L. F., & Almeida, D. M (2012). Conservation of resources theory in the context of multiple roles: An analysis of within- and cross-role mediational pathways. *Community, Work, & Family, 15*, 131–148. https://doi.org/10.1080/13668803.2010.539066

Bragger, J. D., Reeves, S., Toich, M. J., Kutcher, E., Lawlor, A., Knudsen, Q. E., & Simonet, D. (2021). Meaningfulness as a predictor of work-family balance, enrichment, and conflict. *Applied Research in Quality of Life, 16*(3), 1043–1071. https://doi.org/10.1007/s11482-019-09796-z

Braunstein-Bercovitz, H., Frish-Burstein, S., & Benjamin, B. A. (2012). The role of personal resources in work-family conflict: Implications for young mothers' well-being. *Journal of Vocational Behavior, 80*(2), 317–325. https://doi.org/10.1016/j.jvb.2011.10.003

Burr, H., Berthelsen, H., Moncada, S., Nübling, M., Dupret, E., Demiral, Y., Oudyk, J., Kristensen, T. S., Llorens, C., Navarro, A., Lincke, H. J., Bocéréan, C., Sahan, C., Smith, P., & Pohrt, A. (2019). The third version of the copenhagen psychosocial questionnaire. *Safety and Health at Work, 10*(4), 482–503. https://doi.org/10.1016/j.shaw.2019.10.002

Calisir, F., Gumussoy, C. A., & Iskin, I. (2011). Factors affecting intention to quit among IT professionals in Turkey. *Personnel Review, 40*(4), 514–533. https://doi.org/10.1108/00483481111133363

Carlson, D. S., & Kacmar, K. M. (2000). Work–family conflict in the organization: Do life role values make a difference? *Journal of Management, 26*(5), 1031–1054. https://doi.org/10.1016/S0149-2063(00)00067-2

Chalofsky, N. (2003). An emerging construct for meaningful work. *Human Resource Development International, 6*(1), 69–83. https://doi.org/10.1080/1367886022000016785

Chen, Z., Powell, G. N., & Cui, W. T. (2014). Dynamics of the relationships among work and family resource gain and loss, enrichment, and conflict over time. *Journal of Vocational Behavior, 84*(3), 293–302. https://doi.org/10.1016/j.jvb.2014.02.006

DiRenzo, M. S., Greenhaus, J. H., & Weer, C. H. (2011). Job level, demands, and resources as antecedents of work-family conflict. *Journal of Vocational Behavior, 78*(2), 305–314. https://doi.org/10.1016/j.jvb.2010.10.002

Dodanwala, T. C., Santoso, D. S., & Yukongdi, V. (2023). Examining work role stressors, job satisfaction, job stress, and turnover intention of Sri Lanka's construction industry. *International Journal of Construction Management, 23*(15), 2583–2592. https://doi.org/10.1080/15623599.2022.2080931

Edwards, A. B., Zarit, S. H., Stephens, M. A. P., & Townsend, A. (2002). Employed family caregivers of cognitively impaired elderly: An examination of role strain and depressive symptoms. *Aging & Mental Health, 6*(1), 55–61. https://doi.org/10.1080/13607860120101149

Fletcher, L. (2016). How can personal development lead to increased engagement? The roles of meaningfulness and perceived line manager relations. *The International Journal of Human Resource Management, 30*(7), 1203–1226. https://doi.org/10.1080/09585192.2016.1184177

Ford, M. T., Heinen, B. A., & Langkamer, K. L. (2007). Work and family satisfaction and conflict: A meta-analysis of cross-domain relations. *Journal of Applied Psychology, 92*(1), 57–80. https://doi.org/10.1037/0021-9010.92.1.57

Frone, M. R., Russell, M., & Cooper, M. L. (1992). Antecedents and outcomes of work–family conflict: Testing a model of the work–family interface. *Journal of Applied Psychology, 77*, 65–78.

Gorgievski, M. J., & Hobfoll, S. E. (2008). Work can burn us out or fire us up: Conservation of resources in burnout and engagement. *Handbook of Stress and Burnout in Health Care*, 7–22.

Gözükara, İ, & Çolakoğlu, N. (2016). The mediating effect of work family conflict on the relationship between job autonomy and job satisfaction. *Procedia-Social and Behavioral Sciences, 229*, 253–266. https://doi.org/10.1016/j.sbspro.2016.07.136

Grandey, A. A., & Cropanzano, R. (1999). The conservation of resources model applied to work-family conflict and strain. *Journal of Vocational Behavior, 54*(2), 350–370. https://doi.org/10.1006/jvbe.1998.1666

Greenhaus, J. H., & Beutell, N. J. (1985). Sources of conflict between work and family roles. *Academy of Management Review, 10*(1), 76–88. https://doi.org/10.5465/amr.1985.4277352

Greenhaus, J. H., Powell, G. N., Greenhaus, J. H., & Powell, G. N. (2006). When work and family are allies: A theory of work-family enrichment. *The Academy of Management Review, 31*(1), 72–92. https://doi.org/10.5465/amr.2006.19379625

Grönlund, A. (2007). More control, less conflict? Job demand-control, gender and work-family conflict. *Gender, Work and Organization, 14*(5), 476–497. https://doi.org/10.1111/j.1468-0432.2007.00361.x

Guinot, J., Chiva, R., & Roca-Puig, V. (2014). Interpersonal trust, stress and satisfaction at work: An empirical study. *Personnel Review, 43*(1), 96–115. https://doi.org/10.1108/PR-02-2012-0043

Hair, J. F., Hult, G. T. M., Ringle, C. M., & Sarstedt, M. (2017). *A primer on partial least squares structural equation modeling* (2nd ed.). Sage.

Halbesleben, J.R.B., Neveu, J.P., Paustian-Underdahl, S.C., & Westman, M. (2014). Getting to the "COR": Understanding the role of resources in conservation of resources theory. *Journal of Management, 40*(5), 1334–1364. https://doi.org/10.1177/0149206314527130

Henseler, J., Ringle, C. M., & Sarstedt, M. (2015). A new criterion for assessing discriminant validity in variance-based structural equation modeling. *Journal of the Academy of Marketing Science, 43*(1), 115–135. https://doi.org/10.1007/s11747-014-0403-8

Hobfoll, S. E. (1989). Conservation of resources: A new attempt at conceptualizing stress. *American Psychologist, 44*(3), 513–524. https://doi.org/10.1037/0003-066X.44.3.513

Hobfoll, S. E. (2001). The influence of culture, community, and the nested-self in the stress process: Advancing conservation of resources theory. *Applied Psychology, 50*(3), 337–421. https://doi.org/10.1111/1464-0597.00062

Hobfoll, S. E. (2002). Social and psychological resources and adaptation. *Review of General Psychology, 6*(4), 307–324. https://doi.org/10.1037/1089-2680.6.4.307

Johnson, M. J., & Jiang, L. (2017). Reaping the benefits of meaningful work: The mediating versus moderating role of work engagement. *Stress and Health, 33*(3), 288–297. https://doi.org/10.1002/smi.2710

Karatepe, O. M., Sokmen, A., Yavas, U., & Babakus, E. (2010). Work-family conflict and burnout in frontline service jobs: Direct, mediating and moderating effects. *Ekonomie a Management, 13*(4), 61–73.

Kinman, G., & Jones, F. (2005). Lay representations of workplace stress: What do people really mean when they say they are stressed? *Work and Stress, 19*(2), 101–120. https://doi.org/10.1080/02678370500144831

Kossek, E. E., Pichler, S., Bodner, T., & Hammer, L. B. (2011). Workplace social support and work–family conflict: A meta-analysis clarifying the influence of general and work–family-specific supervisor and organizational support. *Personnel Psychology, 64*(2), 289–313. https://doi.org/10.1111/j.1744-6570.2011.01211.x

Lapierre, L. M., & Allen, T. D. (2012). Control at work, control at home, and planning behavior: Implications for work-family conflict. *Journal of Management, 38*(5), 1500–1516. https://doi.org/10.1177/0149206310385868

Le Fevre, M., Matheny, J., & Kolt, G. S. (2003). Eustress, distress, and interpretation in occupational stress. *Journal of Managerial Psychology, 18*(7), 726–744. https://doi.org/10.1108/02683940310502412

Lee, P., & Kleinman, G. (2003). Statistical choices and apparent work outcomes in auditing. *Journal of Managerial Psychology, 18*(2), 105–125. https://doi.org/10.1108/02683940310465018

Li, A., Shaffer, J. A., Wang, Z., & Huang, J. L. (2021). Work-family conflict, perceived control, and health, family, and wealth: A 20-year study. *Journal of Vocational Behavior, 127*(2020), 1–15. https://doi.org/10.1016/j.jvb.2021.103562

Liao, E. Y., Lau, V. P., Hui, R. T. Y., & Kong, K. H. (2019). A resource-based perspective on work–family conflict: meta-analytical findings. *Career Development International, 24*(1), 37–73. https://doi.org/10.1108/CDI-12-2017-0236

Lin, W., Ma, J., Wang, L., & Wang, M. (2015). A double-edged sword: The moderating role of conscientiousness in the relationships between work stressors, psychological strain, and job performance. *Journal of Organizational Behavior, 36*, 94–111. https://doi.org/10.1002/job.1949

Lovibond, P. F., & Lovibond, S. H. (1995). The structure of negative emotional states: Comparison of the depression anxiety stress scales (DASS) with the Beck depression and anxiety inventories.

Behavior Research and Therapy, 33(3), 335–343. https://doi.org/10.1016/0005-7967(94)00075-U

Mansour, S., & Tremblay, D. G. (2018). Work–family conflict/family–work conflict, job stress, burnout and intention to leave in the hotel industry in Quebec (Canada): Moderating role of need for family friendly practices as "resource passageways." *The International Journal of Human Resource Management, 29*(16), 2399–2430. https://doi.org/10.1080/09585192.2016.1239216

Marzocchi, I., Nielsen, K., Di Tecco, C., Vignoli, M., Ghelli, M., Ronchetti, M., & Iavicoli, S. (2024). Job demands and resources and their association with employee well-being in the European healthcare sector: a systematic review and meta-analysis of prospective research. *Work and Stress*, 1–28. https://doi.org/10.1080/02678373.2024.2308812

Mauno, S., Kinnunen, U., & Ruokolainen, M. (2006). Exploring work- and organization-based resources as moderators between work-family conflict, well-being, and job attitudes. *Work and Stress, 20*(3), 210–233. https://doi.org/10.1080/02678370600999969

Meng, L., Du, J., & Lin, X. (2023). Surviving bench stress: Meaningful work as a personal resource in the expanded job demands-resources model. *Current Psychology, 42*(21), 17757–17768. https://doi.org/10.1007/s12144-022-02956-9

Minkkinen, J., Auvinen, E., & Mauno, S. (2020). Meaningful work protects teachers' self-rated health under stressors. *Journal of Positive School Psychology, 4*(2), 140–152.

Mostafa, A. M. S. (2022). The moderating role of self-sacrificing disposition and work meaningfulness on the relationship between work-family conflict and emotional exhaustion. *Journal of Happiness Studies, 23*(4), 1579–1597. https://doi.org/10.1007/s10902-021-00463-5

Pais-Ribeiro, J. L., Honrado, A., & Leal, I. (2004). Contribuição para o estudo da adaptação portuguesa das Escalas de Ansiedade, Depressão e Stress (EADS) de 21 itens de Lovibond e Lovibond. *Psicologia, Saúde & Doenças, 5*(1), 229–239. https://doi.org/10.1002/job.2595

Pal, S., & Saksvik, P. O. (2008). Work-family conflict and psychosocial work environment stressors as predictors of job stress in a cross-cultural study. *International Journal of Stress Management, 15*(1), 22–42.

Park, R., & Jang, S. J. (2017). Family role overload's relationship with stress and satisfaction. *Journal of Managerial Psychology, 32*(1), 61–74. https://doi.org/10.1108/JMP-01-2016-0020

Pasewark, W. R., & Viator, R. E. (2006). Accounting Profession. *Behavioral Research in Accounting, 18*, 147–165. https://doi.org/10.2308/bria.2006.18.1.147

Persellin, J. S., Schmidt, J. J., Vandervelde, S. D., & Wilkins, M. S. (2019). Auditor perceptions of audit workloads, audit quality, and job satisfaction. *Accounting Horizons, 33*(4), 95–117. https://doi.org/10.2308/acch-52488

Podsakoff, P. M., MacKenzie, S. B., Lee, J. Y., & Podsakoff, N. P. (2003). Common method biases in behavioral research: A critical review of the literature and recommended remedies'. *Journal of Applied Psychology, 88*(5), 879–903. https://doi.org/10.1037/0021-9010.88.5.879

Rahim, M. A. (1997). Relationships of stress, locus of control, and social support to psychiatric symptoms and propensity to leave a job: A field study with managers. *Journal of Business and Psychology, 12*(2), 159–174. https://doi.org/10.1023/A:1025018101302

Richardson, H. A., Yang, J., Vandenberg, R. J., DeJoy, D. M., & Wilson, M. G. (2008). Perceived organizational support's role in stressor-strain relationships. *Journal of Managerial Psychology, 23*(7), 789–81

Ringle, C. M., Wende, S., & Becker, J. (2024). *SmartPLS 4*. Bönningstedt: SmartPLS. https://www.smartpls.com

Sánchez-Cardona, I., Vera, M., & Marrero-Centeno, J. (2021). Job resources and employees' intention to stay: The mediating role of meaningful work and work engagement. *Journal of Management and Organization*. https://doi.org/10.1017/jmo.2021.10

Simmons, B. L., & Nelson, D. L. (2007). Eustress at work: Extending the holistic stress model. In C. L. Cooper, D. Nelson (Eds.), *Positive organizational behavior* (pp. 40–54). SAGE Publications.

Singh, R., & Nayak, J. K. (2015). Mediating role of stress between work-family conflict and job satisfaction among the police officials: Moderating role of social support. *Policing: An International Journal of Police Strategies & Management, 38*(4), 738–753. https://doi.org/10.1108/PIJPSM-03-2015-0040

Steger, M. F., & Dik, B. J. (2010). Work as meaning: Individual and organizational benefits of engaging in meaningful work. In P. A. Linley, S. Harrington, & N. Garcea (Eds.), *Oxford handbook of positive psychology and work* (pp. 131–142). Oxford University Press.

Vickovic, S. G., & Morrow, W. J. (2020). Examining the influence of work–family conflict on job stress, job satisfaction, and organizational commitment among correctional officers. *Criminal Justice Review, 45*(1), 5–25. https://doi.org/10.1177/0734016819863099

Wayne, J. H., Grzywacz, J. G., Carlson, D. S., & Kacmar, K. M. (2007). Work–family facilitation: A theoretical explanation and model of primary antecedents and consequences. *Human Resource Management Review, 17*(1), 63–76. https://doi.org/10.1016/j.hrmr.2007.01.002

Zhao, X., Lynch, J. G., & Chen, Q. (2010). Reconsidering Baron and Kenny: Myths and truths about mediation analysis. *Journal of Consumer Research, 37*(2), 197–206. https://doi.org/10.1086/651257

Zheng, C., Molineux, J., Mirshekary, S., & Scarparo, S. (2015). Developing individual and organizational work-life balance strategies to improve employee health and well-being. *Employee Relations, 37*(3), 354–379. https://doi.org/10.1108/ER-10-2013-0142

Part IV
Thematic Area: Organizational Culture and Employee Engagement

Chapter 7
Leadership, Organizational Culture, and Change Management: Navigating Complexity in a Disrupted World"

Paulo Viana and Carolina Feliciana Machado

Abstract Leadership is a multifaceted domain encompassing key leadership styles that significantly impact organizational outcomes, change processes, and cultures. Organizational change is an inevitable and complex process faced by all organizations, regardless of their size or industry. This paper explores the complex relationship between leadership, organizational culture, and change management, examining how diverse change models, leadership styles, and organizational culture dynamics influence the success or failure of change initiatives. The significance of organizational culture as a driving force and an obstacle to effective change is highlighted, emphasizing its impact on employee commitment, adaptation, and performance. Additionally, the paper explores the evolution of organizational culture and the emergence of subcultures in larger organizations, which further complicate the process of change.

Keywords Leadership · Organizational culture · Change management · Navigating complexity · Disrupted world

7.1 Introduction

Research in the field of leadership has traditionally been dedicated to the study of leadership styles and practices that increase effectiveness and productivity as well as employee motivation, job satisfaction, commitment, among other factors (Fosse et al., 2019). Leaders are considered responsible for the success and failure of organizational change initiatives (Bakari et al., 2017) and the style of leadership adopted by them may be influenced by the organizational culture, context and personal traits

P. Viana · C. F. Machado (✉)
School of Economics, Management and Political Science, University of Minho, Braga, Portugal
e-mail: carolina@eeg.uminho.pt

C. F. Machado
Interdisciplinary Centre of Social Sciences (CICS.NOVA.UMinho), University of Minho, Braga, Portugal

C. Machado (ed.), *Human Resource Management in a Disrupted World*, Lecture Notes on Multidisciplinary Industrial Engineering,
https://doi.org/10.1007/978-3-031-89948-5_7

(Mroz et al., 2018; Sousa & Rocha, 2019). It may diverge from context or from "situation to situation or change in a single setting over time" (Mroz et al., 2018, p. 2) and the leader's capacity to adapt himself to these situations, by providing and enhancing flexibility and skills to effectively make changes, is what provides an effective leadership. According to Tran and Gandolfi (2020), regardless of the leadership style's impact may have over time due to the flexibility that a leader should have to comply and adapt to fit the subordinates and the organization's needs, "any leader should acknowledge a thorough understanding of the different styles that effectively empower their leadership" (p. 201).

In today's constant evolution, evolving global landscape, organizations, regardless of their size, age or status, will experience change (Hussain et al., 2018; Tran & Gandolfi, 2020; Venus et al., 2019). Whether for economical, technological, cultural or environmental forces, change will undeniably occur. In fact, it never starts or ends, as it is an incessant set of continuously transforming conditions (Krauss & Vanhove, 2022; Supriharyanti & Sukoco, 2023). The complexity, unpredictability and volatility of such process, often lead to resistance and consequently, failure. Research denotes that between 40 and 70% of change initiatives fail (Yue et al., 2022), primarily due to resistance to change, lack of leadership support, and misalignment with organizational culture (Aitken & von Treuer, 2021; Naeem, 2020a). Due to this intricacy, the role of leadership becomes essential in conducting the change process contributing to overseeing the transition buy also with shaping culture. Hence, denominated culture's victims (Schein, 2009). On the other hand, organizational culture can perform as an enabler and an impediment to change, depending on how it is managed. The interconnection between organizational change and culture is particularly intricate, as culture is one of the most resistant aspects of an organization to transform (Payne et al., 2022). Resistance is not solely a negative force but can be seen as liberation tool for employees to express their concerns and also a valuable and respected source of feedback. Understanding the underlying causes of resistance—be it personal, organizational, or cultural—is essential for leaders who aspire to lead an organization its employees through the complexities of change. A nuanced approach to leadership, one that emphasizes transparency, trust, and communication, can mitigate resistance and foster a culture of openness to change (Jeong et al., 2016; Yue et al., 2019). The true challenge lies in aligning leadership styles with the existing culture, addressing resistance, and fostering an atmosphere where the process of change is not regarded as a disorder but as an opportunity for innovation and growth.

7.1.1 Authentic Leadership

Authentic leadership (AL) has a propensity to consider authenticity, honesty and sincerity as intrinsic attributes of this style and it is highly related to conduce to positivistic outcomes. Subordinates assume greater efforts, connect themselves in organizational citizenship, create new mind-sets along a better attitudinal behaviour, assume a trust enhancement, noticeable higher levels of motivation, the engagement and satisfaction at work and development increases and an outstanding increment on psychological capital is conspicuous (Alvesson & Einola, 2019). It appeals to "mass audiences eager to learn from, be inspired by, or mimic those who are perceived as successful in business"(p. 386). Compactly, "AL promises dramatic benefits" (p. 386). However, given its popularity, it is noticeable that in the last decade, a greater attention is being given to this style due to an inexplicable gist of enthusiasm amongst leadership researchers, as well as for consultants who seek to take advantage from attractiveness's trendy leadership theories, for personal or organizational benefits. Being authentic entails being vulnerable and genuine, allowing putting the self in the centre rather than job function and work eventualities providing a more personal environment. Therefore "the rewards associated with making oneself vulnerable by pursuing authenticity, AL and authentic followership at work will outweigh the risks more often than not" (Gardner et al., 2021, p. 5). Besides being aware of their beliefs and values, having high levels of self-confidence, authentic leaders provide special attention to enhancing follower's strengths rather than their weaknesses, to be in accordance with their true needs and, in a protective way, to fight against followers resource shortfalls (Braun & Nieberle, 2017). Such as other leadership styles (i.e. ethical leadership), authentic leaders take decisions with high transparency and openness involving subordinates in those decisions (decision-making processes) resulting in a reciprocal encouragement that provides trust and reliability (Bakari et al., 2017). Indeed, ethical and authentic leadership "share a common concern for a moral dimension of leadership" (Braun & Nieberle, 2017, p. 2). Leaders generate positive beliefs in subordinate's consciences to generate loyalty, commitment and involvement and adopt non mandatory behaviours (Bakari et al., 2017).

7.1.2 Authoritarian Leadership

Despite the rising popularity of numerous leadership styles such as ethical and servant, the authoritarian leadership (ATL) continues to be a prevalent style that captures the interest for research amongst scholars. This style is highly linked to most Asian, Middle East and numerous Western countries (Siddique et al., 2020). According to Zhang et al. (2020, p. 2) this leadership style is "in line with the respecting the superior notion emphasized by traditional Chinese culture". It is considered as a predominant and efficient style "because of its fit with traditional values" (Wang & Guan, 2018, p. 2). Authoritarian leaders are considered selfish, egocentric, self-centred, stress enhancers and creativity obstructers (Zheng et al., 2021) and inflexible. Due to its objectivity, ATL induces unambiguous and specific goals considered doable to the followers, providing focus and certainty. Thus, authoritarian leaders provide an enhancement of subordinate's sense of identity group participants, motivating them to high level performances (Schaubroeck et al., 2017) and through the exercise or severe control, setting objective instructions, enhancing responsibility and even when distributing punishment and recompenses, subordinates are driven to execute tasks with rigor, professionalism and with an outstanding quality (Wang & Guan, 2018). ATL is considered a style that blocks follower's future and vision. The attitudinal control and persuasion on total obedience acts as a major constraint on subordinates to make long term career paths (Siddique et al., 2020). ATL is also accountable for reducing follower's voluntary efforts through a demotivation process by which leaders infer the ineffectiveness and defenselessness of subordinates (Zheng et al., 2021). According to Sakkar Sudha and Shahnawaz (2020), when negativity occurs, authority, consistency and confidence are displayed by an authoritarian style.

7.1.3 Participative Leadership

Participative leadership (PL) is unique in achieving goals and reducing adverse impacts. It involves followers in decision-making, enhances their sense of ownership and combines their personal goals into organizational ones. Leaders invite subordinates in a participative process of managing by assigning values, organizing reporting among other encouragement strategies (i.e. making jointly decisions) (Wang et al., 2022). PL induces an atmosphere of open communication where subordinates feel valued and encouraged, instigating a sense of commitment towards leader-organization considerably reducing resistance to organizational change that may occur. By involving subordinates in decision making, PL creates "happier and more productive employees" (Lythreatis et al., 2019, p. 637). PL is categorized by "less formalization, an inclination of respect, power sharing, mutual understanding and eradication of reciprocal threats and antagonism" (Bhatti et al., 2019, p. 4). At an individual level, PL provides benefits but equally detriments to subordinates as an improvement of subordinate's well-being is noticeable, but an overreliance on PL by leaders can not only increase work responsibilities generating an onerous feeling to employees, resulting in a psychological stress's increase. (Wang et al., 2022). At an organizational level, although PL has an irrefutable impact on organizational creativity "does not promote team efficiency" (Li et al., 2018, pp. 657–658).

Researchers Lam et al. (2015) posit that some leaders may be disinclined to adopt a PL methodology due to a possible power shrinkage derived from this process. This is because, in the case of employee's failure to achieve the expected goals and objectives, a risk of culpability may arise due to the decision-making involvement. Therefore, rather than a subordinate's total involvement, some leaders prefer to exhibit a reasonable degree of knowledge sharing and decision-making, believing that this down cut is sufficient to improve performance without compromising their status to the organization. Considered an influential style for organization's survival and being considered the type that enhances the most knowledge sharing (Chang et al., 2020) and enhancing decision-making, Chang et al. (2020), argued that if decision-making and knowledge sharing are performance's enhancers, leader's encouragement to provide even more ideas exchange and managerial change, should be explored.

7.1.4 Despotic Leadership

Since its foundation, the domain of leadership has been idealized as the Holy Grail for organizations, highlighting leader's role valuable inputs on followers, but completely ignoring the dark style of leadership (Naseer et al., 2016). According to Nauman et al. (2020), the dark side of leadership is an ongoing pattern of behaviour exhibited by a leader that results in overall negative organizational outcomes based on the interactions between the leader, follower and the environment. Described as a form of supremacy in which a single entity or person rules with absolute power, despotic leadership concerns to leader's behaviours that oppress and are tyrannical. They concentrate themselves on acquiring absolute sovereignty, omnipotence, dominance and are extremely manipulative. It promotes anger and disgust amongst followers and triggers bullying comportments that obstinately and insistently provoke subordinates with negative actions such as exploitation, superiority and dominance (Syed et al., 2020). In collectivistic cultures' countries where poverty and unemployment rates are high, despotic leadership has an enormous influence on followers who are expected to comply without arguing and accepting power imbalances, due to the lack of alternatives from other sources of income (Nauman et al., 2020). Besides offensive and enhancing a demoralizing supervision, cruel, demeaning and aggressive, high self-centred (De Clercq et al., 2021) behaviours can be added (selfish and egocentric) intended to control, manipulate and abuse subordinates for personal advantage. Besides performing unethically and uncooperatively towards followers, it represses and neglects organization's objectives and goals "by engaging in fraudulent and morally incorrect self-serving behaviours" (Naseer et al., 2016, p. 16) and it also decreases performance, social responsibility and creativity (De Clercq et al., 2021; Naseer et al., 2016).

7.1.5 Destructive Leadership

Destructive leadership (DL) can be described as tyrannical, obnoxious supervision, bullying, "severe social stressor" (Fosse et al., 2019, p. 710), providing follower's harassment and humiliation, anger, anxiety and low self-regulation being also held responsible for objectives and goals failure (Collins & Jackson, 2015). According to Brandebo (2020), there are 7 destructive leadership behaviours: Over-controlling, lack of decisiveness, ambiguity, stressed, egocentric, threatens and lack of respect. Thoroughgood et al. (2018) refer to DL as a style that should not be measured solely on personality traits or behaviours. They define DL as "a social, or group, process that involves interactions between flawed, toxic, or ineffective leaders, susceptible followers, and conducive environments. Leader actions or inactions are a part of these processes, but not these processes alone" (p. 633). It is argued that leadership processes should be named destructive by the degree of harm directed to the group's welfare they are meant to serve and not specifically the leader. Fosse et al. (2019)

relate DL with other less harmful styles of leadership such as the laissez-faire. The authors argue that because laissez-faire is a "go with the flow" style, where managers are extremely permissive, avoid making decisions, abdicate responsibility and do not exert the power trusted, results in unwanted outcomes for the subordinates and organization by not complying with what is expected from them.

7.1.6 Ethical Leadership

Considered one of the "exemplary models" (Cullen, 2022, p. 487), ethical leadership (EL) has been widely researched and among all definitions provided by the literature and in accordance with several authors (Bush et al., 2021; Cullen, 2022; Egorov et al., 2020; Lara & Armas, 2019; Martin et al., 2022), stating that leaders play a vital part in the process of influencing subordinates regarding ethical and valuable outcomes concerning the organization. However, throughout the last decade, improvements to Brown et al.'s (2005) definition have been made. Kaptein (2019) adds moral entrepreneur to the two main components of EL (moral person and moral manager). That is, besides of just following current morality, ethical leaders also conduct morality through the creation of new norms. EL is negatively connected to employee's burnout, reducing emotional exhaustion, disengagement (Mostafa et al., 2021) and absenteeism (Eluwole et al., 2022). By enhancing employee's sense of optimism about organization's future and their place within, EL acts as motivator factor to contribute to the organization's success (Badrinarayanan et al., 2019). Hence, encouraging a two-way communication, reinforcing trust and inciting dialogue between members for a relationship of solidarity to each other engaging in helping behaviours, enhances the needed bond between leader and follower. Additionally, by acknowledging their fallibility, particularly when they are not required to, leaders induce a stronger relationship in the leader–follower relationship as honesty is an ethical behaviour that normatively appropriate, enhancing employee's ethical leadership evaluation (Zhang et al., 2024). Ethical leaders have also the power to develop and strengthen relationships, model behaviours (Al Halbusi et al., 2023), focus on procedures and not only results, providing commitment, protecting followers from failures and weak performances. When creating a new team for a task, organizations should rely on EL as a style of leadership that transmits confidence shall adversity arises and does not fall back when environments of uncertainty and doubt occur (Martin et al., 2022). Kalshoven et al. (2020) posits that the more ethical leadership is applied, the more ethical the environment becomes, discouraging employee's unethical behaviours such as knowledge hiding and their negative repercussions (i.e. lying and deception) to leader and the organization (Ali et al., 2024). In fact, EL fosters constructive employee behaviours such as knowledge sharing as it induces a culture of sharing know how and other valuable information (Sun et al., 2024a, 2024b) and by sharing information, enhances their prosocial motivation (Xia & Yang, 2020). EL provides job satisfaction, commitment, self-efficacy, increases performance, psychological wellbeing and reduces stress, triggers moral emotions and consequently avoids the violation of moral conducts and norms (Ahmed & Khan, 2023; Jeong et al., 2024). This also contributes to an emulation of ethical leader standards and conducts by the followers as they feel compelled to reciprocate this leadership style (Badrinarayanan et al., 2019; Sun et al., 2024a, 2024b) due to ethical leaders be seen as trustworthy and reliable role models (Ahmed & Khan, 2023; Ali et al., 2024; Eluwole et al., 2022; Jeong et al., 2024; Zheng et al., 2022). EL is also considered an organizational efficient performance booster as well as organizational culture enhancer (Alkhadra et al., 2023; Zahari et al., 2024) supported by leader's ethical attitude towards the organization. Brooks et al. (2023), posits that is it not even a question of necessity but a matter of survival to ensure organizations act ethically. Due to scandals regarding fraud, corruption in organizations worldwide, pressure from the organizations to leaders to behave ethically has increased significantly (Babalola et al., 2019). Organizations tend to reduce unethical behaviours by implementing a growing number of legal measures (Amory et al., 2024) as an augment of focus on ethics helps to avoid misconduct and can similarly institute a constructive strategy to enhance organizational competitive advantage (Treviño & Nelson, 2019).

7.1.7 Servant Leadership

Besides being considered "still in its early ages" (Chiniara & Bentein, 2018, p. 333) and a way of life rather than a management tool (Liu, 2019), servant leaders sacrifice themselves in favour of subordinates. They exert this style not in search of power and prestige, but with an intrinsic desire to serve (Sun & Shang, 2019) others and organizations in an attempt to "develop (…) employees to become independent and serve others" (Kimakwa et al., 2021, p. 13). Servant leadership (SL) puts the subordinate's needs first, help employees to understand their role in the company and adopt an empowering organizational culture for them to improve organizational performance (Eva et al., 2019; Heine et al., 2022). By exerting ethical and altruistic behaviours, SL positively influences the team and building self-confidence to their subordinates, constructing compassion, emotional support and enhancing feedback so that a strong, positive and long-term relationship with subordinates can be developed. Also, serving individuals distinctively, by putting subordinate's well-being over the leader's self-interest, by reducing the power distance between follower-leader, improving team solidity and collective performance, results in a proven organizational effectiveness (Chiniara & Bentein, 2018), that reduces absenteeism and job burnout as well as, through its ethical components (honesty, transparency), be of great assistance in the change process due to the clarification of organization's goals. Even in times of uncertainty, servant leaders demonstrate ethical integrity to support subordinates being motivated and valuable, helping them to deal with stress derived from the process of change and encouraging them to continue performing for personal and organizational improvement (Heine et al., 2022). Due to the fact that resilience augments work engagement, performance, increases job satisfaction, organizational commitment and reduces turnover intention (Eliot, 2020) and because SL "can have a positive impact on employee resilience at work" (p. 412), organizations are demanding human resources departments to implement training programs and cultivate servant leadership behaviours to enhance "servant leadership development" (Sims, 2018, p. 325). Notwithstanding of SL's innumerable values to the followers and organization and besides the trust, loyalty, engagement and even the transformational side that SL is linked to (Chiniara & Bentein, 2018; Eva et al., 2019; Heine et al., 2022; Kimakwa et al., 2021; Sun & Shang, 2019), employees, rather the empowerment provided, struggled to take action and make decisions during the leader's absenteeism (Liu, 2019).

7.1.8 Transformational Leadership

According to Bass (1990), transformational leadership (TL) is a superior performance leadership style occurring when leaders expand and elevate the interests of their employees, generating awareness and acceptance of the purposes and also when they stimulate their employees to look to the future beyond their own interests for the group's well-being. TL is one of the most effective styles for the organization regarding knowledge, human capital and innovation (Le & Lei, 2019) significantly affecting the ability to innovate by creating a positive climate, being able to praise and emphasize practices for creativity (Lei et al., 2020). According to Kim and Park (2020), this leadership style is also called inspirational because it encourages followers to transcend their own interests in favour of the organization and emphasizes the adoption of innovative practices to deal with situations of relative complexity. According to Jin et al. (2016), "Transformational leaders may be born or made"(p. 81). The authors found that TL is highly related with pleasantness at work and leaders who experience it, are more committed to the organization, hence, the more committed, the more transformational. Rather than a style with a static behavioural pattern, where leaders inspire until a certain level and then remain continuous, Tepper et al. (2018) argue that in the ordinary world, "highly transformational leaders are intermittently transformational" (p. 1345), and dynamically moderate themselves in accordance with the environment they are inserted. However, to motivate subordinates and inventing persuasive ways to convince them to exceed self-interest, chase joint goals and objectives may be exhausting, time consuming and emotional (Lin et al., 2019). Being a charismatic role model to subordinates will undoubtedly extract emotional feelings that can contribute to this exhaustion as well as the continuous visionary speeches (AL-Nawafah et al., 2022), that require extra time and energy. The detrimental part of considering others' needs and worries is that leaders disregard themselves, becoming emotional exhausted causing "thoughts of quitting" (Lin et al., 2019, p. 1560). According to Siangchokyoo et al. (2020), "a leader isn't truly transformational unless followers are transformed" (p. 1). Another assumption is the mixed definitions of TL and transformational leaders. TL's essence is to generate a strong commitment from followers to organization; while transformational leaders tend to transform individual's values, beliefs and mind-sets to perform beyond expectations (Siangchokyoo et al., 2020). Even inside TL, there are two types of leaders: (1) Authentic transformational leaders (role models to motivate followers achieving the extra mile towards the organization); and (2) Pseudo-transformational leaders (focused on their self-interest and status within the organization, authoritarian, and proactivity detraction proving low levels of performance) resulting in followers to perform "selfishly egoistically rather than altruistically" (Lin et al., 2017, p. 181). Controversially, TL has been highly related to the authoritarian style jeopardizing its true veracity and based in the statement "if transformational leadership is not dead, it is surely dying" (Siangchokyoo et al., 2020, p. 14), supports that this style, besides be one the most studied, has been questioned if its credibility is based on the likeness of empathy from followers or simply because it is another positivistic leadership style (Stock et al., 2023).

7.1.9 Transactional Leadership

Transactional leadership (TCL) has been "recognized as the foundation of effective leadership" (Young et al., 2021, p. 2) and acknowledged as a conservatory style due to its objectiveness and nothing extra besides it. That is, transactional leaders preserve stability, monitornig performamnce closely, checking for deviances and errors, leading subordinates to assume that this type of leadership is rigid, inflexible, bureaucatic and centralized. Transactional leaders are only concerned with objectivity and improving efficiency neglecting (unlike other styles of leadership) follower's initiatives and creativity to perform better. The followers, knowing that will not be rewarded for 'bringing out the box', limit themselves to what is expected by them restricting themselves to conformance and performing exactly as they are told, rather than exploring and developing opportunities (Afsar et al., 2017). According to Bass and Bass (2008), TCL is the exchange relationship between the leader and his followers with the aim of satisfying their own interests, entailing managerial functions by defining objectives to be achieved through the recognition of the performance achieved. The needs and desires of subordinates are recognized if they make an incremental effort to perform the task specifying the expected reward and if the result is achieved. This style of leadership comprises the following factors: (1) Contingent Reward–Leader assigns a task or obtains agreement from the follower on what is needed and arranges psychological or material rewards for followers in exchange for satisfactory completion of the task. (2) Management by Exception–If active, the leader monitors the deviations, errors and errors in the performance of the followers and applies corrective actions. If passive, the leader does not take any corrective action before a problem captures their attention by indicating poor follower performance. (3) Laissez-Faire–it implies a scale of non-leadership allowing to avoid taking a position, decision and abdicating of authority. However, besides being considered "the most frequently utilized leadership style in organizations" (Young et al., 2021, p. 16), transactional leaders by increasing obedience and reduce resistance, are considered "change blockers" (Afsar et al., 2017) due to the fact that of "instead of introducing change, trying out new solutions, and envisioning continuous innovations, they seek consistency, stability, status quo, and steadiness" (p. 312).

7.1.10 Organizational Change

For any organization, whether public or private, profitable or non-profitable, at least once during its lifetime, change will happen at a bigger or smaller scale (Hussain et al., 2018; Venus et al., 2019). It is an extremely complex process and failure occurs frequently due to several intrinsic factors that influence the success or failure of this phenomena. Depending on the situation and because of the unpredictable factors related to an organization, not all changes are equal, and the use of different models of change is necessary to comply with the process of change (Galli, 2018). "Change is a factor experienced by all organizations regardless of their size, age, or field" (Tran & Gandolfi, 2020, p. 199). Organizational change (OC) has been defined as an organization that continuously changes its direction to adjust the internal and external changing demands. Whether through the form of new ideas or behaviours, change has been introduced to employees by the organization as a starting point to a new era. According to Sinval et al. (2021), OC is intrinsically connected with behavioural factors corresponding to "a central feature of organizational behaviour, and in turn, in the lives of the individuals that compose them (p. 1). The authors defined OC as a process in which an organization (e.g., an organized, purposeful group, business, or department) alters minor to major structural components to address operational costs, productivity, and/or service quality deficiencies, identify new growth opportunities, or achieve other organizational goals. By its turn, Neill et al. (2020), also defined OC as a planned and structured process that consent organizations to adapt and perform in accordance to adjust new processes, new technologies, values and workforce. OC requires a dis-identification with former organizational strategies, values, culture to a re-identification within the new organization. The major challenge that nowadays organizations face, is, essentially, a constant change always occurring in a "from here to there" process where the "there" process is not a static; it is a flamboyant target—an incessant set of continuously transforming conditions (Supriharyanti & Sukoco, 2023). According to Castillo et al. (2018), every time the process of change starts, it leads to chaos in ordinary tasks, providing change in work routines, generating uncertainty and producing high levels of insecurity augmenting stress and resistance (Castillo et al., 2018). The role of the leaders to persuade followers to accept, commit and avoid resistance is crucial. Whether through sensemaking, storytelling and legitimization, communication is key for a successful organizational change (Shrivastava et al., 2022; Trabucchi et al., 2022). Shall a lack of communication exist and persist, many change efforts become unsuccessful (Shrivastava et al., 2022), generating stress, insecurity and doubt, increasing conflicts between employees and consequently resistance and turnover during or subsequently the process of change (Yue et al., 2022). Therefore, leaders must embrace the process of change as "an inevitable and continuous part or organizational life" (Holten et al., 2020, p. 394). Leaders should be prepared to play a role of increasing employee commitment by effectively managing the change process programme (Yanti & Dahlan, 2017). Success of change is then extremely related with employee organizational commitment. Loyal and faithful relationships from leaders to employees are crucial to engage them when organizational changes are established. The success of organizational changes can vary on several variables such as the kind of change it is, how it is transmitted to employees and how it is managed by leaders of the organization (Aujla & Mclarney, 2020). Despite the extensive literature review on OC, there is a propensity to all attempts fail. More specifically, between 40 and 70% of action concerning OC are considered a failure, whether due to resistance to change, lack of trust and support, low participation, politics, stress, low levels of commitment, absence of rewards, lack of employee and cultural integration and lack of leadership (Aitken & von Treuer, 2021; Naeem, 2020b; Yue et al., 2022). However, due to several factors, internal to an organization's environment, not all changes are equal and applying different models of change depending on the situation and environment is vital (Galli, 2018).

7.1.11 Kurt Lewin's CATS

Kurt Lewin is recognized as the "founding father of change management" (Cummings et al., 2015, p. 34) due to his remarkable ideology 'unfreeze-change-refreeze' also known as CATS (changing as three steps). Lewin believed that the organizations need to accept that change is necessary and the *status quo* needs to be put aside. A time to reflect on change and an organizational involvement analysis prior to "unfreezing" the organization (Galli, 2018, p. 125, 126) is necessary. This first step will undoubtedly generate uncertainty, disequilibrium and confusion but it is essential to change mind-sets for a new beginning and embrace new directions. Besides, a motivational factor must exist, otherwise "change does not occur" (p. 126). Lastly, when changes are gaining shape and mind-sets are equilibrated into "new processes into their routine and discontinued past practices" (p. 126), the organization is ready to stabilize (refreeze). According to Hussain et al. (2018), Lewin's model lacks additional steps that are crucial for the model's effectiveness. That is, the author states that between the processes of "unfreeze" to "change", employee involvement (EI) must be implemented. This factor contributes to raise subordinate's input not only into decisions that involve employee's well-being and the performance of the organization, but also to overcome resistance to change that will arise. Leadership plays a change agent's role in this model to unfreeze the organization. It is considered as "the key factor for change process" (Hussain et al., 2018, p. 126).

7.1.12 Nadler and Tushman Congruence Model

Nadler and Tushman (1980, p. 39) developed "a model that specifies the critical inputs (…) outputs, and the transformation processes that characterize organizational functioning". It is characterized as a model which sees organizations made of inputs, outputs and components and the interdependency between them. The authors stated that these components need to "fit" between them and shall one component plays an outlier role, the model fails. There are four inputs, namely: (1) Environment (markets, suppliers, governmental and regulatory bodies); (2) Resources (employees, technology); (3) History (development over a period of time and the impact of past events); (4) Strategy (to define the organizational outputs). The components are comprised by: (1) Task (how is it done and processed); (2) Individuals (knowledge and skills, experience, education, commitment, expectations and how they like to be rewarded for their work); (3) **Structure** (processes, methods, procedures developed to get employees perform consistently with strategy); (4) Culture (through the leadership style and co-worker's values and beliefs). The outputs are: (1) Organization (how well the objectives are achieved); (2) Team (teams utilizing all available resources); (3) Individual (individual-level outputs and behaviours). When inputs, outputs and components are combined, a transformation process arises (Nadler & Tushman, 1980).

7.1.13 Stace and Dunphy's Model

The contingency model presented by Dunphy and Stace is an extended version of Lewin's CATS. The authors stated that organizations should vary their change strategies in accordance with the environmental changes to achieve what Lewin stated as a perfect "fit". The model focuses in environmental factors as well as leadership styles along with a categorization of change of four different types: (1) Fine tuning (a minimum level of change, improving the "fit" between strategy, people, processes); (2) Incremental adjustment (moderate changes to structure or management procedures); (3) Modular transformation (restructuring and realignment occurs at this level, i.e. restructuring departments, goals changing, changing leaders); (4) Corporate transformation (radical changes ensues at this level, i.e. mission and core values, structural changes, systems and new leaders from outside). When collaborative modes along with incremental strategies are applied, a participative evolution is generated and it should be used when an organization is in "fit" but needs small adjustments and key interest groups are in favour of change. Shall collaborative modes be combined with transformative change strategies (charismatic change strategies), are to be applied if the organization is out of "fit", no time for participation but support for radical change is available. Regarding the coercive mode combined with transformative change strategies (dictatorial transformation), this approach should be used when radical change is vital to organization's survival (Stace & Dunphy, 1991).

7.1.14 Kotter's Model

For a wide range of organizations, "the much bigger challenge is leading change" (Kotter, 2012, p. 30). Kotter's change model is comprised by 8 steps, namely: (1) Establishing a sense of urgency (to change occur, it is crucial to be clear and understandable); (2) Creating the guiding coalition (for effective change, effective leaders must develop into a coalition to construct urgency around the need for change. People must be informed and realize that change is indispensable); (3) Developing a vision and strategy [Change will not be successful without a well- developed strategic vision because the organization does not own a specific layout guide for the process. The employees must comprehend why the change is needed in order to support it (Galli, 2018)]; (4) Communicating the change vision (a vision of change must be shared among employees by team members. Shall this process fails, resistance to change may arise); (5) Empowering employees for broad-based action [involve people to think about the change and how to attain it, and not to look for ways to stop it (Appelbaum et al., 2012)]; (6) Generating short-term wins (change's progress must be shared to everyone in order to keep changeable mind-sets open and focused); (7) Consolidate gains and producing more change [no complacency and continuous progress is imperative. Strengthen people through the process of change and developing them as change agents (Appelbaum et al., 2012)]; (8) Anchoring new approaches in the corporate culture (to institutionalize the change and to anchor it within the organizational culture. If not succeeded, all efforts implied were worthless and a revert process to old habits can occur (Appelbaum et al., 2012; Galli, 2018). Despite its undeniable and recognizable model as an remarkable starting point to implement change, the model "lacks rigorous fundaments" (Appelbaum et al., 2012, p. 765) and "should not be considered as something that guarantees success" (p. 776).

7.1.15 Van De Ven and Poole's Model

Van de Ven and Poole's model of change is based on an extensive literature review based on twenty dissimilar process theories that differ in terminology across disciplines. The authors found that these processes could be resumed into four basic schools of thought: (1) Life-cycle Theory—"change is imminent" (Van De Ven & Poole, 1995, p. 515) and its progression is based on a cumulative and a conjunctive unitary sequence. Similar to other models of change (i.e. Kotter), each step evolves from the previous one and relies on a logical five recognition stages: recognition, research on an issue, elaboration of an idea into a valuable state, commercialization, and diffusion and adoption (Van De Ven & Poole, 1995). (2) Teleological Theory—this approach is purposeful and adaptive and with a constructive mode of change, as it builds a planned end state and implies action to reach ends while monitoring the progress. Distinct of life-cycle theory, although considered an approach that considers planned change, it does not need a sequence to follow (Burton et al.,

2006; Van De Ven & Poole, 1995). (3) Dialectical Theory—Dialectical theory reports that the organization exists in a world filled with a plurality of collisions of events, forces or contradictory values that compete with each other for domination and control. In the process of dialectical theory, stability and change are explained by opposing power references. Forces and accommodations maintain the status quo in producing stability. Change occurs when opposing values, forces, or events gain a sufficient power to control the status quo (Van De Ven & Poole, 1995). (4) Evolutionary Theory—Similar to Kurt Lewin's model, this three-step approach relates to a continuous and repetitive cycle of variation, selection and retention where only the fittest survive. Variation relates to a new organizational process, while selection refers to the competition of scarce resources and retention involves forces (inertia and persistence) that perpetuate and maintain organizational forms (Burton et al., 2006; Van De Ven & Poole, 1995).

7.1.16 Organizational Culture

The interconnectivity amid the process of change and organizational culture (OCT) is, above all definitions, compound, as culture is the most challenging process related to an organization to change (Payne et al., 2022). OCT is considered a "multi-layer construct representing a complex phenomenon, a socially constructed and shared conscious and subconscious that guides employees' behaviour and attitudes" (Byrne et al., 2022, p. 808). According to Krauss and Vanhove (2022), continuous change is incessant, in constant evolution and progressively adaptive. As so, organizations in frequent change develop the continuous change concept. On the other hand, organizations that provide infrequent change represent episodical change which provides revolutionary change triggered by disorderly actions. OCT is a significant factor of whether organizations adopt episodical or continuous change (Albach et al., 2015), on the other hand, Schein and Schein (2016) argue that the more changes an organization suffers (frequent or infrequent), the more influence have on organizational culture. Throughout the last decades, a continuous search of how to make sense of OCT has been considered. According to Martin (1992), OCT can be considered in three perspectives, including integration (only one culture in an organization), differentiation (multiple subcultures) and fragmentation (puts in question if OCT truly exists). Leaders should not take for granted that OCT change will be accepted by all employees especially long-term employees which are traditionally averse to cultural change (Gover et al., 2016). OCT also is highly equated as the most considerable factor that increase or obstruct commitment and readiness to change (Haffar et al., 2023). Although leaders are considered, to some extent, the givers and definers of organizational culture (Boehm et al., 2015), Schein (2009) states that "leaders (...) are victims of culture" (p. xi), and their role is to create cultures and manage cultural issues and "if (leaders) do not manage culture, it will manage (them) and they may not even be aware of the extent to which this is happening" (p. 215).

Edgar Schein believed in a model of organizational culture based on a pyramid framework comprising three different levels: (1) Artifacts—the visible products of the team/group, as the architecture of the space, language, technology, products and style. Schein claimed that this level is "both easy to observe and very difficult to decipher" (Schein & Schein, 2016, p. 29) due to the complexity of different cultures (what makes sense to some, can make no sense to others). Although the importance of this level is acknowledged by Wei and Miraglia (2017), when they affirm that artifacts play a "key role in encouraging and enhancing knowledge transfer practices as continuing endeavour" (p. 578), it is a level that yields little insight into the company's culture and changing it would not result in a significant cultural change. (2) Espoused beliefs and values—company's acknowledged values and norms and these affect how employees behave and interact in the organization. They provide insight into an organization's culture but changing them will not provide a significant change (Burkus, 2014, para. 6). (3) Basic underlying assumptions—Although considered the essence of culture and the bedrock or organizational culture, they are the hardest to notice. They produce an effect on the creativity of the organization's employees. At this level, culture delivers to its employees a sense of identity and self-esteem (Burkus, 2014; Schein & Schein, 2016). Moreover, according to Wei and Miraglia (2017), due to the fact that values and beliefs are considered abstract and even reciprocally ambiguous, understanding the basic assumptions can help to decode and envision people's behaviour more accurately. Organizational culture is undoubtedly a mark of personality and acknowledgement that can be transformed in a lineage of thought throughout the years. OCT not only controls knowledge transfer but also suggests who controls and shares this knowledge and has the power to influence people's beliefs and selections, and "provides terms of reference for what is right or wrong" (Wei & Miraglia, 2017, p. 573). It is a mark of recognition that can provide pride, status and power, not only for the employees but also to the organization. It provides continuity, has a dynamic impact on performance, has the capacity to "influence people on decision making in response to the opportunities and threats that affects an organization" (Low et al., 2015, p. 918) and it is established as the most distinguished feature of successful companies (Arditi et al., 2017). Additionally, it is interpreted as the most imperative "enabler of, and impediment to, the management and transfer or organizational knowledge" (Wei & Miraglia, 2017, p. 573).

Although OCT "defines the way things are done in the organization" (Lee et al., 2018, p. 1179) has an enormous influence on employees to follow a rigorous and objective path to comply with the organization's standards and objectives, indirectly influences mental health problems by maximizing or minimizing the component roles of job design, demands, social relations and rewards, complying with the association of psychological distress via the influence it exerts on work organization (Dextras-Gauthier & Marchand, 2018). OCT's an invisible but extremely powerful social force that influences market-oriented behaviours, financial performance, subordinates attitudes and improves effectiveness in the organization (Hogan & Coote, 2014).

The role of the leaders to maintain, shape, preserve and control organizational culture is demanding. Once the organizational culture is primarily founded, leaders are the main responsible for "managing the evolution of the organization's basic assumptions and for modifying the culture in order to keep up with the changing demands of the environment" (Berson et al., 2008, p. 618). Besides the undeniable influence that organizational culture has to retain and acquire talent (Khan, 2016; Kontoghiorghes, 2016), it is also linked to the success of companies due to fact that culture can decide "whether or not the organization is able to survive under changing conditions" (Chong et al., 2018, p. 975). Culture plays a vital role not only to the organization but also to the process of developing leaders, creating a tremendous part on the organization's milieu and its consequences on leadership's development (Chong et al., 2018). In other words, leadership and culture complement each other as leadership stimuluses culture as much as culture influences leadership (Schein & Schein, 2016).

7.1.17 Organizational Culture's Four Models

Following the well-established framework provided by Cameron and Quinn (2006) suggests that OCT can be defined in four archetypes (clan, adhocracy, market and hierarchy). The clan culture is very similar to an extended family. It creates bounds of trust, commitment, share, mentoring and encourages teamwork (Assens-Serra et al., 2021; Cameron & Quinn, 2006). The market culture is results' dependant, hence, enhances competition and boosts the ideology to lead the market providing high levels of pressure. It pursues stability and focus on external environment elements such as customers, market regulators and suppliers to increase cost-effectiveness and efficiency (Aichouche et al., 2022; Assens-Serra et al., 2021; Strengers et al., 2022). The hierarchy culture provides stability, stimulates "clean lines of decision-making, authority, standardized rules and procedures" (Cameron & Quinn, 2006, p. 37), delivers formalities, focus, and efficiency in conjunction with low cost production. If clear roles are established along with roles and regulation, people perform in accordance (Strengers et al., 2022). If OCT is rules oriented, this type of culture favours the satisfaction of cooperation (Aichouche et al., 2022; Assens-Serra et al., 2021). Such organizations are dominated by strict and inflexible rules that provide to employees a low level of commitment and readiness to change (Haffar et al., 2023). According to Cameron and Quinn (2006), of all the 4 types, adhocracy is the only type which is prepared to face this trending technological, social, and economic boom that has been perceived, studied and researched by several authors and schools of thought. When combined with clan culture, adhocracy is the most desired culture by the employees (Strengers et al., 2022) and therefore allows a considerable number of cases that conduct to theme' attractiveness to scholars. Organizations characterized by an adhocracy culture encourage an individual's inventiveness, which boosts the efficiency of the organization and enhances employees' commitment to change by emphasizing flexibility and encouraging creativity (Cameron & Quinn, 2006; Haffar et al., 2023).

7.1.18 Subcultures' Influence

Culture evolves in an organization at the same pace as it progresses in a society. At the instance that an organization develops from small to medium to large, the established culture becomes complex boosting the arousal of subcultures (Kaul, 2019). Schein and Schein (2016) acknowledges three generic subcultures, namely: the operator subculture (human interaction oriented), the designer and engineering subculture (neglects human beings and relies on machines) and the executive subculture (emphasis on human capital). On the other hand, according to Gooderham et al. (2022), the concept of having a single and unitary organizational culture is skeptical. The authors believe that a dominant organizational culture is present but it is sustained through several local cultures, forming subcultures that can be different from the organization or completely antagonistic. Even inside a subculture, several subunits with divergent cultures may exist for several departments but all "contains common attributes that make up an overarching culture typical of the entire organization" (Cameron & Quinn, 2006, p. 147). Another significant factor is the fact that the bigger the organization, the more likely subcultures arise, becoming more disperse through department levels (Gooderham et al., 2022) and different subcultures are delineated by group leader's past cultural experiences that are transferred almost intrinsically from the previous group to the current one (Kim & Toh, 2019). Despite this and according to Strengers et al. (2022), after a considerable number of employees, it is particularly complex and strenuous to preserve one main dominant culture. The authors assert that this is due to the department's exponential growth, developing groups of interests, and consequently forming subunits and arising subcultures which considers "valuable for organizations and subsequently to understand underlying perspectives and values espoused" (p. 125). Contrariwise, according to Mello and Schloemer (2022), subcultures are groups that do not share the same organizational culture's values and norms allowing members to act independently and contrarily to the instigated existing culture. The more robust, resilient and influential existing subcultures are, the more persuasive it can be to change the direction of the organization "in ways not dictated by the overarching culture" (Mello & Schloemer, 2022, p. 158) monolithically increasing the complexity of change.

7.1.19 Resistance to Change

Resistance is considered a main obstacle to change (Venus et al., 2019), "one of the key sources of change failure" (Neves et al., 2021, p. 872) and as a constraint to the execution and failure of change initiatives (Veloso-Besio et al., 2019). Resistance to chance (RTC) is commonly associated to an impoverishment at an organization level and a motivational decreaser at an individual level (Veloso-Besio et al., 2019; Venus et al., 2019). Whenever experienced with a sudden or a premeditated event, resistance is a natural reaction, nevertheless the influence that change might have. RTC is other regarded as a defensive mechanism's individual asset that is observed when employees perceive change. They experience a reaction process which can be defined in 4 stages: "initial denial, resistance, gradual exploration, and eventual commitment" (Shimoni, 2017, p. 259). If change occurs on a frequent basis, employees face constant stress and discomfort resulting in an extremely difficult adaptation process (Rafferty & Jimmieson, 2017). However, "even when organizations try to introduce a change initiative that is consistent with a societal change that employees already value, employees may resist" (Malhotra et al., 2021, p. 476). Perseverance is a key issue due to the fact that, at whatever phase of the process of change that is undergoing, resistance is vastly high (Schweiger et al., 2018). However, factors such as stress can arise predominantly. Shall stress not be controlled, it generates further resistance obstructing successful change, allowing a continuous and persistent cycle that can annihilate change objectives (Schweiger et al., 2018). RTC can be defined into 2 paradigms. The traditional paradigm, where RTC is seen as a maladaptive anti-change reaction or a threat by defying, contracting, refusing to stop or delay the process of change. The modern paradigm incites that RTC is a valuable resource within the change process. By questioning and hesitating, employees show concern and commitment to understand, adapt and enrich the change process (Schweiger et al., 2018). In accordance with the authors, Galli (2018) argues that RTC can provide positive outcomes as it "…forces management to check their vision or roadmap to help identify problem areas (…) providing management with information about the intensity of an employee's emotions…" (p. 129). According to Waldman and O'Reilly (2018), when the process of change begins, only one third will embrace the change, one third are unconvinced and unsure and one third will actually resist to change divided by four main categories (Warrick, 2022): (1) Personal reasons—when people interests are threatened, uncertainty about what can derive from change; (2) Organizational reasons—In an organization managed with skills, trust and respected leaders who truly have concerns for the organizations and its employees, RTC tends to diminish in comparison with organizations lacking the same values. (3) Change agents—the influence and preponderance of change agents through their status and leadership style [considered a major reason that activates followers to resist change (Chatterjee et al., 2022)] can negatively influence followers and therefore increase resistance; (4) The way change is managed—If well managed, change can decrease resistance and even increase support for change. On the other hand, defectively change is the most significant basis to enhance RTC. Shall organizations recurrently disregard change, an organization's memory concerning change arises, and also upsurges RTC (Warrick, 2022).

7.1.20 *Openness to Change*

Openness to change (OTC) has been defined has "the extent to which employees are willing to support a change, and their positive affect about the potential consequences of the change" (Augustsson et al., 2017, p. 351). Employees are active participants of change initiatives instead of passive recipients, implying that their beliefs and behaviours considerably contribute (positively or negatively) to the change phenomena. Ertürk (2008) believes that employees openness to change is "the most important element of organizational culture that leaders should look to build" (p. 463) to help leaders to successfully manage the process of change. OTC is "characterized by innovative actions, free thinking, a readiness for new experiences and taking-risks" (Kruse et al., 2019, p. 7). Openness, honesty and encouragement at administrative levels lead to a greater employee receptiveness enhancing their commitment, trust and consequently their readiness to change (Jeong et al., 2016; Sinval et al., 2021). Transformational leadership is "significantly related to less resistance from followers to change and more commitment to change" (Jeong et al., 2016, p. 498) and it is a style that along with effective and transparent communication, induces to employee's organizational trust controlling uncertainty and psychological stress, leading to employee's openness to change (Yue et al., 2019).

7.1.21 *Leadership's Impact on Change and Culture's Influence*

Throughout an organization's lifetime, change will undeniably occur. Whether for economical, socio-political, environmental, technological or cultural factors, change is inevitable (Appelbaum et al., 2012; Neves et al., 2021; Venus et al., 2019). Organizations, in order to face the constant globalization, market demands and constant stock exchange fluctuations, have the "obligation" to constantly reinvent or modulate themselves. A sense of trust is prompted on leaders that have the role to perform as a balance scorecard between the members of the organization and the organization per se. According to Hussain et al. (2018), a fundamental aspect of leadership is its styles. Notwithstanding, some authors believe that depending on the environment and (i.e. radical change), although context is a key factor for influencing leadership styles (Wang et al., 2022) and variables (i.e. stress), some leadership styles are underrated but extremely effective and necessary. Rather concerning with positive leaders transforming and preparing follower's minds, beliefs and practices to change (Waheed et al., 2019), organizations should be aware that dark leadership styles such as destructive (although considered by some authors as "goal blockage" (Collins & Jackson, 2015, p. 386)) and "directive or coercive change leadership may be essential" (Stace & Dunphy, 1992, p. 205) during standardized processes of organizational change are necessary and extremely effective under extreme situations (Neves & Schyns, 2018).

At a cultural level, leaders act as breadwinners and contributors of culture (Srimulyani & Hermanto, 2022). Organizations, also denominated as cultures (Ehrhart et al., 2014), can also "shape" leadership's behaviours and traits influencing culture perspective (Srimulyani & Hermanto, 2022). According to Wei and Clegg (2018), when members of an organization realize that their organization will be acquired by other entity, a wave of stress, discontentment and resistance arises, originating a "sense of discontinuity in the identity of the pre-acquisition organization (becoming) detrimental to identity post-acquisition" (p. 337). Leaders act as "sensegivers and the providers of culture" (Srimulyani & Hermanto, 2022, p. 2) and their behaviours such as showing respect and appreciation for subordinates, along with effectiveness (Srimulyani & Hermanto, 2022) encourages and supports innovative behaviours (Hogan & Coote, 2014) to comply with organizational culture (OC) being considered primal to organizational effectiveness and leaders contemplated as truly important members in establishing culture. This reciprocal link is indispensable for organizations to perform effectively (Vito, 2020). According to Ehrhart et al. (2014), "organizations are cultures" (p. 125) and leaders, through their personality traits, influence the culture of the organization. Leadership and OC are "interdependent because every aspect of leadership practice eventually shapes OC" (Srimulyani & Hermanto, 2022, p. 5). Notwithstanding, their behaviour can also be shaped by the organizational culture, providing structure and meaning but constraining leadership's practices. Organizational outcomes are influenced by leadership and these outcomes are organizational culture dependents, therefore, the importance of OC and leadership in determining the success of the organization, is undeniable (Lokaj & Sadrija, 2020).

Although there is a reciprocal influence between OC and leadership, the effectiveness of both depends upon the leadership style that is being practiced. For instance, both despotic and destructive leadership completely ignore culture, channeling its focus on leader's self-interests and organizational objectives. Servant leadership (SL) besides their uniqueness in serving subordinates, does not provide richness to OC, whereas styles such as ethical, transformational and authentic, do focus on culture and on the continuity of it by retaining talent, enhancing reciprocal interactions between leader and follower, and therefore transmitting cultural values and beliefs of the organization. According to Azanza et al. (2013), authentic leadership is vital to OC as it "flows through to the followers (becoming) part of the fabric of the organizational culture" (p. 49).

7.2 Conclusion

Throughout the last decades, a vast majority of researchers offered particular interest to several leadership styles that are more suitable to attend the organization and follower's needs. Research has shown that the transformational and transactional styles are primordial, however there are alternative styles that deserve mention, such as the authentic style which provides transparency and openness (Bakari et al., 2017) but augments vulnerability (Gardner et al., 2021). The authoritarian style enhances the follower's sense of identity (Wang & Guan, 2018) nevertheless, at the same instance, blocks their future and vision (Siddique et al., 2020). The participative style induces respect and shares power (Bhatti et al., 2019; Wang et al., 2022). However, this power sharing can diminish leader's power, providing a sense of vulnerability (Lam et al., 2015). The similarity of the dark leadership styles characterized by despotic and dark is remarkable, however some discrepancies can be mentioned. Indeed, despite the despotic style be considered cruel, unethical, repressive and dominant (De Clercq et al., 2021; Naseer et al., 2016; Syed et al., 2020), according to De Clercq et al. (2021), these leaders are easible manipulated by followers by acquiring their confidence to benefit from less harmful punishments and higher benefits. On the other hand, the destructive style is contemplated as tyrannical, obnoxious and high levels of anger developer (Brandebo, 2020; Collins & Jackson, 2015; Schmid et al., 2018). However, some authors argue that culture and followers can influence and dictate this leadership style and it is responsible to avoid conflict between parties, providing more focus and motivating followers to comply with organization's objectives (Thoroughgood et al., 2018). Honesty, reliability and concern for follower's well-being (Egorov et al., 2020) are intrinsic to ethical leadership. However it is not suitable to all followers as it requires leader dependent and low esteem's followers to perform accordingly, as the less dependent, the less required this leadership style is needed (Velez & Neves, 2018). The servant leadership is considered a job satisfaction and organizational improvement and commitment increaser (Eliot, 2020; Heine et al., 2022), however it is leader dependent, obstructing followers to perform without leader's direct supervision (Liu, 2019). The transformational style is charismatic (Nassif et al., 2021), provides a proactive philosophy in optimizing the development and innovation process (Bass & Bass, 2008) and inspirational (AL-Nawafah et al., 2022; Tepper et al., 2018), but can be exhausting, emotional and time consuming to keep motivating followers to exceed themselves (Lin et al., 2019). In turn, besides transactional leaders be considered as obstructors of change, rigid, inflexible, bureaucratic and centralized (Afsar et al., 2017), this style is recognized as effective (Young et al., 2021) and particularly objective (Bass & Bass, 2008).

The true complexity that a process of organizational change has and the number of variables it is dependent of has been immensely researched. Due to this intricacy, any organizational change becomes diffuse, uncertain and unpredictable. In accordance to these variables, the use of different models of change is indispensable to comply with change (Galli, 2018). However, depending on variables such as culture and leadership, a careful model's choice has to be applied. Indeed, for change to occur, the organizational culture must not be discarded as it represents continuity (Low et al., 2015) through a set of philosophies transmitted from generations (Khan, 2016) and a true influencer for an episodical or a continuous change (Albach et al., 2015). Despite this, organizational culture is contemplated as the most considerable factor that increase or obstruct commitment and readiness to change (Haffar et al., 2023), can truly influence leadership (Schein, 2009) and decide the future of the organizations under a process of change (Chong et al., 2018). An impact of the new integration approach is perceived as well as enhancing forms of resistance (Wei &

Clegg, 2018). Considered a main obstacle to change, a primordial factor of change failure (Neves et al., 2021; Velez & Neves, 2018; Venus et al., 2019), a defensive mechanism's individual asset (Shimoni, 2017) and a motivational decrease factor (Veloso-Besio et al., 2019; Venus et al., 2019), resistance is crucial for a successful organizational change. Although is commonly seen as a maladaptive anti-change reaction (traditional paradigm), it is also regarded as a valuable resource within the change process (modern paradigm) as followers, through the process of questioning, show concern understand, adapt and enrich within the change process (Galli, 2018; Schweiger et al., 2018). Conversely, openness to change should be regarded as the main focus by leaders depositing efforts to preserve the existing organizational culture (Ertürk, 2008), as it allows free thinking and innovative actions (Kruse et al., 2019).

References

Afsar, B., Badir, Y. F., Saeed, B. B., & Hafeez, S. (2017). Transformational and transactional leadership and employee's entrepreneurial behavior in knowledge–intensive industries. *International Journal of Human Resource Management, 28*(2), 307–332. https://doi.org/10.1080/09585192.2016.1244893

Ahmed, M., & Khan, M. I. (2023). Beyond the universal perception: Unveiling the paradoxical impact of ethical leadership on employees' unethical pro-organizational behavior. *Heliyon, 9*(11), 1–16. https://doi.org/10.1016/j.heliyon.2023.e21618

Aichouche, R., Chergui, K., Brika, S. K. M., El Mezher, M., Musa, A., & Laamari, A. (2022). Exploring the relationship between organizational culture types and knowledge management processes: A meta-analytic path analysis. *Frontiers in Psychology, 13*, 1–12. https://doi.org/10.3389/fpsyg.2022.856234

Aitken, K., & von Treuer, K. (2021). Leadership behaviours that foster organisational identification during change. *Journal of Organizational Change Management, 34*(2), 311–326. https://doi.org/10.1108/JOCM-01-2020-0029

Al Halbusi, H., Ruiz-Palomino, P., & Williams, K. A. (2023). Ethical leadership, subordinates' moral identity and self-control: Two- and three-way interaction effect on subordinates' ethical behavior. *Journal of Business Research, 165*, 1–18. https://doi.org/10.1016/j.jbusres.2023.114044

Albach, H., Meffert, H., Pinkwart, A., & Reichwald, R. (2015). *Management of permanent change* (1st ed.). Springer Gabler.

Ali, M., Usman, M., Khan, M. A. S., Shafique, I., & Mughal, F. (2024). "Articulating cognizance about what to hide what not": Insights into why and when ethical leadership regulates employee knowledge-hiding behaviors. *Journal of Business Ethics, 190*(4), 885–895. https://doi.org/10.1007/s10551-023-05426-9

Alkhadra, W. A., Khawaldeh, S., & Aldehayyat, J. (2023). Relationship of ethical leadership, organizational culture, corporate social responsibility and organizational performance: A test of two mediation models. *International Journal of Ethics and Systems, 39*(4), 737–760. https://doi.org/10.1108/IJOES-05-2022-0092

AL-Nawafah, S. S., Al-Amaeara, A. F., & Alkhawaldah, R. A. (2022). Applying transformational leadership in Jordan governmental institutions. *Journal of Management Information & Decision Sciences, 25*, 1–13.

Alvesson, M., & Einola, K. (2019). Warning for excessive positivity: Authentic leadership and other traps in leadership studies. *The Leadership Quarterly, 30*(4), 383–395. https://doi.org/10.1016/j.leaqua.2019.04.001

Amory, J., Wille, B., Wiernik, B. M., & Dupre, S. (2024). Ethical leadership on the rise? A cross-temporal and cross-cultural meta-analysis of its means, variability, and relationships with follower outcomes across 15 years. *Journal of Business Ethics*, 1–29. https://doi.org/10.1007/s10551-024-05632-z

Appelbaum, S. H., Habashy, S., Malo, J.-L., & Shafiq, H. (2012). Back to the future: Revisiting Kotter's 1996 change model. *Journal of Management Development, 31*(8), 764–782.

Arditi, D., Nayak, S., & Damci, A. (2017). Effect of organizational culture on delay in construction. *International Journal of Project Management, 35*(2), 136–147. https://doi.org/10.1016/j.ijproman.2016.10.018

Assens-Serra, J., Boada-Cuerva, M., Serrano-Fernandez, M.-J., & Agullo-Tomas, E. (2021). Gaining a better understanding of the types of organizational culture to manage suffering at work. *Frontiers in Psychology, 12*, 1–12. https://doi.org/10.3389/fpsyg.2021.782488

Augustsson, H., Richter, A., Hasson, H., & von Thiele Schwarz, U. (2017). The need for dual openness to change: A longitudinal study evaluating the impact of employees' openness to organizational change content and process on intervention outcomes. *The Journal of Applied Behavioral Science, 53*(3), 349–368. https://doi.org/10.1177/0021886317691930

Aujla, S., & Mclarney, C. (2020). The effects of organizational change on employee commitment. *IUP Journal of Organizational Behavior, 19*(1), 7–22.

Azanza, G., Moriano, J. A., & Molero, F. (2013). Authentic leadership and organizational culture as drivers of employees' job satisfaction. *Journal of Work and Organizational Psychology, 29*(2), 45–50. https://doi.org/10.5093/tr2013a7

Babalola, M. T., Bligh, M. C., Ogunfowora, B., Guo, L., & Garba, O. A. (2019). The mind is willing, but the situation constrains: Why and when leader conscientiousness relates to ethical leadership. *Journal of Business Ethics, 155*(1), 75–89. https://doi.org/10.1007/s10551-017-3524-4

Badrinarayanan, V., Ramachandran, I., & Madhavaram, S. (2019). Mirroring the Boss: Ethical leadership, emulation intentions, and salesperson performance. *Journal of Business Ethics, 159*(3), 897–912. https://doi.org/10.1007/s10551-018-3842-1

Bakari, H., Hunjra, A. I., & Niazi, G. S. K. (2017). How does authentic leadership influence planned organizational change? The role of employees' perceptions: Integration of theory of planned behavior and Lewin's three step model. *Journal of Change Management, 17*(2), 155–187.

Bass, B. M. (1990). From transactional to transformational leadership: learning to share the vision. *Organizational Dynamics, 18*(3), 19–31.

Bass, B., & Bass, R. (2008). *The bass handbook of leadership : Theory, research, and managerial applications* (4th ed.). Free Press.

Berson, Y., Oreg, S., & Dvir, T. (2008). CEO values, organizational culture and firm outcomes. *Journal of Organizational Behavior, 29*(5), 615–633.

Bhatti, M. H., Ju, Y., Akram, U., Bhatti, M. H., Akram, Z., & Bilal, M. (2019). Impact of participative leadership on organizational citizenship behavior: Mediating role of trust and moderating role of continuance commitment: evidence from the pakistan hotel industry. *Sustainability, 11*(4). 1170. https://doi.org/10.3390/su11041170

Boehm, S. A., Dwertmann, D. J. G., Bruch, H., & Shamir, B. (2015). The missing link? Investigating organizational identity strength and transformational leadership climate as mechanisms that connect CEO charisma with firm performance. *The Leadership Quarterly, 26*(2), 156–171. https://doi.org/10.1016/j.leaqua.2014.07.012

Brandebo, M. F. (2020). Destructive leadership in crisis management. *Leadership and Organization Development Journal, 41*(4), 567–580. https://doi.org/10.1108/LODJ-02-2019-0089

Braun, S., & Nieberle, K. W. A. M. (2017). Authentic leadership extends beyond work: A multilevel model of work-family conflict and enrichment. *Leadership Quarterly, 28*(6), 1–18. https://doi.org/10.1016/j.leaqua.2017.04.003

Brooks, S., Richmond, J., & Blenkinsopp, J. (2023). Applying a lens of temporality to better understand voice about unethical behaviour. *Journal of Business Ethics, 188*(4), 681–692. https://doi.org/10.1007/s10551-023-05505-x

Burkus, D. (2014). *How to tell if your company has a creative culture.* Harvard Business Review. https://hbr.org/2014/12/how-to-tell-if-your-company-has-a-creative-culture
Burton, R. M., Eriksen, B., Hakonsson, D., & Snow, C. (2006). *Organization design: The evolving state-of-the-art* (vol. 6). Springer.
Bush, J. T., Welsh, D. T., Baer, M. D., & Waldman, D. (2021). Discouraging unethicality versus encouraging ethicality: Unraveling the differential effects of prevention- and promotion-focused ethical leadership. *Personnel Psychology, 74*(1), 29–54.
Byrne, Z. S., Cave, K. A., & Raymer, S. D. (2022). Using a generalizable photo-coding methodology for assessing organizational culture artifacts. *Journal of Business and Psychology, 37*(4), 797–811. https://doi.org/10.1007/s10869-021-09773-0
Cameron, K. S., & Quinn, R. E. (2006). *Diagnosing and changing organizational culture : Based on the competing values framework* (Revised Ed). Jossey-Bass.
Castillo, C., Fernandez, V., & Sallan, J. M. (2018). The six emotional stages of organizational change. *Journal of Organizational Change Management, 31*(3), 468–493. https://doi.org/10.1108/JOCM-05-2016-0084
Chang, Y. Y., Chang, C. Y., Seih, Y. T., Chen, Y. C. K., & Chang, S.-Y. (2020). Participative leadership and unit performance: evidence for intermediate linkages. *Knowledge Management Research and Practice*, 1–15. https://doi.org/10.1080/14778238.2020.1755208
Chatterjee, S., Chaudhuri, R., Vrontis, D., & Jabeen, F. (2022). Digital transformation of organization using AI-CRM: From microfoundational perspective with leadership support. *Journal of Business Research, 153*, 46–58. https://doi.org/10.1016/j.jbusres.2022.08.019
Chiniara, M., & Bentein, K. (2018). The servant leadership advantage: When perceiving low differentiation in leader-member relationship quality influences team cohesion, team task performance and service OCB. *The Leadership Quarterly, 29*(2), 333–345. https://doi.org/10.1016/j.leaqua.2017.05.002
Chong, M. P. M., Shang, Y., Richards, M., & Zhu, X. (2018). Two sides of the same coin? Leadership and organizational culture. *Leadership and Organization Development Journal, 39*(8), 975–994. https://doi.org/10.1108/LODJ-05-2017-0122
Collins, M. D., & Jackson, C. J. (2015). A process model of self-regulation and leadership: How attentional resource capacity and negative emotions influence constructive and destructive leadership. *The Leadership Quarterly, 26*(3), 386–401. https://doi.org/10.1016/j.leaqua.2015.02.005
Cullen, J. G. (2022). Moral recovery and ethical leadership. *Journal of Business Ethics, 175*(3), 485–497. https://doi.org/10.1007/s10551-020-04658-3
Cummings, S., Bridgman, T., & Brown, K. G. (2015). Unfreezing change as three steps: Rethinking Kurt Lewin's legacy for change management. *Human Relations, 69*(1), 33–60. https://doi.org/10.1177/0018726715577707
De Clercq, D., Fatima, T., & Jahanzeb, S. (2021). Ingratiating with despotic leaders to gain status: The role of power distance orientation and self-enhancement motive. *Journal of Business Ethics, 171*(1), 157–174.
de Lara, P. Z. M., & Armas, M. V. (2019). Does ethical leadership motivate followers to participate in delivering compassion? *Journal of Business Ethics, 154*(1), 195–210.
Dextras-Gauthier, J., & Marchand, A. (2018). Does organizational culture play a role in the development of psychological distress? *International Journal of Human Resource Management, 29*(12), 1920–1949. https://doi.org/10.1080/09585192.2016.1216874
Egorov, M., Kalshoven, K., Pircher Verdorfer, A., & Peus, C. (2020). It's a match: Moralization and the effects of moral foundations congruence on ethical and unethical leadership perception. *Journal of Business Ethics, 167*(4), 707–723.
Ehrhart, M. G., Schneider, B., & Macey, W. H. (2014). *Organizational climate and culture: An introduction to theory, research, and practice*. Routledge.
Eliot, J. L. (2020). Resilient leadership: The impact of a servant leader on the resilience of their followers. *Advances in Developing Human Resources, 22*(4), 404–418.

Eluwole, K. K., Karatepe, O. M., & Avci, T. (2022). Ethical leadership, trust in organization and their impacts on critical hotel employee outcomes. *International Journal of Hospitality Management, 102*, 1–10. https://doi.org/10.1016/j.ijhm.2022.103153

Ertürk, A. (2008). A trust-based approach to promote employees' openness to organizational change in Turkey. *International Journal of Manpower, 29*(5), 462–483. https://doi.org/10.1108/01437720810888580

Eva, N., Robin, M., Sendjaya, S., van Dierendonck, D., & Liden, R. C. (2019). Servant leadership: A systematic review and call for future research. *Leadership Quarterly, 30*(1), 111–132. https://doi.org/10.1016/j.leaqua.2018.07.004

Fosse, T. H., Martinussen, M., Skogstad, A., & Einarsen, S. V. (2019). Active and passive forms of destructive leadership in a military context: A systematic review and meta-analysis. *European Journal of Work and Organizational Psychology, 28*(5), 708–722. https://doi.org/10.1080/1359432X.2019.1634550

Galli, B. J. (2018). Change management models: A comparative analysis and concerns. *IEEE Engineering Management Review, 46*(3), 124–132. https://doi.org/10.1109/EMR.2018.2866860

Gardner, W. L., Karam, E. P., Alvesson, M., & Einola, K. (2021). Authentic leadership theory: The case for and against. *The Leadership Quarterly, 32*(6), 101495. https://doi.org/10.1016/j.leaqua.2021.101495

Gooderham, P. N., Pedersen, T., Sandvik, A. M., Dasí, À., Elter, F., & Hildrum, J. (2022). Contextualizing AMO explanations of knowledge sharing in MNEs: The role of organizational and national culture. *Management International Review, 62*(6), 859–884. https://doi.org/10.1007/s11575-022-00483-0

Gover, L., Halinski, M., & Duxbury, L. (2016). Is it just me? Exploring perceptions of organizational culture change. *British Journal of Management, 27*(3), 567–582. https://doi.org/10.1111/1467-8551.12117

Haffar, M., Al-Karaghouli, W., Djebarni, R., Al-Hyari, K., Gbadamosi, G., Oster, F., Alaya, A., & Ahmed, A. (2023). Organizational culture and affective commitment to e-learning' changes during COVID-19 pandemic: The underlying effects of readiness for change. *Journal of Business Research, 155*, 1–11. https://doi.org/10.1016/j.jbusres.2022.113396

Heine, E. C. E., Stouten, J., & Liden, R. C. (2022b). Providing service during a merger: The role of organizational goal clarity and servant leadership. *Journal of Business Ethics*, 1–22. https://doi.org/10.1007/s10551-022-05162-6

Hogan, S. J., & Coote, L. V. (2014). Organizational culture, innovation, and performance: A test of Schein's model. *Journal of Business Research, 67*(8), 1609–1621. https://doi.org/10.1016/j.jbusres.2013.09.007

Holten, A.-L., Hancock, G. R., & Bøllingtoft, A. (2020). Studying the importance of change leadership and change management in layoffs, mergers, and closures. *Management Decision, 58*(3), 393–409. https://doi.org/10.1108/MD-03-2017-0278

Hussain, S. T., Lei, S., Akram, T., Haider, M. J., Hussain, S. H., & Ali, M. (2018). Kurt Lewin's change model: A critical review of the role of leadership and employee involvement in organizational change. *Journal of Innovation & Knowledge, 3*(3), 123–127. https://doi.org/10.1016/j.jik.2016.07.002

Jeong, S., Hsiao, Y., Song, J. H., Kim, J., & Bae, S. H. (2016). The moderating role of transformational leadership on work engagement: The influences of professionalism and openness to change. *Human Resource Development Quarterly, 27*(4), 489–516.

Jeong, S., Sun, C. (Timothy), & Yam, K. C. (2024). The affective processes of ethical leadership: The role of moral emotions. *Journal of Business Ethics*, 1–19. https://doi.org/10.1007/s10551-024-05718-8

Jin, S., Seo, M.-G., & Shapiro, D. L. (2016). Do happy leaders lead better? Affective and attitudinal antecedents of transformational leadership. *The Leadership Quarterly, 27*(1), 64–84.

Kalshoven, K., Den Hartog, D. N., & De Hoogh, A. H. B. (2020). Ethical leadership at work questionnaire (ELW): Development and validation of a multidimensional measure. *Leadership Quarterly, 22*(1), 51–69. https://doi.org/10.1016/j.leaqua.2010.12.007

Kaptein, M. (2019). The moral entrepreneur: A New component of ethical leadership. *Journal of Business Ethics, 156*(4), 1135–1150. https://doi.org/10.1007/s10551-017-3641-0

Kaul, A. (2019). Culture vs strategy: Which to precede, which to align? *Journal of Strategy and Management, 12*(1), 116–136. https://doi.org/10.1108/JSMA-04-2018-0036

Khan, S. R. K. (2016). Influence of organizational culture on teacher's work engagement: An empirical study. *IUP Journal of Organizational Behavior, 15*(4), 27–45.

Kim, Y. J., & Toh, S. M. (2019). Stuck in the past? The influence of a leader's past cultural experience on group culture and positive and negative group deviance. *Academy of Management Journal, 62*(3), 944–969. https://doi.org/10.5465/amj.2016.1322

Kim, E. J., & Park, S. (2020). Transformational leadership, knowledge sharing, organizational climate and learning: An empirical study. *Leadership and Organization Development Journal, 41*(6), 761–775. https://doi.org/10.1108/LODJ-12-2018-0455

Kimakwa, S., Gonzalez, J. A., & Kaynak, H. (2021). Social entrepreneur servant leadership and social venture performance: How are they related? *Journal of Business Ethics*, 1–25. https://doi.org/10.1007/s10551-021-04986-y

Kontoghiorghes, C. (2016). Linking high performance organizational culture and talent management: Satisfaction/motivation and organizational commitment as mediators. *International Journal of Human Resource Management, 27*(16), 1833–1853. https://doi.org/10.1080/09585192.2015.1075572

Kotter, J. P. (2012). *Leading change, with a new preface by the author* (1st ed.). Harvard Business Review Press.

Krauss, J., & Vanhove, A. J. (2022). Organizational culture perceptions and change frequency: The moderating effect of members' hierarchical level in the organization. *Leadership & Organization Development Journal, 43*(2), 302–314. https://doi.org/10.1108/LODJ-10-2021-0464

Kruse, P., Wach, D., Costa, S., & Moriano, J. A. (2019). Values matter, don't they?–Combining theory of planned behavior and personal values as predictors of social entrepreneurial intention. *Journal of Social Entrepreneurship, 10*(1), 55–83. https://doi.org/10.1080/19420676.2018.1541003

Lam, C. K., Huang, X. U., & Chan, S. C. H. (2015). The threshold effect of participative leadership and the role of leader information sharing. *Academy of Management Journal, 58*(3), 836–855.

Le, P. B., & Lei, H. (2019). Determinants of innovation capability: The roles of transformational leadership, knowledge sharing and perceived organizational support. *Journal of Knowledge Management, 23*(3), 527–547. https://doi.org/10.1108/JKM-09-2018-0568

Lee, J., Chiang, F. F. T., van Esch, E., & Cai, Z. (2018). Why and when organizational culture fosters affective commitment among knowledge workers: The mediating role of perceived psychological contract fulfilment and moderating role of organizational tenure. *International Journal of Human Resource Management, 29*(6), 1178–1207. https://doi.org/10.1080/09585192.2016.1194870

Lei, H., Leaungkhamma, L., & Le, P. B. (2020). How transformational leadership facilitates innovation capability: The mediating role of employees' psychological capital. *Leadership and Organization Development Journal, 41*(4), 481–499. https://doi.org/10.1108/LODJ-06-2019-0245

Li, G., Liu, H., & Luo, Y. (2018). Directive versus participative leadership: Dispositional antecedents and team consequences. *Journal of Occupational & Organizational Psychology, 91*(3), 645–664.

Lin, C., Huang, P. C., Chen, S. J., & Huang, L. C. (2017). Pseudo-transformational leadership is in the eyes of the subordinates. *Journal of Business Ethics, 141*(1), 179–190.

Lin, S. H., Scott, B. A., & Matta, F. K. (2019). The dark side of transformational leader behaviors for leaders themselves: A conservation of resources perspective. *Academy of Management Journal, 62*(5), 1556–1582. https://doi.org/10.5465/amj.2016.1255

Liu, H. (2019). Just the servant: An intersectional critique of servant leadership. *Journal of Business Ethics, 156*(4), 1099–1112. https://doi.org/10.1007/s10551-017-3633-0

Lokaj, A. S., & Sadrija, T. L. (2020). Organizational culture influenced by leadership styles: The case of private businesses in Kosovo. *Problems and Perspectives in Management, 18*(3), 306–314.

Low, W. W., Abdul-Rahman, H., & Zakaria, N. (2015). The impact of organizational culture on international bidding decisions: Malaysia context. *International Journal of Project Management, 33*(4), 917–931. https://doi.org/10.1016/j.ijproman.2014.10.010

Lythreatis, S., Mostafa, A. M. S., & Wang, X. (2019). Participative leadership and organizational identification in SMEs in the MENA region: Testing the roles of CSR perceptions and pride in membership. *Journal of Business Ethics, 156*(3), 635–650.

Malhotra, N., Zietsma, C., Morris, T., & Smets, M. (2021). Handling resistance to change when societal and workplace logics conflict. *Administrative Science Quarterly, 66*(2), 475–520.

Martin, S. R., Emich, K. J., McClean, E. J., & Woodruff, C. T. (2022). Keeping teams together: How ethical leadership moderates the effects of performance on team efficacy and social integration. *Journal of Business Ethics, 176*(1), 127–139.

Martin, J. (1992). *Cultures in organizations : Three perspectives*. Oxford University Press.

Mello, J. E., & Schloemer, H. (2022). Do organizational subcultures matter? A case study of logistics and supply chain management. *The International Journal of Logistics Management, 33*(1), 141–164. https://doi.org/10.1108/IJLM-04-2021-0230

Mostafa, A. M. S., Farley, S., & Zaharie, M. (2021). Examining the boundaries of ethical leadership: The harmful effect of co-worker social undermining on disengagement and employee attitudes. *Journal of Business Ethics, 174*(2), 355–368.

Mroz, J. E., Yoerger, M., & Allen, J. A. (2018). Leadership in workplace meetings: The intersection of leadership styles and follower gender. *Journal of Leadership and Organizational Studies, 25*(3), 309–322. https://doi.org/10.1177/1548051817750542

Nadler, D. A., & Tushman, M. L. (1980). A model for diagnosing organizational behavior. *Organizational Dynamics, 9*(2), 35–51.

Naeem, M. (2020a). Understanding the role of social media in organizational change implementation. *Management Research Review, 43*(9), 1097–1116. https://doi.org/10.1108/MRR-07-2019-0336

Naeem, M. (2020b). Using social networking applications to facilitate change implementation processes: Insights from organizational change stakeholders. *Business Process Management Journal, 26*(7), 1979–1998. https://doi.org/10.1108/BPMJ-07-2019-0310

Naseer, S., Raja, U., Syed, F., Donia, M. B. L., & Darr, W. (2016). Perils of being close to a bad leader in a bad environment: Exploring the combined effects of despotic leadership, leader member exchange, and perceived organizational politics on behaviors. *Leadership Quarterly, 27*(1), 14–33. https://doi.org/10.1016/j.leaqua.2015.09.005

Nassif, A. G., Hackett, R. D., & Wang, G. (2021). Ethical, virtuous, and charismatic leadership: An examination of differential relationships with follower and leader outcomes. *Journal of Business Ethics, 172*(3), 581–603.

Nauman, S., Zheng, C., & Basit, A. A. (2020). How despotic leadership jeopardizes employees' performance: the roles of quality of work life and work withdrawal. *Leadership & Organization Development Journal*, 1–16. https://doi.org/10.1108/LODJ-11-2019-0476

Neill, M. S., Men, L. R., & Yue, C. A. (2020). How communication climate and organizational identification impact change. *Corporate Communications: An International Journal, 25*(2), 281–298. https://doi.org/10.1108/CCIJ-06-2019-0063

Neves, P., & Schyns, B. (2018). With the bad comes what change? The interplay between destructive leadership and organizational change. *Journal of Change Management, 18*(2), 91–95.

Neves, P., Pires, D., & Costa, S. (2021). Empowering to reduce intentions to resist future change: Organization-based self-esteem as a boundary condition. *British Journal of Management, 32*(3), 872–891.

Payne, D., Trumbach, C., & Soharu, R. (2022). The values change management cycle: ethical change management. *Journal of Business Ethics*, 1–12. https://doi.org/10.1007/s10551-022-05306-8

Rafferty, A. E., & Jimmieson, N. L. (2017). Subjective perceptions of organizational change and employee resistance to change: Direct and mediated relationships with employee well-being. *British Journal of Management, 28*(2), 248–264.

Sakkar Sudha, K., & Shahnawaz, M. G. (2020). Narcissism personality trait and performance: Task-oriented leadership and authoritarian styles as mediators. *Leadership and Organization Development Journal, 41*(2), 280–293. https://doi.org/10.1108/LODJ-09-2019-0399

Schaubroeck, J. M., Shen, Y., & Chong, S. (2017). A dual-stage moderated mediation model linking authoritarian leadership to follower outcomes. *Journal of Applied Psychology, 102*(2), 203–214. https://doi.org/10.1037/apl0000165

Schein, E. H., & Schein, P. A. (2016). *Organizational culture and leadership* (5th Ed.). Wiley.

Schein, E. H. (2009). *The corporate culture survival guide* (2nd ed.). Jossey-Bass.

Schmid, E. A., Pircher Verdorfer, A., & Peus, C. V. (2018). Different shades-different effects? Consequences of different types of destructive leadership. *Frontiers in Psychology, 9*, 1289. https://doi.org/10.3389/fpsyg.2018.01289

Schweiger, S., Stouten, H., & Bleijenbergh, I. L. (2018). A system dynamics model of resistance to organizational change: The role of participatory strategies. *Systems Research & Behavioral Science, 35*(6), 658–674. https://doi.org/10.1002/sres.2509

Shimoni, B. (2017). What is resistance to change? A habitus-oriented approach. *Academy of Management Perspectives, 31*(4), 257–270.

Shrivastava, S., Pazzaglia, F., Sonpar, K., & McLoughlin, D. (2022). Effective communication during organizational change: A cross-cultural perspective. *Cross Cultural & Strategic Management, 29*(3), 675–697. https://doi.org/10.1108/CCSM-08-2021-0144

Siangchokyoo, N., Klinger, R. L., & Campion, E. D. (2020). Follower transformation as the linchpin of transformational leadership theory: A systematic review and future research agenda. *The Leadership Quarterly, 31*(1), 1–18. https://doi.org/10.1016/j.leaqua.2019.101341

Siddique, C. M., Siddique, H. F., & Siddique, S. U. (2020). Linking authoritarian leadership to employee organizational embeddedness, LMX and performance in a high-power distance culture: A mediation-moderated analysis. *Journal of Strategy and Management, 13*(3), 393–411. https://doi.org/10.1108/JSMA-10-2019-0185

Sims, C. M. (2018). The diversity intelligent servant leader: Developing leaders to meet the needs of a diverse workforce. *Advances in Developing Human Resources, 20*(3), 313–330. https://doi.org/10.1177/1523422318778009

Sinval, J., Miller, V., & Marôco, J. (2021). Openness toward organizational change scale (OTOCS): Validity evidence from Brazil and Portugal. *PLoS ONE, 16*(4), 1–22.

Sousa, M. J., & Rocha, Á. (2019). Leadership styles and skills developed through game-based learning. *Journal of Business Research, 94*, 360–366. https://doi.org/10.1016/j.jbusres.2018.01.057

Srimulyani, V. A., & Hermanto, Y. B. (2022). Organizational culture as a mediator of credible leadership influence on work engagement: empirical studies in private hospitals in East Java, Indonesia. *Humanities & Social Sciences Communications, 9*(1).

Stace, D. A., & Dunphy, D. C. (1991). Beyond traditional paternalistic and developmental approaches to organizational change and human resource strategies. *International Journal of Human Resource Management, 2*(3), 263–283.

Stace, D. A., & Dunphy, D. C. (1992). Translating business strategies into action: Managing strategic change. *Journal of Strategic Change, 1*(4), 203–216.

Stock, G., Banks, G. C., Voss, E. N., Tonidandel, S., & Woznyj, H. (2023). Putting leader (follower) behavior back into transformational leadership: A theoretical and empirical course correction. *The Leadership Quarterly, 34*(6), 1–17. https://doi.org/10.1016/j.leaqua.2022.101632

Strengers, J., Mutsaers, L., van Rossum, L., & Graamans, E. (2022). The organizational culture of scale-ups and performance. *Journal of Organizational Change Management, 35*(8), 115–130. https://doi.org/10.1108/JOCM-09-2021-0268

Sun, P., & Shang, S. (2019). Personality traits and personal values of servant leaders. *Leadership and Organization Development Journal, 40*(2), 177–192. https://doi.org/10.1108/LODJ-11-2018-0406

Sun, U., Park, H., & Yun, S. (2024). Ethically treated yet closely monitored: Ethical leadership, leaders' close monitoring, employees' uncertainty, and employees' organizational citizenship behavior. *Journal of Organizational Behavior, 45*(5), 702–719. https://doi.org/10.1002/job.2760

Sun, U. Y., Xu, H., Kluemper, D. H., McLarty, B. D., & Yun, S. (2024). Ethical leadership and knowledge sharing: A social cognitive approach investigating the role of self-efficacy as a key mechanism. *Journal of Business Research, 174*, 1–15. https://doi.org/10.1016/j.jbusres.2024.114531

Supriharyanti, E., & Sukoco, B. M. (2023). Organizational change capability: A systematic review and future research directions. *Management Research Review, 46*(1), 46–81. https://doi.org/10.1108/MRR-01-2021-0039

Syed, F., Akhtar, M. W., Muhammad, K., Asrar-ul-Haq, M., Ul ain, Q., Husnain, M., & Aslam Kashif, M. (2020). When leader is morally corrupt: interplay of despotic leadership and self-concordance on moral emotions and bullying behavior. *Leadership & Organization Development Journal*, 1–18. https://doi.org/10.1108/LODJ-11-2019-0476

Tepper, B. J., Dimotakis, N., Lambert, L. S., Koopman, J., Matta, F. K., Man Park, H., & Goo, W. (2018). Examining follower responses to transformational leadership from a dynamic, person-environment fit perspective. *Academy of Management Journal, 61*(4), 1343–1368. https://doi.org/10.5465/amj.2014.0163

Thoroughgood, C. N., Sawyer, K. B., Padilla, A., & Lunsford, L. (2018). Destructive leadership: A critique of leader-centric perspectives and toward a more holistic definition. *Journal of Business Ethics, 151*(3), 627–649.

Trabucchi, D., Buganza, T., Bellis, P., Magnanini, S., Press, J., Verganti, R., & Zasa, F. P. (2022). Story-making to nurture change: Creating a journey to make transformation happen. *Journal of Knowledge Management, 26*(11), 427–460. https://doi.org/10.1108/JKM-07-2022-0582

Tran, T. T., & Gandolfi, F. (2020). Implementing Lewin's change theory for institutional improvements: A Vietnamese case study. *Journal of Management Research, 20*(4), 199–210.

Treviño, L., & Nelson, K. (2019). Managing business ethics: Straight talk about how to do it right. *Personnel Psychology, 72*(4), 625–626. https://doi.org/10.1111/peps.12360

Van De Ven, A. H., & Poole, M. S. (1995). Explaining development and change in organizations. *Academy of Management Review, 20*(3), 510–540.

Velez, M. J., & Neves, P. (2018). Shaping emotional reactions to ethical behaviors: Proactive personality as a substitute for ethical leadership. *The Leadership Quarterly, 29*(6), 663–673. https://doi.org/10.1016/j.leaqua.2018.06.004

Veloso-Besio, C. B., Cuadra-Peralta, A., Ponce-Correa, F., Gil-Rodríguez, F., & Söberg-Tapia, O. (2019). Effectiveness of training, based on positive psychology and social skills, applied to supervisors, to face resistance to organizational change. *Journal of Organizational Change Management, 32*(2), 251–265. https://doi.org/10.1108/JOCM-04-2018-0099

Venus, M., Stam, D., & van Knippenberg, D. (2019). Visions of change as visions of continuity. *Academy of Management Journal, 62*(3), 667–690.

Vito, R. (2020). Key variations in organizational culture and leadership influence: A comparison between three children's mental health and child welfare agencies. *Children and Youth Services Review, 108*, 1–8. https://doi.org/10.1016/j.childyouth.2019.104600

Waheed, Z., Hussin, S., Khan, M. I., Ghavifekr, S., & Bahadur, W. (2019). Ethical leadership and change: A qualitative comparative case study in selected malaysian transformed schools. *Educational Management Administration & Leadership, 47*(4), 624–639.

Waldman, D. A., & O'Reilly, C. (2018). *Leadership for organizations* (1st ed.). Sage Publications, Inc.

Wang, Q., Hou, H., & Li, Z. (2022). Participative leadership: A literature review and prospects for future research. *Frontiers in Psychology, 13*, 1–12.

Wang, H., & Guan, B. (2018). The positive effect of authoritarian leadership on employee performance: The moderating role of power distance. *Frontiers in Psychology, 9*. https://doi.org/10.3389/fpsyg.2018.00357

Warrick, D. D. (2022). Revisiting resistance to change and how to manage it: What has been learned and what organizations need to do. *Business Horizons*, 1–19. https://doi.org/10.1016/j.bushor.2022.09.001

Wei, T., & Clegg, J. (2018). Effect of organizational identity change on integration approaches in acquisitions: Role of organizational dominance. *British Journal of Management, 29*(2), 337–355.

Wei, Y., & Miraglia, S. (2017). Organizational culture and knowledge transfer in project-based organizations: Theoretical insights from a Chinese construction firm. *International Journal of Project Management, 35*(4), 571–585. https://doi.org/10.1016/j.ijproman.2017.02.010

Xia, Z., & Yang, F. (2020). Ethical leadership and knowledge sharing: The impacts of prosocial motivation and two facets of conscientiousness. *Frontiers in Psychology, 11*, 1–10. https://doi.org/10.3389/fpsyg.2020.581236

Yanti, S., & Dahlan, J. A. (2017). The effects of organizational culture, leadership behavior, and job satisfaction on employee organizational commitment. *Journal of Positive Management, 8*(4), 80–96.

Young, H. R., Glerum, D. R., Joseph, D. L., & McCord, M. A. (2021). A meta-analysis of transactional leadership and follower performance: Double-edged effects of LMX and empowerment. *Journal of Management, 47*(5), 1–26.

Yue, C. A., Men, L. R., & Ferguson, M. A. (2019). Bridging transformational leadership, transparent communication, and employee openness to change: The mediating role of trust. *Public Relations Review, 45*(3), 1–13. https://doi.org/10.1016/j.pubrev.2019.04.012

Yue, C. A., Thelen, P. D., & Walden, J. (2022). How empathetic leadership communication mitigates employees' turnover intention during COVID-19-related organizational change. *Management Decision*, 1–21. https://doi.org/10.1108/MD-01-2022-0011

Zahari, A. I., Said, J., Muhamad, N., & Ramly, S. M. (2024). Ethical culture and leadership for sustainability and governance in public sector organisations within the ESG framework. *Journal of Open Innovation: Technology, Market, and Complexity, 10*(1), 1–11. https://doi.org/10.1016/j.joitmc.2024.100219

Zhang, Y., Lu, L., Gao, P., & Zhang, J. (2020). Effect of authoritarian leadership on user resistance to change: Evidence from IS project implementation in China. *Industrial Management and Data Systems, 120*(10), 1813–1834. https://doi.org/10.1108/IMDS-11-2019-0619

Zhang, K., Zhao, B., & Yin, K. (2024). When leaders acknowledge their own errors, will employees follow suit? A social learning perspective. *Journal of Business Ethics, 189*(2), 403–421. https://doi.org/10.1007/s10551-023-05329-9

Zheng, Y., Graham, L., Farh, J.-L., & Huang, X. (2021). The impact of authoritarian leadership on ethical voice: A moderated mediation model of felt uncertainty and leader Benevolence. *Journal of Business Ethics, 170*(1), 133–146. https://doi.org/10.1007/s10551-019-04261-1

Zheng, Y., Epitropaki, O., Graham, L., & Caveney, N. (2022). Ethical leadership and ethical voice: The mediating mechanisms of value internalization and integrity identity. *Journal of Management, 48*(4), 973–1002. https://doi.org/10.1177/01492063211002611

Part V
Thematic Area: Talent Development and Continuous Learning

Chapter 8
Online Recruitment: A Focus Group Study with Portuguese Recruiters

Ana Sofia Freitas and Catarina Brandão

Abstract Technological development has influenced recruitment and selection (R&S) processes, with the increasing use of online recruitment (OR). Despite this development, the literature on this process is scarce, especially regarding the perspectives and practices of Portuguese recruiters. This study explores and characterises OR in the Portuguese context from the perspective of recruiters, and addresses the operationalisation of this process. Two focus groups were conducted with 10 recruiters. A reflexive thematic analysis supported the definition of five main themes. Technology permeates the R&S process from the outset, ranging from management software and job boards to social media, whose role goes beyond the dissemination of vacancies. The results also reveal the transmission of organisational values by recruiters during the process, the unchanged profile of the desired candidate in OR, and the growing normalisation of OR, which is perceived as the future of R&S, while preserving its traditional fundamental goals. This study contributes to a more in-depth understanding of the use of OR in Portugal and how it is perceived by recruiters and supports the understanding of this contemporary R&S method.

Keywords Online Recruitment · Recruitment and Selection Practices · Human Resource Management · Digital Transformation in HR · Recruiter Perspectives

---

A. S. Freitas
Faculty of Psychology and Educational Sciences, University of Porto, Porto, Portugal

C. Brandão (✉)
Center for Psychology at University of Porto; Faculty of Psychology and Education Sciences, University of Porto, Porto, Portugal
e-mail: catarina@fpce.up.pt

C. Machado (ed.), *Human Resource Management in a Disrupted World*, Lecture Notes on Multidisciplinary Industrial Engineering,
https://doi.org/10.1007/978-3-031-89948-5_9

8.1 Introduction

The growth of the information technology (IT) sector and the pandemic have brought a high level of complexity to the strategic area of recruitment and selection (R&S) (Blacksmith & Poeppelman, 2014; Halpern et al., 2021). This lends significant importance and relevance to human resources (HR) departments, as innovation in HR processes and strategies is essential to attract talent. Recruiters need to be creative, flexible and online (Halová & Müller, 2021; Halpern et al., 2021). The use of the Internet to carry out processes, especially R&S, and digitisation, are essential for the sustainability of organisations (Brandão et al., 2022; Tutar & Ay, 2021).

Despite being an increasingly used method, the literature on online recruitment (OR) or E-recruitment is still limited (Brandão et al., 2019). The existing literature is fragmented (Blacksmith & Poeppelman, 2014), and current studies tend to lack a broader perspective on the variables related to and influencing OR. It is even more relevant to understand the potential of OR given the unprecedented technological advances that occurred during the COVID-19 pandemic (Fernandes & Machado, 2022). Furthermore, there is limited information on the Internet as a tool for R&S in the Portuguese context. The aim of this study is to contribute to the understanding of OR in Portugal, using a comprehensive and in-depth approach.

8.1.1 Human Resource Management and Technology

Information technologies (IT) include "(…) systems of information, Internet, information and communication-related technologies, and their infrastructure including computer software, networks, and hardware, which process or transmit information to enhance the effectiveness of individuals and organizations" (Onn & Sorooshian, 2013, p. 139). IT is widely used and has become essential to many organisations and functions. The field of HR is no exception: technology underpins various functions performed in HR departments, including recruitment and selection (R&S). Much of the administrative work performed by HR departments is now delegated to specific programs, such as CV screening. This allows HR professionals to devote more time to other aspects of their roles and reduces the demand for physical HR resources (Monteiro & Alturas, 2012). The literature also refers to the use of advanced technologies applied to HR processes, in particular virtual reality (e.g., Blacksmith & Poeppelman, 2014; Kaul, 2021), artificial intelligence (Kaul, 2021), and predictive technologies (González Iglesias, 2020).

8.1.2 Online Recruitment and Selection Process

R&S involves several stages, from advertising a job vacancy to attracting and maintaining candidates' interest, and influencing their decisions, making it a complex and dynamic process (Breaugh, 2013). Specifically, recruitment involves attracting a desired pool of candidates, followed by the selection phase (Brandão et al., 2019), which involves analysing, evaluating, and selecting candidates based on the specifications required for the position (Chiavenato, 1999). These two processes are commonly referred to together as the R&S process.

While some advocate a restricted view of OR (or E-recruitment), understanding it as the recruitment process carried out via the Internet and/or electronic/digital tools (Rosoiu & Popescu, 2016), others (e.g., Brandão et al., 2019; Costa, 2002; Wolfswinkel et al., 2010), argue that OR goes beyond the mere posting of a job vacancy online, as it can include technological tools for CV screening, candidate monitoring, and even conducting remote interviews and interacting with candidates. Thus, OR is not limited to the online posting of job vacancies (i.e., recruitment), but also includes elements of the selection process.

The different functionalities and structures of IT mentioned above are essential for conducting an OR process, as technology serves as the cornerstone of OR, enabling different stages of the process. To carry out an OR process, recruiters can rely on different platforms, such as the company website, job-specific websites, emails, and even social media (Brandão et al., 2019; Halpern et al., 2021).

Several authors (e.g., Bartram, 2000; Cappelli, 2001; Peter, 2001) predicted that the Internet would revolutionise recruitment and that many companies would adopt OR. However, the pandemic has accelerated digitalisation and various technological advances. According to a study by McKinsey and Company (2020), the organisational world advanced five years in eight weeks in terms of technological adaptation because of the coronavirus. The use of new Internet-based methods and technologies is expected to increase in the coming years (Halová & Müller, 2021).

Social media platforms are increasingly being used to strategically recruit candidates (Hedenus et al., 2021). In addition to posting job vacancies, recruiters use social media as a source of supplementary information to assess candidates' skills and personal qualities (Brown & Vaughn, 2011; Morgan, n.d.). Social media has also led to a paradigm shift: recruiters are increasingly taking the initiative to find talented individuals (passive candidates), rather than candidates actively seeking new opportunities (Blacksmith & Poeppelman, 2014).

The candidate's profile is no longer limited to what is presented in their CV and the presentation of an individual's online profile is increasingly influential in R&S decisions. As a result, how a candidate presents themselves online is of critical importance (Rosoiu & Popescu, 2016). In addition, the recruiter must consider the candidate's network of contacts to be authentic and "trustworthy" (Hedenus et al., 2021). According to Campos et al. (2018), candidates who use the Internet tend to have higher levels of education and characteristics associated with better professional outcomes.

As an online-driven process, OR tends to attract and be used by younger people (Rosoiu & Popescu, 2016). Millennials and Generation Z entering the job market are comfortable with technology, which means they are more informed and technologically competent (Brandão et al., 2017; Halpern et al., 2021). Therefore, it is increasingly likely that recruitment processes will be conducted via the Internet (Brandão et al., 2019).

In addition, the behaviour and attitudes of recruiters influence R&S, including the likelihood of a candidate accepting an offer (Blacksmith & Poeppelman, 2014; Chapman et al., 2005). Today, recruiters use marketing and advertising strategies, and assume the role of experts in the selection of individuals (González Iglesias, 2020).

8.1.3 Advantages and Disadvantages of Online Recruitment

Several advantages have been attributed to OR. One of the main advantages is its convenience. For companies, conducting the process via the Internet reduces time and costs (Monteiro & Alturas, 2012; Rosoiu & Popescu, 2016). The Internet allows for real-time interaction, and speeds up all stages of the process, making OR on average about 70% faster than traditional R&S (Rosoiu & Popescu, 2016). OR also allows greater flexibility and adaptability (Monteiro & Alturas, 2012). Through online platforms, recruiters can stay informed about the labour market and have access to diverse candidate databases and quality information (Rosoiu & Popescu, 2016). As it is an online process, the pool of potential candidates with access to job vacancies increases, and geographical constraints are eliminated (if possible) (Brandão et al., 2019). It has also been observed that many people use OR tools regularly, even if they are employed or have found a new job (Monteiro & Alturas, 2012). In addition, employers can strengthen their corporate image and reputation by using OR (Monteiro & Alturas, 2012).

Online application processes offer advantages for candidates, including lower costs and less time and effort (Monteiro & Alturas, 2012; Tutar & Ay, 2021). There is no need to travel to companies, there is greater flexibility in scheduling interviews, and it is easier to submit CVs and necessary information online.

OR can also present challenges that are almost the opposite of its benefits. As it is technology-based, OR is susceptible to technical and operational constraints (Monteiro & Alturas, 2012), which can disrupt the entire process when complications arise. From a company perspective, one disadvantage is the high volume of applications received (Brandão et al., 2019), which may include unsuitable candidates for the position (Autor, 2001), slowing down the process. Another criticism of this method is its impersonal nature (Brandão et al., 2019; Monteiro & Alturas, 2012). Some stages of R&S require interpersonal contact, which can be beneficial for both candidates and recruiters.

There are also ethical concerns regarding OR, with the most pressing issue being the privacy of candidate data (Monteiro & Alturas, 2012). Nowadays, especially with social media, recruiters can have access to information that they would not have in traditional R&S processes. Furthermore, OR can be discriminatory, excluding those who do not use the Internet or have a lower level of digital literacy (Brandão et al., 2019). Another critical issue is the possibility of fraud, which can be committed by both recruiters (Brandão et al., 2019) and candidates (Monteiro & Alturas, 2012).

8.1.4 The Portuguese Context

Little is known about OR in Portugal, with only a few studies focusing on this context. In 2017, Brandão et al. found that HR professionals considered OR as "the future" and predicted that more organisations would adopt online tools. In a recent study by Lopes et al. (2022), with recruiters and HR managers from 38 companies operating in the northern and central regions of Portugal (mainly in the IT sector), participants reported familiarity with widely used platforms such as LinkedIn, Indeed, and Facebook. The study also showed that conducting online interviews via platforms such as Microsoft Teams has become common practice. However, participants were not particularly receptive to the use of more advanced technologies such as artificial intelligence.

In another study focusing on the Portuguese context, Brandão et al. (2019) found that managers believed that OR facilitated access, was fast and practical, was more cost-effective than traditional R&S, was more impersonal than traditional R&S, despite having a specific relational dimension, and that traditional R&S methods provided strategic access to some candidates. Overall, although the challenges associated with OR were acknowledged, the data suggested that the assessment was overwhelmingly positive. At the time, however, organisations were not yet exploiting the full potential of OR.

The subject of OR is of great importance, especially in an increasingly technological and globalised world. The use of technology in R&S has revolutionised HR departments, enabling them to be more efficient and to gain insights into the labour market that were previously unattainable. R&S is no longer constrained by geographical or temporal factors, and encompasses a wide range of candidates worldwide. OR has experienced a rapid boom in a short period of time. Particularly in the Portuguese context, little is known about this process, especially from the perspective of recruiters. This research contributes to a deeper understanding of the unique challenges and opportunities in the online context, generating insights, advancing knowledge in the field, and promoting organisational sustainability (especially in HRM) (Tutar & Ay, 2021). This opens the door to a more in-depth discussion of OR in a context that still offers room for analysis—the Portuguese context.

Table 8.1 Specific Research Objectives and Research Questions

Research objectives	Research questions
1. Identify the stages of the OR process and the specific stages at which technologies are used.	RQ1: How does the OR process unfold?
2. Explore how technologies are impacting the OR process.	RQ2: What is the role of technology in the OR process?
3. Identify the social media platforms being used in OR and how.	RQ3: How are social media platforms used in OR?
4. Describe how recruiters have adapted to the technological and remote changes in R&S, which has become increasingly virtual and technology-dependent.	RQ4: What behaviours do recruiters adopt during (direct or indirect) interactions with candidates in OR?
5. Identify the characteristics and behaviours that recruiters value in candidates.	RQ5: Which candidate profiles are most valued by recruiters in OR?

8.2 Objectives of the Study and Research Questions

The aim of this study is to explore and characterise OR within the Portuguese context, focusing on recruiters' perspectives. Table 8.1 presents the specific research objectives of the study and the related research questions (RQ), which were formulated in order to gain a more thorough understanding of the topic under investigation.

8.3 Method

A cross-sectional design (Kesmodel, 2018) with a qualitative approach (Flick & Barbour, 2009) was adopted for the study. Due to its broad scope, the qualitative methodology is particularly suitable for studies in which the aim is to collect very specific data and analyse them in a detailed and interpretative way, with an emphasis on understanding participants' perspectives, contexts and behaviours (Maxwell, 2008).

Considering the proposed objectives and the literature review, the research strategy adopted proved to be an appropriate strategy to provide a detailed and holistic picture of the current state of OR in the Portuguese context. The study was approved by the Ethics Committee of the FPCEUP (approval 2023/08-01).

8.3.1 Recruitment

Participants were eligible if they had a professional role as a recruiter and were working in the Portuguese labour market. Participants had to be actively involved in and knowledgeable about the national labour market to ensure that the study provided meaningful insights into the current state of OR in Portugal. Recruitment was carried out by contacting the HR departments of two organisations on a convenience basis (Miles & Huberman, 1994). Once participants met the inclusion criteria and were identified in each organisation, they were formally invited to participate by the research team.

8.3.2 *Data Collection*

Two online focus groups were conducted in 2023. Focus groups allow discussions with a depth that is not as feasible with other methods, and allow the experiences of a particular group to be explored (How to Run Focus Groups, 2015). In this case, convening professionals with extensive knowledge of OR allowed for detailed insights into the process and a comprehensive understanding of the process and the dimensions under study. A pilot focus group was conducted to validate both the focus group guide and the facilitation of the focus groups via an online platform.

When scheduling the focus groups, participants were provided with a document explaining the background of the study, an informed consent form, and a short sociodemographic questionnaire.

Given the physical dispersion of the participants, the focus groups were conducted online and synchronously via a video conferencing platform (Zoom), and lasted between 35 and 50 min. Each group had a moderator and an assistant moderator, following Falter et al. (2022). There were five participants in each group. At the end of each focus group, a summary of the conversation was produced and participants were asked to confirm that the summary effectively reflected their contributions.

8.3.3 *Participants*

Data were collected from professionals with experience in R&S from two organisations operating in Portugal: a recruitment consultancy specialising in headhunting (Company A) and a retail organisation (Company B).

A total of 10 recruiters, ages 23–61 ($M = 37.7$, $SD = 10.8$), participated in the focus groups. Only one participant was male, 30% had a master's degree, and 70% had a bachelor's degree. Their educational backgrounds ranged from psychology (50%) and human resource management (40%) to adult education and training (10%). The participants' experience as recruiters ranged from 1 to 34 years ($M = 9.9$, $SD = 10.5$).

All participants reported using various technology tools in R&S processes, including job boards (90%), social media platforms (70%), email (80%), CV screening software (80%), applicant tracking systems (40%), and video conferencing tools (80%). None of the respondents reported using artificial intelligence tools.

8.3.4 Data Analysis

The analysis began with the transcription of the focus group discussions, which were then checked against the video and audio recordings to ensure accuracy (Braun & Clarke, 2006). The transcriptions were also sent to the participants to confirm their content, and no participants requested changes or raised objections to the transcriptions. Analysis was conducted on verbatim transcriptions of the focus groups, using NVivo 14®.

The method of analysis was reflexive thematic analysis (RTA) (Braun & Clarke, 2022). Braun and Clarke (2022) describe RTA as a theoretically flexible method with no standardised format. The approach to data analysis is not exclusive, allowing both semantic and latent meanings to be captured and data to be analysed both inductively and deductively, as if along a continuum (Braun & Clarke, 2022; Braun et al., 2019; Byrne, 2021). Another notable aspect of this type of analysis is its emphasis on the subjectivity and reflexivity of the researcher, making it an inherently interpretive practice (Braun & Clarke, 2022). This technique offers several advantages, the most notable being its flexibility throughout the data analysis process, and its ability to generate numerous unanticipated insights (Braun & Clarke, 2006).

In keeping with the qualitative nature and philosophy of RTA, a constructionist epistemological stance was adopted in the study. In this approach, the meaning and significance of the data is recognised as the central criterion for the coding process, as opposed to an essentialist perspective, where the primary criterion is the recurrence of themes in the data, thereby losing the interpretive potential (Braun et al., 2016; Byrne, 2021). Codes were generated both deductively and inductively. Finally, the data were coded semantically and latently, to capture the meanings conveyed by the participants and the data while allowing space for reflection on and interpretation of these meanings (Byrne, 2021). The process of data analysis using RTA followed a non-linear trajectory, requiring revisiting, re-analysing, and rethinking the data and its meanings. Braun and Clarke (2020) emphasise that coding and theme development are flexible and organic, often evolving throughout the analysis process, as was the case in this study. The process was recursive and iterative (Braun & Clarke, 2013, 2020).

8.4 Findings and Discussion

The presentation and discussion of the findings is organised according to the defined themes (Fig. 8.1), which are inherently linked to the research questions proposed in the study.

The themes are presented throughout the discussion in order to provide a clearer picture of how the study objectives were addressed. Participants' extracts are identified by the letter corresponding to their organisation (A or B) and a number assigned according to the order of their contribution during data collection (Px).

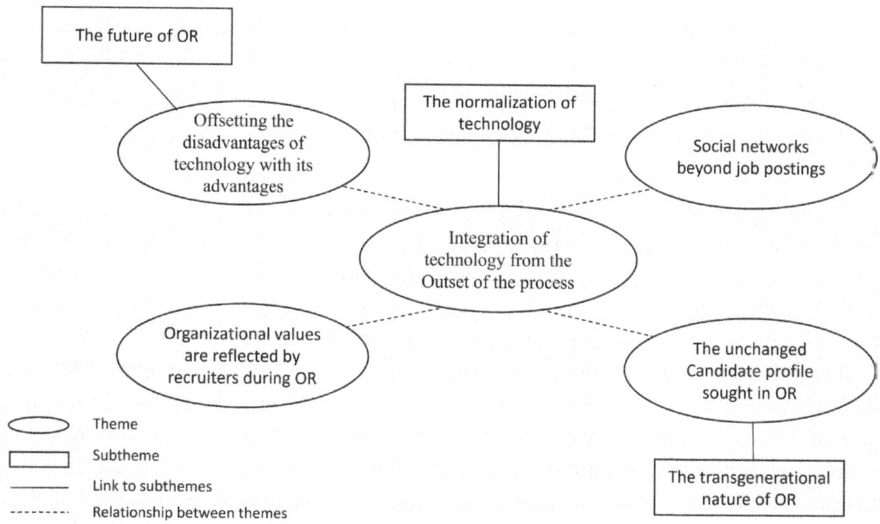

Fig. 8.1 Final thematic map

8.4.1 *Integrating Technology into R&S from the Outset*

This first theme addresses RQ1: *How does the OR process unfold?*

The use of technology in both organisations appears to be diverse. However, in both cases, technology plays a particularly important role at the start of the process. Both companies use online platforms such as LinkedIn and Indeed to attract candidates, with each company's selection criteria determining the appropriate platform at any given time. Company A also searches for candidates directly on LinkedIn. The application process is facilitated online, with both organisations receiving CVs in digital format and Company B using e-forms. Technology plays a key role in the screening of CVs—Company A uses dedicated software (similar to Excel) and Company B conducts the screening process online, using e-forms submitted by candidates. Company B's use of robotic software facilitates the generation of automated responses to candidates, freeing up recruiters for other tasks. Occasionally, Company A also uses tests and reference checks, which are usually conducted online. The process then moves to the client side, where candidate follow-up continues, albeit with variations depending on the client. In Company B, on the other hand, once the CVs have been screened, suitable candidates are contacted by telephone. This is followed by face-to-face contact—group dynamics and individual interviews.

> (A, P6) "The first [phase] (…) is the direct search on LinkedIn, which happens at the same time as the CV screening when we publish the job announcement. The second phase is a telephone screening phase when we speak to candidates who are interested or whose profile, we also consider makes sense for the project. And the third phase, on our side, is a company interview, which we are currently doing mainly online (…). From here, you go to the client phase (…)."

(B, P5) "(...) we are doing a hybrid recruitment process, but mainly in person. When necessary, we do some part of the process online; however, we have been doing dynamics and interviews in person. (…) we conduct an initial screening of the applications, then telephone contact, which can also serve as an initial screening (...). Next, we schedule group dynamics whenever possible. (…) and then do a first and second interview… as many interviews as necessary."

In both organisations, technology plays a supporting role, assisting only in administrative and managerial tasks—in the screening stage—as Nikolaou (2021) observed. For example, neither company uses software that screens CVs without human intervention. Both companies use email and other services—such as WhatsApp—for text and video communication with candidates (the latter for interviews).

In certain recruitment contexts, such as those focused on diversity and inclusion, adaptations in the use of technology, or even its absence, may be required. Specifically, in this area, Company B conducts its R&S processes entirely in person, without reliance on technological support. This particular context of R&S, characterised by the necessity for close and meticulous candidate monitoring, underscores the conclusion that the utilisation of technology might not be universally appropriate for all R&S processes. In this particular instance, technology appears to act as an impediment to the establishment of connections and communication with candidates.

(B, P13) "The recruitment process is 100% in person. (…) People, as a rule, come from very different backgrounds. There is a training and preparation process that the institutions develop, so there is also different monitoring. Here we are talking about things that are perhaps a little different from a normal recruitment process (…)"

It is evident that, despite the pivotal role of technology in the preliminary stages of the R&S process, human intervention remains a critical component in the analysis of applications. Technology is regarded as an effective tool for attracting candidates and assisting with CV screening; however, it is not yet sufficiently advanced to replace human judgement and automate the early stages of recruitment. This finding aligns with the extant literature, which suggests that technology can optimise, but not necessarily replace, the fundamental role of recruiters in the R&S process (e.g., García, 2020).

Throughout the discussions, participants made comments that alluded to both the normalisation of technology in R&S and to their expectations for the future of OR:

(A, P10) "If we think about it, email communication is also online. Sometimes I even forget about these aspects."

This extract highlights the ease with which certain technologies, in this case, online communication (email), are currently viewed in the context of R&S. It suggests that the Internet and related technologies have become such an integral part of recruiters' daily lives that they are no longer seen as specific or unique tools within what is considered an OR process. Instead, it reflects the normalisation of the online environment and digital tools in R&S, where online or technology-mediated processes have become routine.

The same participant from Company A noted that she does not perceive a distinction between traditional recruitment and online recruitment, reinforcing the idea that the two forms of R&S are similar and already commonly used, thus to some extent trivialising the notion of OR:

> (A, P10) "To this day, I don't think we do online recruitment. We do recruitment, period."

8.4.2 *Offsetting the Disadvantages of Technology with Its Advantages*

This theme addresses RQ2: *What is the role of technology in the OR process?*

There are advantages and disadvantages to the use of technology in R&S, and in most situations, technology is seen as a useful tool. With regard to the software used to manage the process, in particular that used by Company A, although compared to Excel, it is specifically designed for R&S and has several features that help the recruiters:

> (A, P10) "(…) the software is more robust than Excel. Firstly, it allows us to be online all at the same time and (…) to be in a network, Porto and Lisbon, and any consultant who now works from home. So it allows us to speed up the process. It allows us to manage the curricula and save the curricula from one process to another and to save all the data of the candidates from one process to another. So there are several advantages (…)."

> (A, P7) "(...) given the volume of candidates and the volume of projects we have at the moment, the system ends up being much more agile to use than Excel."

Video calling tools are also seen as beneficial for recruiters and candidates, particularly in terms of time management, speeding up R&S processes:

> (A, P10) "(…) doing an interview via video call (…) makes things a lot easier in terms of the agenda, whether it's the candidates' agenda (…) or our own agenda, to be able to coordinate. So, I think it makes the process faster, more agile."

When it comes to job boards (and the like), recruiters believe there are "two sides of the same coin". In other words, the advantages that technology brings to R&S can also have disadvantages. One of the main benefits highlighted by all participants is the scope offered by the use of websites, as well as the simplification and facilitation of the job search and offer process:

> (A, P10) "I think that when we publicise, we reach an audience, we reach a greater number of people, or we can reach a greater number of people, and especially nowadays, we can reach a greater number of interested, motivated people."

> (B, P1) "(...) nowadays it is very easy for me to apply for a job from home, and I think that also helps a lot. (…) Something which was unthinkable a few years ago. (…) I think this has really made it much easier for companies, both for those who apply and for companies"

The use of job sites represents a significant change in R&S processes. They offer greater reach and a more proactive approach, making the process more efficient and affordable. These changes reflect the evolution of technology and the resulting evolution and approach to the R&S process itself, for candidates and organisations (Halpern et al., 2021; Peter, 2001; Rosoiu & Popescu, 2016). However, recruiters say that the convenient (and therefore advantageous) scope that the Internet offers in attracting candidates is also one of the biggest disadvantages of advertising vacancies online, as mentioned in the literature (e.g., Brandão et al., 2019):

> (A, P10) "(…) placing an advertisement, on the one hand, you reach more motivated and interested people who may be in line with your offer, which is spectacular, but on the other hand, you have a huge number of applications here (…) that does not interest us (…). It's the thing about (…) online, which is that you always have to deal with the good and the bad of both (…)"

In terms of interaction, recruiters surprisingly emphasise that technology brings recruiters and candidates closer together. Technology is often seen as an obstacle to the relationship between recruiters and candidates due to its impersonality and lack of face-to-face contact (e.g., Brandão et al., 2019; Nikolaou, 2011; Nikolaou & Judge, 2007). Our focus group participants believe the opposite:

> (B, P5) "By being formal, [emails] end up creating a little more proximity based on the responses, that is, when you start exchanging responses, you also end up creating a certain proximity that we didn't have before with the candidates."

> (B, P3) "And this closeness sometimes has to do with the response time, because now… Now, it's been a while… since we use our phone number that the company gives us, SMS ends up being faster or WhatsApp, right? And this proximity is sometimes created by the speed of the response"

> (A, P10) "(…) we have no difficulty in creating a closer relationship here, more of a "you" relationship, of addressing people as "you" and not as "[formal] you",[1] and not using titles. We can also convey this in an email, for example, in a LinkedIn message when we approach candidates, in this close way (…)"

These transcripts show that, for these recruiters, technology does not undermine the relationship established with a candidate, compared to the relationship that would develop through face-to-face contact. On the contrary, according to our analysis, the current closeness and trust has been facilitated by the introduction of technology in R&S. As the impersonality of the OR is repeatedly mentioned in the literature (e.g., Brandão et al., 2019; Monteiro & Alturas, 2012), the experience of these recruiters suggests that one of the main disadvantages identified in the use of technology is perhaps negligible, or that ways to overcome it are being developed.

Participants recognise that despite the limitations associated with remote R&S processes, OR is the future, with technology potentially playing an increasingly important role.

> (B, P5) "[Those limitations] will certainly have to be resolved because if that [OR] is the future, there's no alternative."

> (A, P10) "And that's where technology might help us over time. There could be more robust software for this filtering process. At present, the software we use, with only the filters it provides, is very rudimentary."

> (B, P3) "(…) in the future, we'll see the impact that robots and artificial intelligence will have on our performance as professionals in the field. Clearly, we're still in an embryonic

[1] In Portuguese, the words "tu" and "você" both translate to "you" in English, but carry different levels of formality. "Tu" is an informal form of address, whereas "você" is a formal or neutral form.

phase, but we've already started to take some steps, and [so] in time, there might be some impact (…)."

Participants show an understanding of the inevitability of OR as a leading method for R&S and demonstrate a certain degree of openness to the introduction of new technologies into the process. In other words, they are willing to adapt to keep pace with progress (Kaul, 2021; Lopes et al., 2022). OR is increasingly becoming an integral part of the hiring landscape as the technology evolves and its potential applications within the context of R&S expand (Peter, 2001). The normalisation of OR is evident as its limitations and challenges are overcome, paving the way for a future where technology will play an increasingly important role in talent selection processes.

8.4.3 *Social Media Beyond Job Advertising*

This theme refers to RQ 3: *How are social media platforms used in OR?*

The analysis supports the fact that social media, in particular LinkedIn, Facebook, and Instagram, are used not only to advertise job vacancies but also to share content and news. In terms of job postings, LinkedIn emerges as the central resource, particularly for Company A, given its focus on headhunting (Blacksmith & Poeppelman, 2014). The platforms' ability to provide access to a pool of specialised candidates is particularly relevant in this context.

(A, P7) "(…) our work, being in headhunting, (…) is very proactive in the direct search for candidates, and nowadays LinkedIn is a key tool because most people end up using it. This makes it easier because we can quickly identify candidates who match the requirements, contact them directly through messages, and gauge their interest in a particular project."

However, when recruiting for roles that do not require a high level of skill, LinkedIn presents some challenges.

(B, P5) "I think the main challenge is that most candidates on LinkedIn tend to have a more skilled profile. This creates challenges, particularly in managing the expectations of those who contact us, as they are often expecting more senior roles. Meanwhile, the roles we need to fill are typically in sales, supermarkets, or catering, which means we often attract candidates whose expectations exceed our day-to-day requirements"

(B, P3) "(…) in [region in Portugal], LinkedIn has a completely different impact. The success rate is much higher because of the industries and the positioning there, with our headquarters in [region in Portugal]. (…) [here] LinkedIn was necessary for the occasional hiring of middle management positions. In [region in Portugal] it has a significant impact, but not here"

LinkedIn is also widely used to share content and news, making it an important tool for employer branding.

(B, P3) "(…) everything related to the activities carried out by the company, not only matters related to recruitment, but everything that the company wants to communicate, right? From a partnership agreement signed with an institution to any event in which the company participates, such as the representation of a director, this platform [LinkedIn] is often used to share

> such content (…). In essence, this creates value for the company itself, which can lead people to identify with the company's policies and apply."

As technology and R&S practices evolve, the effective use of social media is an essential component of a successful talent acquisition strategy that goes beyond job posting. However, as with job boards, social media platforms present a certain duality: the benefits they bring to the R&S process often come with challenges for recruiters, such as unqualified applications and managing expectations. LinkedIn stands out as the platform of choice for recruiters, largely because of its professional focus. It facilitates the identification of qualified talent, and engagement with passive candidates, and enhances employer branding efforts (Nikolaou, 2021). However, the effectiveness of social media in OR requires strategies to be adapted to local realities. Ultimately, while social media can be a powerful recruitment tool, using it effectively requires a thorough understanding of the nuances and specific challenges of each platform.

8.4.4 Organisational Values Are Reflected by Recruiters During OR

This theme relates to RQ4: *What behaviours do recruiters adopt during (direct or indirect) interactions with candidates in OR?*

Participants focus on the importance of aligning their actions with the organisation's values, regardless of whether the R&S process takes place online or in person.

> (A, P9) "I think that the behaviour, at least the way I try to position myself, is very much in line with our values of closeness, cordiality, and transparency with people throughout all stages of the process (...)"

> (A, P10) "(...) the idea is to maintain consistency in the way we interact, whether in person or online. And indeed, the idea is that the candidate, from the first contact to the last, will always receive a message from us, whatever the medium, with the principles (...) that guide us, which are our values (...)"

> (B, P5) "I have the same attitude online and in person. That is, a slightly formal attitude because we are a company with that characteristic, with a certain formality in dealing with candidates. However, as a recruiter, in moments when it is necessary to be more informal or closer, I do so both digitally and in person"

Organisational values are reflected in a tangible—and intentional—way in the way recruiters interact with candidates, as recruiters appear to be aware of the behaviours and attitudes they should adopt in order to convey their organisation's values. This intentionality is crucial, as recruiters play a vital role in representing the company's culture and values of the organisation and communicating the identity of the organisation to candidates (Chapman et al., 2005). Participants communicate that their approach is independent of whether the R&S is conducted online or in person and during direct and indirect interactions with candidates (e.g., through messaging).

8.4.5 *The Unchanged Candidate Profile Sought in Online Recruitment*

This theme relates to RQ5: *Which candidate profiles are most valued by recruiters in OR?*

The candidate profile valued in OR processes does not appear to differ from that valued in traditional R&S processes, i.e., it is not method dependent. Attributes such as resilience, adaptability, attention to presentation, and teamwork remain highly valued, regardless of the R&S format. The remote nature of OR does not change the essence of the desired requirements, demonstrating consistency in talent acquisition.

> (A, P10) "Let me tell you, they are no different from those [behaviours] we look for in a non-online process. (…) The competencies are exactly the same as those we look for in a non-online candidate"

> (B, P2) "(…) you expect candidates to show the same commitment and motivation in an online interview as in a face-to-face interview, don't you? So, we always look for that—the picture should be as appropriate as if it were in person, and the effort throughout the interview should be either very similar or identical to what it would be in person (…)."

However, it can be more difficult to make an accurate assessment of a candidate online.

> (A, P10) "That there are other challenges—this will be the next point—that there are other challenges in assessing competencies [online], I would say 'yes'."

> (B, P5) "(…) an important aspect of online recruitment (…) is that you can make more mistakes, mainly because what you see is just this, what we see here on the screen, isn't it? (…) those mistakes may affect the hiring decision because a person may struggle to do certain tasks, (…) and that is something that we don't see in an online interview."

On the other hand, OR allows recruiters to observe candidates in a different environment, once again highlighting a duality in online candidate assessment:

> (B, P3) "(…) generally, the person is at home, in their comfort zone, right? And they may behave in a completely different way than they would in a company interview. They are more relaxed, more confident, aren't they? This can help us later in the integration process to understand whether the person is going to be more reserved or someone who, after a short time, is going to interact with everyone in a very casual manner. (…) So, although it can be a disadvantage, it can also be an advantage, because it helps us analyse the person's characteristics."

Thus, OR seems to blur the boundaries between personal and professional spaces, affecting both candidates' privacy and their attitude and behaviour during the process, which recruiters perceive as more relaxed and potentially less professional. Participation in OR requires candidates to draw these boundaries, as there is no physical corporate environment to set the tone of professionalism that traditional R&S naturally imposes. The way candidates behave in the online environment can give an indication of their level of preparation and professionalism. Some participants shared experiences of online interviews that would not occur in a face-to-face setting:

(A, P6) "Or when they type on the computer, for example."

(B, P3) "I remember doing interviews [online] and (…) two situations stand out: one was a cat walking across the table during the interview (…). And the other was someone (…) doing the interview while sitting on their bed."

(B, P5) "(…) when they leave the microphone on and there is background noise. We've even had cases where the person was eating during a group dynamic session."

While OR can make certain aspects of candidate assessment more challenging, it also enables a different type of assessment that can ultimately help recruiters make more informed and accurate hiring decisions. Reflecting on the extracts highlights the continuity of certain assessment standards, regardless of the R&S method used. Recruiters value certain attitudes, behaviours, and core competencies, as well as the demonstration of professional etiquette, which are consistent across all types and forms of R&S. Despite the difficulties associated with assessing candidates in OR, these challenges are offset by new opportunities to analyse a candidate's profile.

8.5 Conclusion

The analysis allows us to understand that OR represents a remarkable convergence between technology and traditional R&S practices, highlighting how technological innovation has profoundly shaped the field (Halová & Müller, 2021). For recruiters, technology seems to be not just a tool, but a fundamental ally in the search for talent.

Looking at traditional R&S and OR, it is clear that at their core they share the same goal: to identify qualified candidates to fill specific positions. Traditional R&S relies on face-to-face contact, while OR takes advantage of various digital and online platforms. However, the essence of the process remains the same—finding talent.

In many cases, the disadvantages attributed to OR do not appear to be significant. Issues such as impersonality do not seem to exist, and other limitations commonly cited for this method are offset by advantages such as greater efficiency in the process and new perspectives in evaluating candidates that did not previously exist. The paradigm seems to indicate that the disadvantages do not invalidate this form of recruitment, but rather encourage adaptations and adjustments (Fraij, 2021; Halová & Müller, 2021; Lopes et al., 2022) that improve and consolidate OR as an increasingly popular method and one that stands out as the preferred choice for contemporary needs.

Technology is intrinsically linked to R&S and this relationship will intensify as society advances technologically (Brandão et al., 2019; Morgan, n.d.). The normalisation of OR reflects this evolution, indicating that technology is not just an auxiliary tool, but an inevitable part of the R&S scenario. In the future, even more thorough technological integration can be foreseen, where digital tools are not limited to optimising the process, but become an essential and inalienable part of it, as already pointed out by Peter (2001).

This study contributes to the field of human resource management, especially to the field of R&S, which has been significantly influenced by contemporary technological advances. The focus of the study is on the experiences and meanings that Portuguese recruiters ascribe to OR, an increasingly relevant modality, although a more in-depth understanding of it is still lacking. The analysis reveals different technologies used since the beginning of the R&S process, each with different purposes. Recruiters from both organisations highlight both the advantages and disadvantages associated with the use of technology in talent sourcing, while recognising the significant potential of OR. Contrary to expectations derived from the existing literature, the results of this study show discrepancies. Some of the disadvantages traditionally associated with this form of R&S were not relevant to the participants and OR allows access to new perspectives in the R&S process.

This study highlights the need for continuous adaptation and customisation of OR processes (Halová & Müller, 2021), and shows that different companies from different sectors and even different areas under the jurisdiction of HR departments have specific R&S needs. Therefore, in order to further improve the process, it would be beneficial for organisations to take the different types of technology and other factors addressed in this study into account. Organisations would benefit from a more detailed analysis of the platforms used, to understand how they can impact R&S processes. Another aspect to consider is the assessment of candidates through digital means. Candidate behaviour may change depending on the medium used, so new approaches to assessment may be beneficial. Recruiters can emphasise the observation of non-verbal behaviour and the assessment of candidates' soft skills in the online environment. This study also opens the door to a more thorough analysis of the ethical issues related to OR, which are becoming increasingly relevant as its use becomes more widespread.

Future research may consider organisations that use different technologies in their processes. The study and analysis of future trends in OR is also relevant to help organisations adapt their R&S methodologies. In addition, the study of organisations that have successfully implemented the use of OR throughout the R&S process may be relevant to identify key practices and aspects of its implementation. It is also relevant to understand whether organisations where the process is essentially online implement more stages in the R&S process. The predictive power of OR should also be investigated.

Like all studies, this one has limitations that need to be considered. The difficulty in recruiting diverse participants, due to scheduling constraints, limited the representation of different industries and organisations and the breadth of the analysis, as well as the conduct of more focus groups on this topic. Nevertheless, these two organisations illustrate different operationalisations of OR in Portugal, allowing us to understand that Portuguese organisations are at different stages of implementing information technology in R&S. However, the use of technology is still limited.

This study extends the existing knowledge on OR and provides a demystified view of the topic from the perspective of one of the main actors in the process, the recruiter. It is intended to contribute to the ongoing development of the field of organisational psychology, particularly in relation to R&S and it is expected that it will inspire future research to further expand the understanding of OR, especially in the Portuguese context.

Although OR represents a change in the R&S approach through its transition to the digital environment, its essence seems to remain the same. The symbiosis between technology and human practice will continue to evolve, outlining a new paradigm through a new way of conducting processes, but potentially without changing the core and objective of the R&S process.

References

Autor, D. (2001). Wiring the labor market. *Journal of Economics Perspectives, 15*(1), 25–40. https://doi.org/10.1257/jep.15.1.25

Bartram, D. (2000). Internet recruitment and selection: Kissing frogs to find princes. *International Journal of Selection and Assessment, 8*(4), 261–274. https://doi.org/10.1111/1468-2389.00155

Blacksmith, N., & Poeppelman, T. (2014). Three ways social media and technology have changed recruitment. *The Industrial Organizational Psychologist, 52*(1), 114–121. https://doi.org/10.1037/e570072013-008

Brandão, C., Silva, R., & dos Santos, J. (2019). Online recruitment in Portugal: Theories and candidate profiles. *Journal of Business Research, 94*, 273–279. https://doi.org/10.1016/j.jbusres.2018.04.011

Brandão, C., Morais, C., Dias, S., Silva, A. R., & Mário, R. (2017). Using online recruitment: Implicit theories and candidates' profile. In *Recent advances in information systems and technologies* (Vol. 35, pp. 293–301). Springer International Publishing. https://doi.org/10.1007/978-3-319-56541-5_30

Brandão, C., Veloso, A., Vieira dos Santos, J., Gonçalves, S.P., Silva, I., & Ribeiro, R. (2022). Work changes due to COVID-19: Exploring Portuguese workers' expectations using IRaMuTeQ. In A.P. Costa, A. Moreira, M. C. Sánchez-Gómez, & S. Wa-Mbaleka (Eds.), *Computer supported qualitative research. WCQR 2022. Lecture Notes in Networks and Systems* (vol. 466). Springer, Cham. https://doi.org/10.1007/978-3-031-04680-3_19

Braun, V., & Clarke, V. (2013). *Successful qualitative research: A practical guide for beginners.* SAGE Publications.

Braun, V., & Clarke, V. (2006). Using thematic analysis in psychology. *Qualitative Research in Psychology, 3*(2), 77–101. https://doi.org/10.1191/1478088706qp063oa

Braun, V., & Clarke, V. (2020). One size fits all? What counts as quality practice in (reflexive) thematic analysis? *Qualitative Research in Psychology, 18*(3), 328–352. https://doi.org/10.1080/14780887.2020.1769238

Braun, V., & Clarke, V. (2022). Toward good practice in thematic analysis: Avoiding common problems and be(com)ing a *knowing* researcher. *International Journal of Transgender Health, 24*(1), 1–6. https://doi.org/10.1080/26895269.2022.2129597

Braun, V., Clarke, V., Hayfield, N., & Terry, G. (2019). Understanding ta. *Thematic Analysis.* https://www.thematicanalysis.net/understanding-ta/

Breaugh, J. A. (2013). Employee recruitment. *Annual Review of Psychology, 64*, 389–416. https://doi.org/10.1146/annurev-psych-113011-143757

Brown, V. R., & Vaughn, E. D. (2011). The writing on the (Facebook) wall: The use of social networking sites in hiring decisions. *Journal of Business and Psychology 26*, 219–225. https://doi.org/10.1007/s10869-011-9221-x

Byrne, D. (2021). A worked example of Braun and Clarke's approach to reflexive thematic analysis. *Quality & Quantity, 56*(3), 1391–1412. https://doi.org/10.1007/s11135-021-01182-y

Campos, R., Arrazola, M., & de Hevia, J. (2018). Finding the right employee online: Determinants of internet recruitment in Spanish firms. *Applied Economics, 50*(1), 79–93. https://doi.org/10.1080/00036846.2017.1319560

Cappelli, P. (2001). Making the most of on-line recruiting. *Harvard Business Review, 79*(3), 139–148.

Chapman, D. S., Uggerslev, K. L., Carroll, S. A., Piasentin, K. A., & Jones, D. A. (2005). Applicant attraction to organizations and job choice: A meta-analytic review of the correlates of recruiting outcomes. *Journal of Applied Psychology, 90*(5), 928–944. https://doi.org/10.1037/0021-9010.90.5.928

Chiavenato, I. (1999). *Gestão de Pessoas: o novo papel dos recursos humanos nas organizações* [People Management: the new role of human resources in organizations]. Campus.

Costa, T. (2002). E-Rh: O impacto da tecnologia para a gestão competitiva de recursos humanos. Universidade de São Paulo. https://docplayer.com.br/1675205-E-rh-o-impacto-da-tecnologia-para-a-gestao-competitiva-de-recursos-humanos.html

Falter, M., Arenas, A. A., Maples, G. W., Smith, C. T., Lamb, L. J., Anderson, M. G., Uzzell, E. M., Jacobs, L. E., Cason, X. L., Griffis, T. A. N., Polzin, M., & Wafa, N. Z. (2022). Making room for zoom in focus group methods: Opportunities and challenges for novice researchers (during and beyond COVID-19). *Forum Qualitative Sozialforschung Forum: Qualitative Social Research, 23*(1). https://doi.org/10.17169/fqs-23.1.3768

Fernandes, A. B. A., & Machado, C. F. (2022). E-recruitment and the impact of digital age on recruitment: A critical literature review. *Organizational Innovation in the Digital Age*. https://doi.org/10.1007/978-3-030-98183-9_8

Flick, U., & Barbour, R. (2009). *Grupos focais* [Focus Groups]. Artmed Editora.

Fraij, J. (2021). E-HRM to overcome HRM challenges in the pandemic. *SEA–Practical Application of Science, 9*(25), 41–49.

García, J. (2020). Technology doesn't kill the human touch. *Capital Humano, 349*, 25–28. https://capitalhumano.wolterskluwer.es/ch/2020/01/01/technology-doesn-t-kill-the-human-touch

González Iglesias, C. (2020). Reclutamiento responsable: cuando la urgencia debe convertirse en técnica. *Capital Humano, 349*, 37–41. https://capitalhumano.wolterskluwer.es/ch/2020/01/01/reclutamiento-responsable-cuando-la-urgencia-debe-convertirse-en-tecnica

Halová, D., & Müller, M. (2021). Innovative approaches to recruiting: Using social media to become the employer of choice for generation Z. In *16th European conference on innovation and entrepreneurship*, pp. 1135–1143. ISCTE Business School, Instituto Universitário de Lisboa, Portugal.

Halpern, C., Halpern, B., & Cerchiaro, I. (2021). As vagas vagas: Impedimentos no recrutamento online de estagiários. *Revista De Administração Unimep, 19*(5), 152–175.

Hedenus, A., Backman, C., & Håkansson, P. (2021). Whom do you know? Recruiters' motives for assessing jobseekers' online networks. *The International Journal of Human Resource Management, 32*(8), 1754–1777. https://doi.org/10.1080/09585192.2019.1579245

Kaul, A. (2021). *Methods to transform, innovate and evolve recruitment in 2021*. https://www.hr.com/en/magazines/talent_acquisition/september_2021_talent_acquisition/methods-to-transform-innovate-and-evolve-recruitme_ktmcqpcl.html

Kesmodel, U. (2018). Cross-sectional studies—what are they good for? *Acta Obstetricia Et Gynecologica Scandinavica, 97*(4), 388–393. https://doi.org/10.1111/aogs.13331

Lopes, A., Yanchuk, A., Morais, E., Moniz, S., Walter, C. E., & Au-Yong-Oliveira, M. (2022). Remote work and the challenges of the global search for talent: How companies in Portugal are adapting to a new reality. In *2022 17th Iberian conference on information systems and technologies (CISTI)* (pp. 1–6). IEEE. https://doi.org/10.23919/CISTI54924.2022.9820100

Maxwell, J. A. (2008). Designing a qualitative study. *The SAGE Handbook of Applied Social Research Methods*. https://doi.org/10.4135/9781483348858.n7

McKinsey & Company. (2020). How COVID-19 has pushed companies over the technology tipping point—and transformed business forever. *McKinsey Digital and Strategy & Corporate Finance Practices*. https://www.mckinsey.com/business-functions/strategy-and-corporate-finance/our-insights/how-covid-19-has-pushed-companies-over-the-technology-tipping-point-and-transformed-business-forever

Miles, M. B., & Huberman, A. M. (1994). *Qualitative data analysis: An expanded sourcebook* (2ª). Sage Publications.

Monteiro, D., & B. Alturas (2012). The adoption of e-Recruitment: The Portuguese case: Study of limitations and possibilities by the point of view from candidates and from recruiters. In *7th Iberian conference on information systems and technologies (CISTI)* (pp. 1–6). IEEE.

Morgan, H. (n.d.). How social recruiting impacts job search. *Career Planning & Adult Development Journal: A Quarterly Publication of the Career Planning and Adult Development Network, 32*(2), 28–33.

Nikolaou, I. (2011). Core processes and applicant reactions to the employment interview: An exploratory study in Greece. *The International Journal of Human Resource Management, 22*(10), 2185–2201. https://doi.org/10.1080/09585192.2011.580187

Nikolaou, I., & Judge, T. A. (2007). Fairness reactions to personnel selection techniques in Greece: The role of core self-evaluations. *International Journal of Selection and Assessment, 15*(2), 206–219. https://doi.org/10.1111/j.1468-2389.2007.00382.x

Nikolaou, I. (2021). What is the role of technology in recruitment and selection? *The Spanish Journal of Psychology, 24*(2). https://doi.org/10.1017/SJP.2021.6

Onn, C. W., & Sorooshian, S. (2013). Mini literature analysis on information technology definition. *Information and Knowledge Management, 3*(2), 139–140.

Peter, C. (2001). Making the most of on-line recruiting. *Harvard Business*, 139–146.

Rosoiu, O., & Popescu, C. (2016). E-recruiting platforms: Features that influence the efficiency of online recruitment systems. *Informatica Economica, 20*(2/2016), 46–55. https://doi.org/10.12948/issn14531305/20.2.2016.05

Tutar, H., & Ay, G. (2021). Investigation of selected companies' websites for online recruitment and sustainability indicators. *Cumhuriyet University Journal of Economics and Administrative Sciences, 22*(2), 456–479. https://doi.org/10.37880/cumuiibf.994540

Wolfswinkel, J., Furtmueller, E., & Wilderom, C. (2010). Reflecting on e-recruiting research using grounded theory. In *18th European conference on information systems, ECIS 2010 proceedings*, 52. http://aisel.aisnet.org/ecis2010/52

Chapter 9
Executive Coaching as a Mental Health Intervention Instrument Integrated in Human Resources Management

Manuel Salvador Gomes de Araújo
and Vanda Gisela Conde Pereira Branco Afonso

Abstract The focus on mental health (MH) represents one of the biggest challenges for Human Resource Management (HRM) in a disruptive world. It is estimated that 12 billion working days are lost every year due to the psychosocial risks of the twenty-first century, requiring immediate help, according to the (World Health Organization and International Labour Organization in Crucial changes needed to protect workers' health while teleworking, 2022). Although executive coaching (EC) has other main purposes in the HRM strategy, it is also important to analyze other side effects (on people and contexts) derived from the choice of this intervention tool. For this reason, the aim of this study was to evaluate the perceptions of different stakeholders in EC, in the workplace, with reference to its relationship with MH in organizations. Quantitative and cross-sectional methodology was used, with descriptive, correlational, predictive and comparative analyses. The data was collected through an online questionnaire on the Google Forms platform, where 268 answers were obtained. The data collected was analyzed using IBM-SPSS (version 28.0). The results demonstrated that EC is a process with a socially positive representation in the organizational culture, with evidence not only at a behavioral and organizational level, but also in terms of psychological development, making it possible to reduce stress and anxiety, with a positive impact on MH. In addition, using the General Health Questionnaire (GHQ-12), it was found that the MH of HR who had participated in EC processes was better than those who had not participated in EC.

Keywords Coaching · Executive coaching · Mental health · HRM

M. S. G. de Araújo (✉) · V. G. C. P. B. Afonso
Porto Accounting and Business School, Polytechnic Institute of Porto, São Mamede de Infesta, Portugal
e-mail: manuelsalvador@iscap.ipp.pt

C. Machado (ed.), *Human Resource Management in a Disrupted World*, Lecture Notes on Multidisciplinary Industrial Engineering,
https://doi.org/10.1007/978-3-031-89948-5_10

9.1 Introduction

Today's dynamic and complex labour market, which is increasingly competitive and requires people in organizations to perform with flexibility in order to deal with vertiginous organizational changes, is conducive to the incidence of MH problems on a global scale, with repercussions on well-being, low productivity, absence, presenteeism (RANDE Europe, 2022) and many other occupational problems (OPP, 2023; Toh et al., 2023). According to the WHO (2022b), this is an increasingly alarming scenario, given that "About one in eight people in the world live with a mental disorder (p. xv)."

According to the Portuguese Health Regulatory Authority (ERS, 2023), "In Europe, Portugal is the country with the highest prevalence rate (23%) of symptoms associated with psychological problems" (p. 3), which is why MH and emotional distress occupy first place among problems related to the well-being of the workforce, followed by burnout (AON, 2022).

A study by the Portuguese Laboratory for Healthy Working Environments (LABPATS—Gaspar et al., 2023) concluded that around 80% of Portuguese workers have at least one symptom of burnout and that 63% of workers have at least three symptoms. These indicators show that Portugal is a country with a high risk of burnout and depression related to professional activity (Peralta, 2023), which must be taken into account as a priority for HRM, since for the WHO (2022a, 2022d), depression and anxiety are the main cause of incapacity for work.

Based on national and international evidence, it is necessary to integrate MH policies in the workplace, in line with The Sustainable Development Goals of the UN Agenda 2030 (2015), in line with the guidelines stipulated by several organizations (AEST, 2023; EU, 2021; ICOH, 2022; OECD, 2021).

It is extremely important for leaders and managers to become aware of this imperative and urgent need and to become stakeholders in order to promote strategies that encourage employees' MH. Coaching, although its main objective is not to promote the MH of coachees, can be one of these strategies because it is a process that facilitates personal or organizational support and is considered effective in promoting positive changes in individuals and organizations (Korotov, 2023; Passarelli et al., 2022; Wang et al., 2022).

In line with the concept of MH defined by the WHO (2022c), which is inherent to the individual's state of well-being, allowing them to manage the usual stresses of everyday life, we are focusing on the study of EC as an intervention tool, in a non-clinical population (Cavanagh, 2005; Grant, 2008), distinct from psychotherapy and other organizational support resources, with benefits for psychological well-being evidenced by scientific research (Brooks et al., 2023; Spence & Grant, 2007; Theeboom et al., 2013; Wang et al., 2022). In addition, coaching is now seen "as a workplace learning and development (L&D) activity of choice" (Wang et al., 2022, p. 78).

According to Oliveira-Silva et al. (2018), EC consists of "a growing professional practice" (p. 363) whose purpose "is to provide more humanized organizational environments" (Sarsur & Parente, 2019, p. 139), which contributes to encouraging personal and professional growth, improving talent management and optimization (Albizu Gallastegui et al., 2019).

As Boyatzis et al. (2022) mention, "If coaching is to continue to help create and support a better society worldwide, it is crucial to study, develop and improve the practice of coaching." (p. 203). There has been an increase in empirical research attesting to the effectiveness of EC in general, but there is still a lot of scope for scientific research into the discipline (Albizu Gallastegui et al., 2019), particularly in terms of its impact and effectiveness, the circumstances in which it is most effective (De Haan, 2021), as well as health outcomes (Bachkirova & Borrington, 2020; Boyatzis et al., 2022; Brooks et al., 2023).

This justifies the choice of the theme due to its particular importance in contemporary HRM, as the coaching methodology has been increasingly recognized as a potential strategic and successful resource for personal, professional and organizational development (Ali et al., 2020; AMA, 2019; De Haan, 2021; Grant, 2020; Grover & Furnham, 2016; McCarthy & Milner, 2020; Sarsur & Parente, 2019; Spence & Grant, 2007; Salvadorinho & Teixeira, 2022; Theeboom et al., 2013; Wang et al., 2022). In this way, the primary objective of this study is to understand the perception of the contribution of EC to improving the MH of employees, who are the backbone of the success of organizations (Salvadorinho & Teixeira, 2022). And here comes the central research question: How is the activity of Executive Coaching perceived in terms of its relationship and impact on improving mental health?

9.2 Theoretical Background

9.2.1 Mental Health in the Workplace

Once the concept of MH was confined to the absence of mental illness, this paradigm, in line with the evolution of society, has changed to other, more comprehensive meanings that integrate multiple biological, psychological, sociocultural, socioeconomic and environmental factors, with an impact on individual and collective well-being (Mental Health Europe, 2022; WHO, 2018).

Nowadays, the concept of MH refers to "a state of mental well-being that enables people to cope with the stresses of life, realize their abilities, learn well and work well, and contribute to their community." (WHO, 2022c).

MH is therefore a fundamental human right, which conditions our ability to make decisions, to relate and to interact actively in the world in which we live. As such, it involves a state of balance between the person/family/organization, which allows them to be able to earn a living and simultaneously enjoy it, based on well-being and quality of life, develop their own potential, be able to deal with the normal challenges of life, adapt to new realities and work constructively and effectively in the community (WHO, 2022b).

The literature has highlighted that MH at work is directly related to productivity, commitment and talent retention (de Oliveira et al., 2023). Thus, proactive HRM intervention strategies that promote these indices are crucial for organizational sustainability.

9.2.2 Executive Coaching as an HRM Strategy

EC, an emerging and complex practice, which is expanding and whose initial focus was on developing leaders, now extends across hierarchical levels, bringing benefits to the entire organization. According to several international studies, EC is one of the strategies for resistance to change and the promotion of an agile mentality, favoring MH, with a substantially positive impact on the workplace (Adi et al., 2022; Geraghty, 2021; Grover & Furnham, 2016; Nicolau & Constantin, 2021). Despite the ideological separation between the two contexts (EC and MH), imposed by the Professional Code of Ethics for Coaching, the intersection between the two is seen as mutually influential and beneficial (Boyatzis et al., 2023), although studies on the correlation between EC and MH are still limited (Gerathy, 2021). That's why we can consider our study to be a pioneer in Portugal, whose contribution to the development of the theory and practice of EC, within the scope of HRM, was intended to contribute, in some way, to the gap mentioned in the literature regarding the need to investigate the impact on employees' MH statistics when monitored in a EC process.

9.3 Methodology

9.3.1 Study Design

This study mobilized quantitative and transversal methodology, with descriptive, correlational, predictive and comparative analyses. Based on the literature analyzed, we aimed to verify the influence that EC has on MH, and the following hypotheses were formulated.

Hypothesis 1: EC is a process with a positive social representation in Portuguese organizational culture and with an impact on MH.

Hypothesis 2: EC is an organizational support tool with positive evidence in terms of psychological development.

Hypothesis 3: The participants in this study show adequate literacy to conceptualize EC.

Hypothesis 4: Individuals who have already participated in EC processes have lower scores on the GHQ-12.

Hypothesis 5: The perception of absenteeism and presenteeism is due to MH reasons and work-related stress factors.

Exploratory: There are significant differences in the answers to the GHQ-12 between coaches and HR professionals.

9.3.2 Data Collection Instruments

The data was collected through an online questionnaire on the Google Forms platform, designed ad hoc, whose conception was guided by the theoretical framework analyzed and supported by previous reflections and research, organized into three parts:

Part I—Sociodemographic characterization

Part II—Perceptions of EC in organizations

Part III—Perceptions of the relationship between EC and MH. This section ends with the inclusion of Goldberg's GHQ-12 (1978), adapted to Portugal by McIntyre et al. (1999), which is one of the best known and most used instruments due to its brevity and effectiveness "as a one-dimensional indicator of psychological morbidity, i.e. an indicator of symptoms that require clinical attention, as indicated by the authors" (Araújo, 2009, p. 211). This questionnaire contains 12 items, which are answered on a Likert-scale with four categories: 1 = Better, 2 = Same, 3 = Worse, 4 = Much Worse, with high scores indicating worse MH.

9.3.3 Statistical Procedures and Operationalization of Variables

After obtaining the answers, we extracted them into an Excel file which was analyzed and filtered, taking into consideration the information considered necessary for the study. Subsequently, the data was uploaded to the Statistical Package for the Social Sciences (SPSS) analysis software, IBM SPSS program (version 28.0) for descriptive and exploratory analysis, where we made the necessary updates to be processed, using the statistical measures and graphs, hypothesis tests, confidence intervals, and simple linear regression that were necessary to reach some conclusions.

The age class variable was operationalized a posteriori into six categories with an interclass range of ten years (16–25, 26–35, 36–45, 46–55, 56–65 and 66–75 years). The GHQ-12 scale is a reduced version of the original GHQ-28 by Goldberg (1978) adapted by McIntyre and Redondo (1999). This scale was operationalized according to the authors' criteria and, in the end, a variable was created, resulting from the sum of the scores for each of the 12 items on the scale. The score ranged from 0 to 36 points, considering that low values for the GHQ-12 scale translate into better MH. A score between 0 and 24 corresponds to a better state of health and between 25 and 36 is considered pathological, i.e. a worse state of MH. We also operationalized the "EC Literacy" variables, resulting from the sum of question 10 for coaches and question

53 for HR, and the "EC Literacy" variable, resulting from the sum of question 9 for coaches and question 52 for HR.

For statistical treatment, we applied descriptive and inferential statistics. With regard to descriptive statistics, we determined: absolute and relative frequencies, measures of central tendency were only applied to continuous numerical variables (means, medians and modes) and measures of dispersion and variability (standard deviation, minimum and maximum, asymmetry and flatness). The nominal and ordinal qualitative variables were evaluated taking into account the modal class. With regard to inferential statistics, simple linear regressions were used to analyze the variables "EC Representation and Impact on MH" and "Psychological Evidence of EC" as a function of "Position in which they filled in this questionnaire". The analysis of the results of the simple linear regressions was done taking into account Cohen's classification (1988), indicating that correlation values between 0.10 and 0.29 indicate a non-existent or small correlation; values between 0.30 and 0.49 indicate that there is a medium correlation and values between 0.50 and 1 can be interpreted as signs of a high correlation.

In addition, T-tests were conducted for Hypotheses 3, 4 and the Exploratory Hypothesis. In the first phase, the equality of variances was considered using Levene's test. If the p-value was less than 0.05, the variances were assumed to be non-equal. If it was higher, they were assumed to be equal. For the T-test for equality of means, it was assumed that the means were not statistically equal whenever the bilateral p-values were less than 0.05.

9.3.4 Sample

Due to the triadic dimension of EC, represented by the organization's HRM service, the coaches and the coachees, we considered it pertinent to know the perspective of all those involved in the process. In addition, we wanted to reinforce the triangulation of data in this study with the perspective of other individuals who knew people in the coaching process or who had already completed it, in order to get a more reliable picture of reality and a more complete understanding of the phenomenon under study.

The sample selection complied with the inclusion criteria aligned with the purpose of the study:

- Being an HR professional
- Being a coaching professional
- Being or having been a coaching client
- Knowing someone who is or has been in a coaching process.

We received 268 responses. The maximum possible representativeness was achieved with 134 coaches, followed by 122 HR, 10 coachee knowers and only 2 coachees, the remaining two figures not being able to be extrapolated because they make statistical analysis impossible. However, considering that professionals in HR departments, who are supposed to be strategic in the EC process, should be familiar with its implementation, monitoring and benefits for the organization (Sarsur & Parente, 2019), and considering coaches "as the main instrument of coaching" (Bachkirova, 2016), this study only allows us to validate the most representative indicators, i.e. in the coaches (Group I) and HR professionals (Group II) categories, and the rest were excluded because they were not sufficiently powerful to demonstrate the effectiveness of EC. Thus, the "definitive" sample consisted of 256 individuals, the majority of whom were women (67.2%, n = 172).

9.4 Results and Discussion

Group I

We began by checking the coaches' level of literacy about the philosophy of EC, as shown in Fig. 9.1, in which we analyze the answers' interpretation of the concept of EC. Given their level of training and accreditation, we are not surprised that the majority of coaches (94%) are familiar with and properly aware that it is a "process of personal, professional and organizational development" as referenced in the literature (De Haan, 2021; Grant, 2020; McCarthy & Milner, 2020; Salvadorinho & Teixeira, 2022; Wang et al., 2022).

When coaches were asked about the importance that organizations, in general, give to people's MH, 47% considered it insufficient, 29.9% said it was sufficient, 17.2% perceived it as good, 3.7% said it was bad and 2.2% said it was very good. Critically examining the data in Fig. 9.2, we can also identify a lack of attention on the part of organizations to MH, which has been persistently analyzed in the literature (Bachkirova & Borrington, 2020), and which international organizations have been

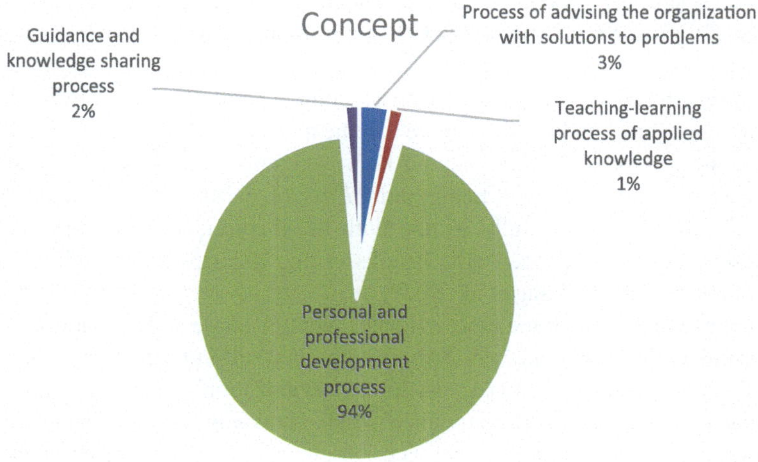

Fig. 9.1 EC concept

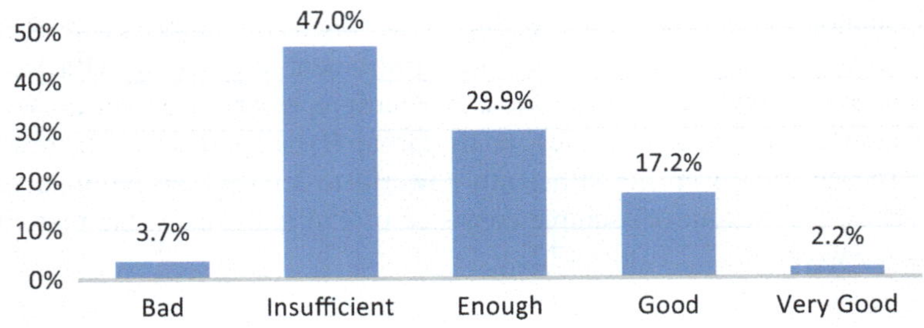

Fig. 9.2 Coaches' perception of the importance that organizations attribute to MH

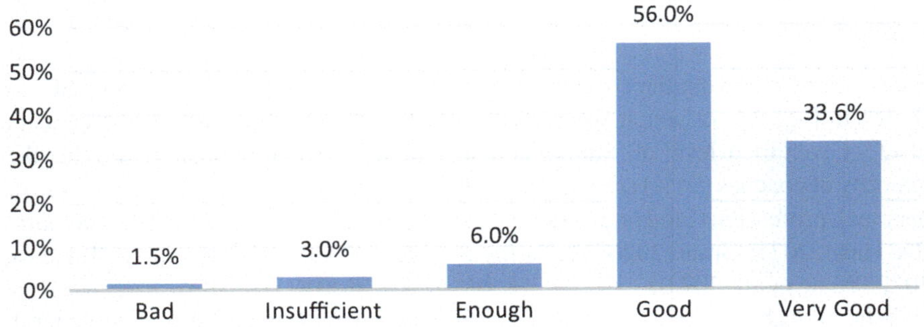

Fig. 9.3 Evaluation of EC results in the promotion of MH

calling for greater investment in well-being, embodied in the promotion of MH as a human right of basic social justice.

Regarding the evaluation of the EC's results in promoting MH, 56% said it was good, 33.6% said very good, 6% said sufficient, 3% said insufficient and 1.5% said bad. These figures reveal a strong belief in the impact of EC on MH, as shown in Fig. 9.3.

Analyzing the evidence considered most significant of the impact of EC on MH, at a psychological level, in Fig. 9.4, we see that the majority (84.3%) report a reduction in stress and anxiety; 73.1% registered improved self-esteem; 69.4% reported an increase in psychological well-being and 27.8% of participants indicated "Other evidence", all of which is fully in line with the theoretical framework analyzed (Athanasopoulou & Dopson, 2018; Brooks et al., 2023; Spence & Grant, 2007; Theeboom et al., 2013; Wang et al., 2022).

We believe that some of the other evidence mentioned are no less important, such as: "learning to deal with bad days through emotional self-control tools"; "increased personal happiness resulting in success for the organization"; "increased motivation and initiative"; "increased self-confidence"; "self-knowledge and self-management"; "self-awareness and increased coherence"; "self-worth"; "discovering priorities"; "empowerment"; "balance between personal and professional life". These conclusions support previous studies (Ballesteros-Sánchez et al., 2019; Geraghty, 2021;

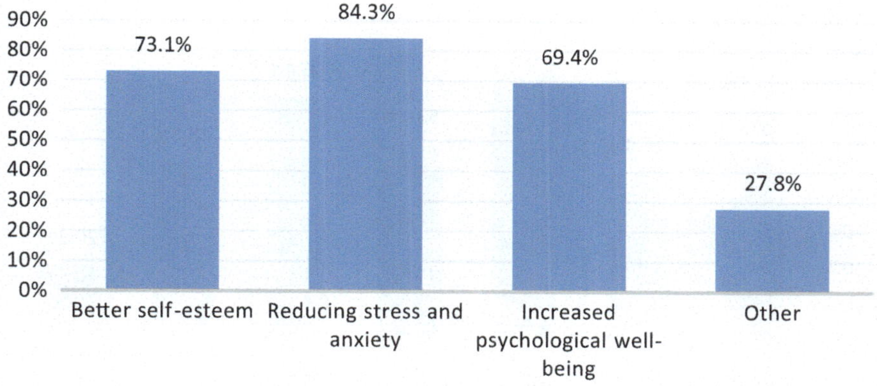

Fig. 9.4 Evidence of the impact of EC on MH, at a psychological level

Grover & Furnham, 2016; Nicolau & Constantin, 2021) and reinforce the benefits of EC with an impact on MH.

Finally, we consider it pertinent to report this response "identification of cases that need psychological/psychiatric support and appropriate clinical referral", which denotes a clear awareness that coaches, despite inevitably facing significant MH problems in some clients, do not deal with them and, therefore, they already have a relevant role in detecting cases and cooperating with the respective clinical areas, according to Grant (2007).

Moving on to an analysis of the most significant evidence of the impact of EC on MH at a behavioral level, Fig. 9.5 shows that the following stand out in descending order: better self-efficacy in leadership and management (78.4); greater empathy (66.4%); better performance (63.4%); greater assertiveness (53.7%); greater security (48.5%); greater flexibility (46.3%); more empathy (66.4%); better performance (63.4%); more assertiveness (53.7%); more security (48.5%) and more flexibility (46.3%), benefits that have also been proven in various studies (Albizu Gallastegui et al., 2019; Ballesteros-Sánchez et al., 2019; De Haan, 2021; Trevisan et al., 2021; Vesoloski & Garcia, 2022).

Other evidence (20.3%) includes: "better management of emotions", "self-care", "development of emotional intelligence", "development of self-leadership", "greater assertiveness", "greater attention to others", "more ability to influence changes in the organizational environment"; "greater ability to challenge perceptions and relativize"; "better stress management"; "better relationships"; "better decisions"; "healthy and collaborative relationships". These behavioral changes are emphasized by researchers (Albizu Gallastegui et al., 2019; Boyatzis et al., 2022; Grover & Furnham, 2016; Sarsur & Parente, 2019) as by products of EC that promote more humanized work climates.

At an organizational level, the greatest impact of EC on MH (see Fig. 9.6), according to this sample, is the increase in confidence, support, recognition and

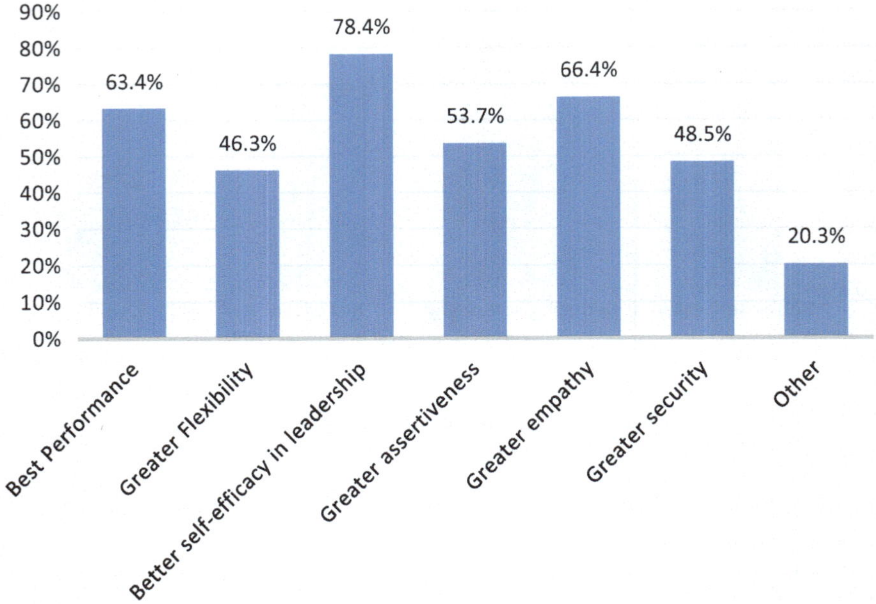

Fig. 9.5 Evidence of the impact of EC on MH, at the behavioral level

feedback (85.8%), certainly because they are rewarding for self-esteem, commitment and a sense of belonging, according to Inceoglu et al. (2018), cited by Zhao and Liu (2020). In second place comes the reference to a better, more motivating, connected and participatory climate (59.7%), followed by an increase in productivity (49.6%), corroborating the theoretical framework analyzed (Boyatzis et al., 2023; De Haan, 2021; Nicolau & Constantin, 2021).

The remaining evidence (11.9%) follows the same line of thought, indicating to "greater value for the company due to more and better results"; "better work life balance; 'greater self-knowledge of talents and potential, allowing the 'right person for the right position' to be identified, generating more engagement and impact on the business". With regards to the coaches' perception of the durability of the positive results of EC, as shown in Fig. 9.7, the majority (87.3%) confirmed this, with 2.2% rejecting it. A percentage (10.4%) of the respondents chose the option "Other", expressing their lack of knowledge about the duration, while the rest believe that the results can be long-lasting if certain constraints contribute to their consistency, namely the "need for continuity", "depending on the person and the support of peers and above all the top management of the organization"; being guaranteed if the support is provided "with consistency and commitment". These constraints are in line with the meta-analytical results carried out by Nicolau et al. (2023), which proved its stability over time, indicating that EC is an effective tool that mainly results in changes in individuals' behaviors, especially cognitive behaviors, without

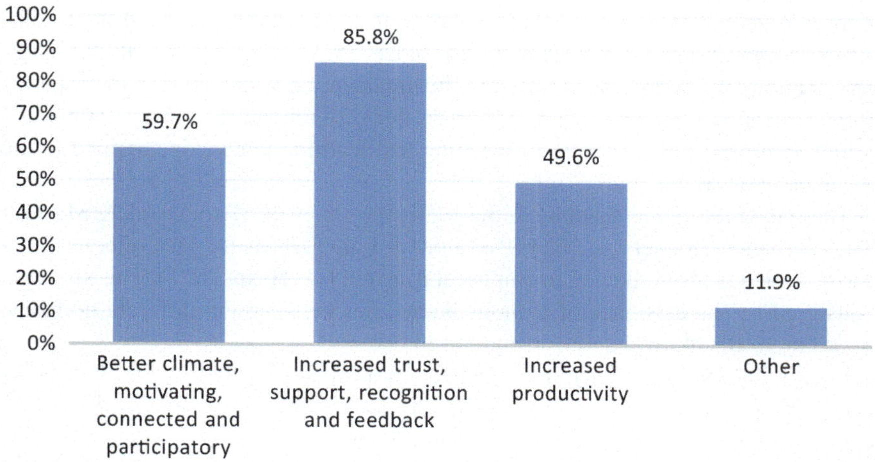

Fig. 9.6 Evidence of the impact of EC on MH, at the organizational level

any conditional effect of the number of sessions or the duration of the coaching program.

Group II

In Fig. 9.8, we analyzed the concept of EC from the HR perspective and we concluded that the majority (81.1%) are well informed, seeing it as a process of personal, professional and organizational development, which is commonly defined in the literature, as it is intended "to help people live better lives and feel more fulfilled at work" (Bachkirova & Borrington, 2020). However, we also found in this group that the remaining individuals are still mistaken, showing confusion with other types of

Fig. 9.7 Durability of positive EC results

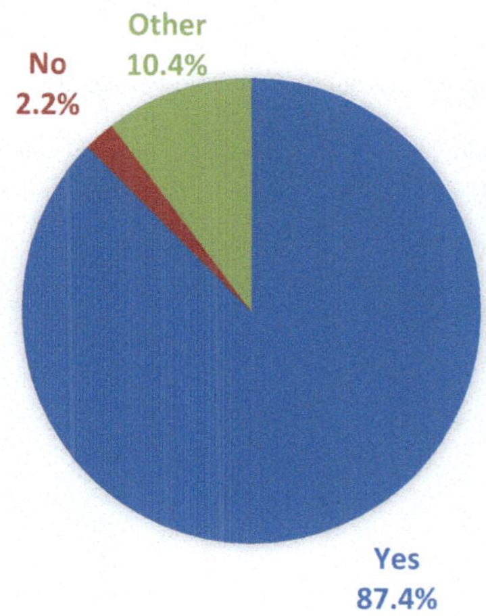

support, namely "Mentoring" when (8.2%) understand that it is a process of guidance and knowledge sharing (ICF, 2020); others (5.7%) confuse it with "Consulting", thinking it is a process of advising the organization with solutions to problems (Boysen et al., 2021) and some (4.9%) do not differentiate EC from "Training", as they interpret it as a process of teaching and learning knowledge applied to the organization (Boysen et al., 2021).

These data remind us that one of the controversies of EC (Brockbank & MacGill 2012; Jordan & Henderson, 2024; Passmore & Lai, 2020) is the lack of conceptual clarity, which is also reflected in empirical research (De Haan, 2021). This situation is justified by the existence of tenuous boundaries between the different domains, according to Bachkirova and Borrington (2019).

Looking at the importance the organization attributes to the EC, Fig. 9.9 shows that 16.7% consider it to be "Very Good", 53.7% consider it to be "Good"; 20.4% report "Enough" and only 9.3% report "Insufficient".

We conclude that EC deserves positive social recognition, whose visibility justifies the main reasons that lead organizations to implement it, in order of relevance (see Table 9.1): the development of professional skills (61.1%); the development of

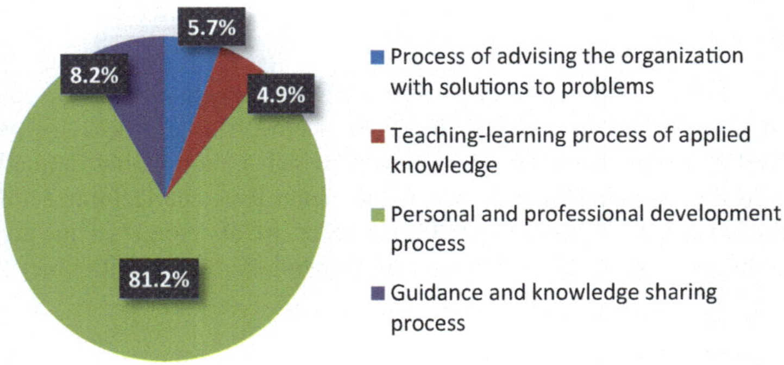

Fig. 9.8 Executive coaching concept

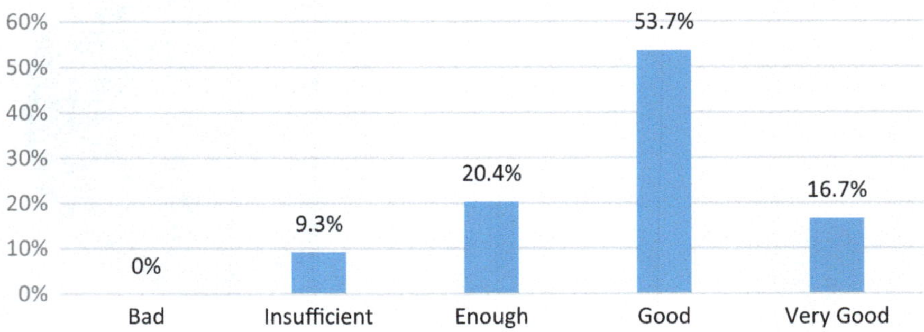

Fig. 9.9 Importance attributed to EC by organizations

Table 9.1 Reasons that lead organizations to use EC

Reasons	%
Development of professional skills	61.1
Development of personal behaviors	57.4
Development of organizational strategies	44.4
Another reason	7.4

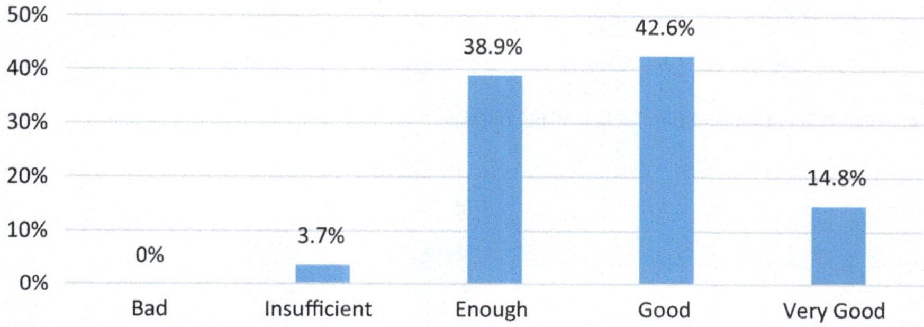

Fig. 9.10 Opinion of people in the organization about EC

personal behaviors (57.4%); the development of organizational strategies (44.4%) and other reasons are pointed out (7.4%), namely leadership development, which, by the way, is the one that is currently being intensively emphasized, according to the Executive Coaching Survey (Sacha, 2020).

When HR was asked about their perception of the opinion of people in the organization regarding the EC (see Fig. 9.10), we found that it is well regarded, with 42.6% of the respondents considering it to be "Good"; 14.8% believe it to be "Very Good"; 38.9% rate it as "Enough" and only 3.7% attribute it to "Insufficient".

Regarding the influence of EC in mitigating MH risks in organizations, we found a clearly positive perception on the HR side, with 21.3%, 51.6% and 8.2% attributing it to "Enough", "Good" and "Very Good" respectively. Some (18%) consider it to be "Insufficient" and the rest (0.8%) say it is "Bad", as shown in Fig. 9.11. Several authors refer to EC as an "anchor" for getting through difficult circumstances, reducing crises of stress and anxiety with repercussions on improving MH (Albizu Gallastegui et al., 2019; Ballesteros-Sánchez et al., 2019; Gerathy, 2021).

In relation to the evaluation of the results of EC in promoting mental health, Fig. 9.12 clearly shows the significant impact reported by 80.1% of the respondents; there were 15.6% of HR who gave it an "Insufficient" rating and only 1.6% gave it a "Bad" evaluation. These results are in line with research conducted with coachees, whose participation in EC programs was associated with better MH, quality of life and achievement of goals (Brooks et al., 2023; De Haan, 2021; Jarosz, 2021; Trevisan et al., 2021).

Analyzing the most significant evidence of the impact of EC on MH, at a psychological level, the participants reaffirm the theoretical framework (Greif & Palmer,

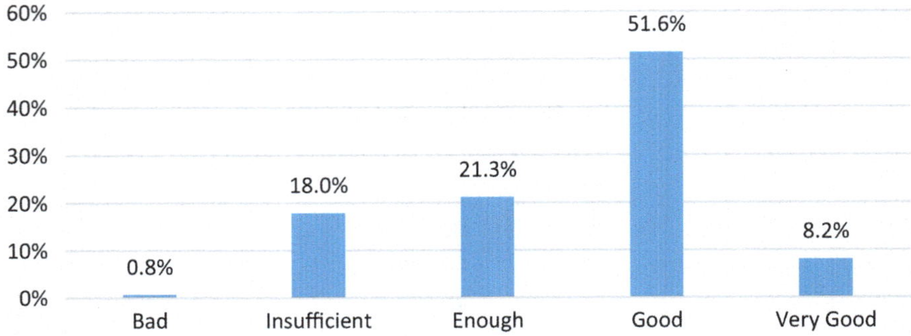

Fig. 9.11 Perception of the influence of EC on MH

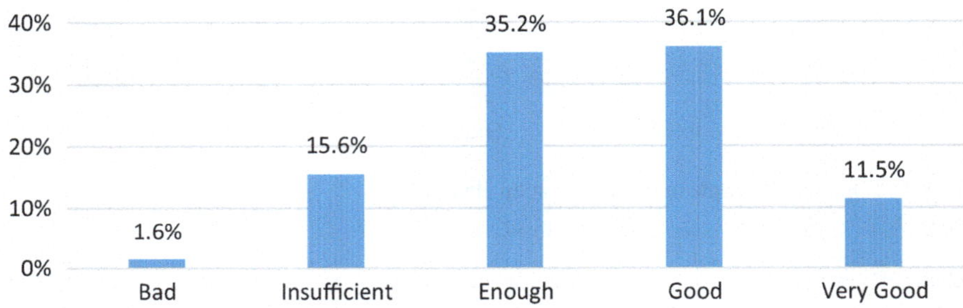

Fig. 9.12 Evaluation of EC results in the promotion of MH

2022), such as the increase in psychological well-being reported by 65.6% of the HR, followed by the reduction of stress and anxiety in the perception of 62.3% of the participants, as well as the improvement of self-esteem in 36.9% of the participants (Fig. 9.13).

We think it is important to highlight that 9.8% of the participants also see other evidences related to "self-knowledge", the "ability to organize and prioritize" and

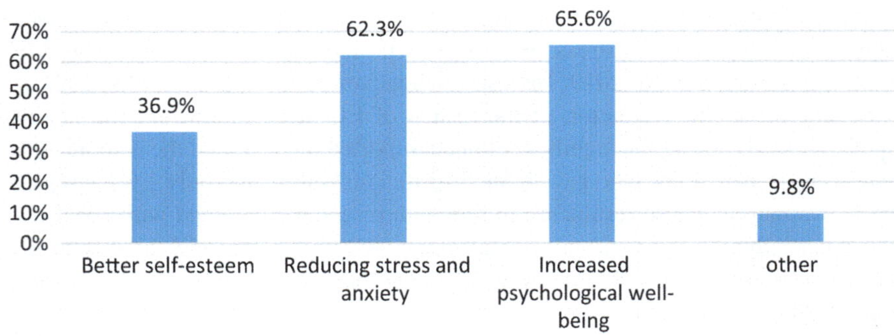

Fig. 9.13 Evidence of the impact of EC on MH, at a psychological level

the fact that they feel "more seen and welcomed" as contributing to personal success. It's impossible not to transcribe the following opinions: "It breaks down that idea that you have to be one at work and the other at home and in your personal life."; "I see EC as much more about professional development than health; I see little focus from coaches on MH, but the situation should be reversed. However, indirectly and in the short term, I see effective results." This perception, in line with some theoretical frameworks, sees EC as a multidimensional construct, where the personal and professional sides cannot be disconnected, i.e. socio-professional development must go hand in hand with human emotional development (Cavanagh & Grant, 2004). As for the most significant evidence of the impact of EC on MH, at a behavioral level, shown in Fig. 9.14, the following are highlighted: "Better self-efficacy in leadership/management" (62.3%); "Best Performance" (60.7%); "Greater Assertiveness" (51.6%); "Greater Flexibility" (30.3%). Some HR (5.7%) added other evidence such as "Better Self-Management and Emotional Balance".

In the last analysis, looking at the most significant evidence of the impact of EC on MH at an organizational level, the most perceptible indicators from the HR perspective are mentioned, such as: "Increased trust, support, recognition and feedback" (71.2%); "Better climate, more motivating, connected and participative" (56.6%); "Increased productivity" (37.7%) and 4.1% of respondents said they had no data to answer, as shown in Fig. 9.15.

When HR professionals were asked about their perception of the long-term effectiveness of the positive results of EC (Fig. 9.16), 66.4% agreed with the affirmative view of Nicolau et al. (2023) and 13.9% responded negatively, the remaining 19. 7% said "I believe that in order for them to be durable, they need to be regular throughout life"; "it depends on the coachee's openness to the process"; "it depends

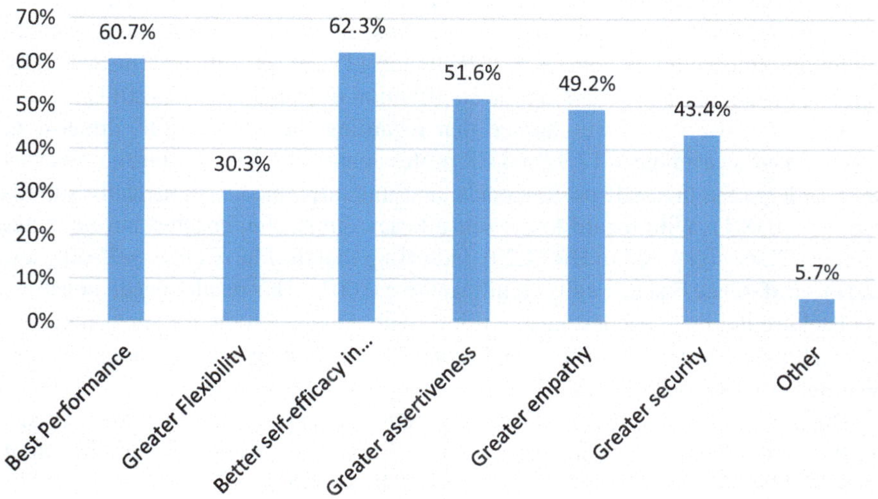

Fig. 9.14 Evidence of the impact of EC on MH, at the behavioral level

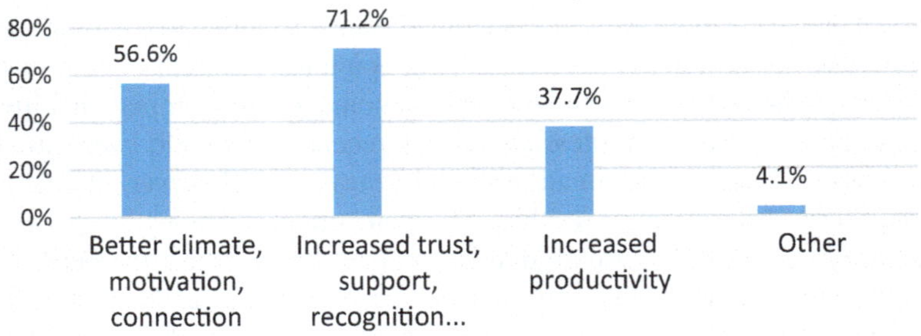

Fig. 9.15 Evidence of the impact of EC on MH, at the organizational level

Fig. 9.16 Durability of positive EC results

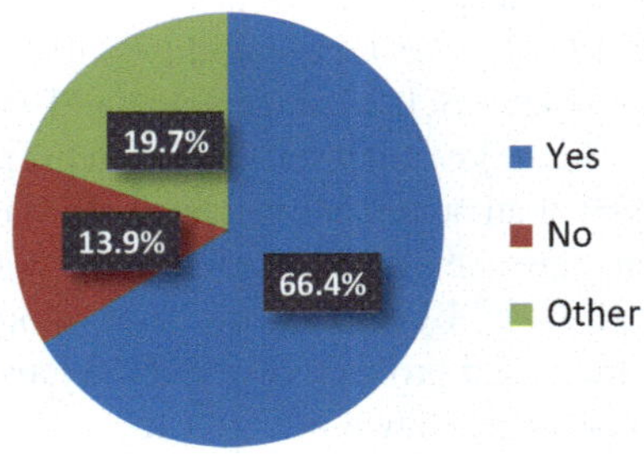

on the coachee's circumstances and MH"; "it depends on the coachee and the business context"; "it depends on if the professional has acquired the tools to continue on their own".

We end by presenting the results of the combined analysis of the GHQ-12 administered to the two groups of coaches and HR professionals in the sample, as shown in Tables 9.2 and 9.3. For this analysis, we chose to show two dimensions, namely depression (measured by items 2, 5, 6, 9, 10 and 11) and self-efficacy (measured by items 1, 3, 4, 8 and 12), following the contribution of Damásio et al. (2011).

Analyzing the items, we can see that regarding the "depression" dimension, coaches have an average of 12.92 and HR professionals 11.84, indicating that coaches are a little worse than HR professionals, and this difference is statistically significant ($p < 0.001$). With regard to the self-efficacy dimension, coaches have a better average (2.56) compared to HR (4.20), indicating that they have better self-efficacy, and this difference is statistically significant ($p < 0.001$). This result may demonstrate the background that coaches have, a priori, with adequate tools for stress management, allowing them to more effectively control states of mental or emotional distress resulting from several factors.

Thus, it can be seen that coaches have better general health (since the average for this category is 15.49) compared to HR professionals (average general health of 16.04), although this difference is not statistically significant.

We conclude by highlighting the understanding of the OPP (2023): "Workers experience well-being when they demonstrate their competencies and feel their own effectiveness or collective effectiveness as a consequence of their behaviors." (p.11).

Table 9.2 Means and SD of GHQ-12 in coaches (n = 134)

Item	Depression	Self-efficacy	Health General
	Mean (DP)	Mean (DP)	Mean (DP)
2 (worries, loss of sleep)	2.227 (±0.627)	–	2.227 (±0.627)
5 (under pressure)	2.24 (±0.603)	–	2.24 (±0.603)
6 (unable to overcome difficulties)	0.57 (±0.581)	–	0.57 (±0.581)
7 (pleasure in activities)	0.54 (±0.608)	–	0.54 (±0.608)
9 (sad and depressed)	2.35 (±0.628)	–	2.35 (±0.628)
10 (loss of confidence)	2.42 (±0.579)	–	2.42 (±0.579)
11 (self-worth)	2.54 (±0.71)	–	2.54 (±0.571)
1 (concentration)	–	0.66 (±0.561)	0.66 (±0.561)
3 (important role)	–	0.42 (±0.539)	0.42 (±0.539)
4 (able to make decisions)	–	0.45 (±0.514)	0.45 (±0.514)
8 (able to face problems)	–	0.51 (±0.558)	0.51 (±0.558)
12 (feeling happy)	–	0.52 (±0.597)	0.52 (±0.597)
Total	12.92* (±1.77)	2.56* (±2.27)	15.49 (±1.99)

Table 9.3 Means and SD of GHQ-12 in HRM (n = 122)

Item	Depression	Self-efficacy	Health general
	Mean (DP)	Mean (DP)	Mean (DP)
2 (worries, loss of sleep)	1.89 (±0.784)	–	1.89 (±0.784)
5 (under pressure)	1.89 (±0.791)	–	1.89 (±0.791)
6 (unable to overcome difficulties)	0.91 (±0.643)	–	0.91 (±0.643)
7 (pleasure in activities)	0.99 (±0.848)	–	0.99 (±0.848)
9 (sad and depressed)	1.98 (±0.823)	–	1.98 (±0.823)
10 (loss of confidence)	2.02 (±0.749)	–	2.02 (±0.749)
11 (self-worth)	2.16 (±0.754)	–	2.16 (±0.754)
1 (concentration)	–	0.96 (±0.648)	0.96 (±0.648)
3 (important role)	–	0.86 (±0.775)	0.86 (±0.775)
4 (able to make decisions)	–	0.79 (±0.683)	0.79 (±0.683)
8 (able to face problems)	–	0.79 (±0.718)	0.79 (±0.718)
12 (feeling happy)		0.81 (±0.696)	0.81 (±0.696)
Total	11.84* (±2.29)	4.20* (±2.76)	16.04 (±2.78)

9.4.1 Hypothesis Testing

With the necessary information obtained, we are now in a position to analyze the results in order to test the hypotheses:

Hypothesis 1 claims that “EC is a process with a socially positive representation in organizational culture and with an impact on MH”. In order to test this hypothesis, a linear regression was performed between the variable “EC Representation and Impact on MH” and the variable “Position in which they found themselves when completing this questionnaire”. The “EC Representation and Impact on MH” variable was operationalized by adding up questions 19 and 21 for the Coaches and 62 and 73 for the HR Professionals, giving a total of 10 points. Table 9.4 shows a simple regression, with “EC Representation and Impact on MH” as the dependent variable and “Position in which they filled in this questionnaire” as the independent variable.

The score for “EC Representation and the Impact on MH” is better in coaches than in HR, which means a statistically significant difference. The independent variable, explained by 7.3% for a significance of $p < 0.001$ and seen in isolation, is not a predictor of the dependent variable although the linear relationship is statistically significant ($F = 15.729$, $p < 0.001$), the correlation is not very strong, and most of the variability in the dependent variable is explained by other factors that are not part of this simple linear model. The null hypothesis was excluded, since the data found confirmed the hypothesis that EC in fact a process with a socially positive representation in organizational culture and with an impact on MH.

Hypothesis 2 affirms that “EC is an organizational support tool, with positive evidence also in terms of psychological development”. To test this hypothesis, a linear regression was performed between the dependent variable “Psychological Evidence of the EC” and the independent variable “Position in which they found themselves when completing this questionnaire”. The variable “Psychological Evidence of EC” was operationalized through the sum of questions 22 from the Coaches and 74 from the HR Professionals.

In Table 9.5, a simple regression is presented, considering the variables mentioned above. The independent variable “Position in which they found themselves when filling in this questionnaire” explains 16.4% for a significance of $p < 0.001$ and a moderate correlation ($r = 0.404$).

The score for the “Psychological Evidence of EC” obtained by the coaches was higher than that of the HR people, resulting in a statistically significant difference, which leads to the conclusion that the coaches attach more importance to it than the HR people. The null hypothesis was excluded, since the evidence found confirms the hypothesis that EC is an organizational support tool, implemented by HRM, with positive evidence also in terms of psychological development.

Table 9.4 Simple linear regression of the variable EC representation and the impact on MH on the variable position ($n = 256$)

Position						
Variable	n	p	F	r	r^2	β
Impact of EC on MH	188	0.000*	15.729	0.279	0.073	−0.444

* $p < 0.001$

Table 9.5 Simple linear regression of the variable "Psychological Evidence of the EC" on the Position ($n = 256$)

Position						
Variable	n	p	F	r	r^2	β
Psychological evidence	256	0.000*	49.684	0.404	0.164	0.399

* $p < 0.001$

Hypothesis 3 affirms that "The participants in this study show adequate literacy in the conceptualization of EC". For this hypothesis, the variable "EC literacy" and the variable "ECC literacy" were tested against the "Position in which they found themselves when completing this questionnaire". In this sense, Hypothesis 3 was subdivided into:

- Hypothesis 3.1: "EC literacy" is influenced by "the position they are in when they filled in this questionnaire".

To test this sub-hypothesis, a T-test was performed, as shown in Table 9.6. Looking at the table, we see that the result of the T-test p is < 0.001, meaning that the "EC Literacy" averages between the two groups are statistically significant. In other words, the results of the T-test show that there is a significant difference in "EC Literacy" scores between the "As a Coach" and "As an HR Professional" groups. Coaches report significantly higher "EC Literacy" scores than HR professionals.

- Hypothesis 3.2: "ECC literacy" is influenced by "the position they are in when completing this questionnaire". To test this sub-hypothesis, we proceeded in the same way, i.e. we performed a T-test, as shown in Table 9.7.

Table 9.6 T-Test for "EC Literacy" as a function of "Position"

Position						
Variable		n	Mean	t	df	P (bilateral)
EC literacy	Coach	134	2.2687	9.007	234.492	0.000
	HRM	122	1.4180			

*$p < 0.001$

Table 9.7 T-Test for "ECC Literacy" as a function of position

Position						
Variable		n	Mean	t	df	P (bilateral)
ECC literacy	Coach	134	5.2090	7.229	230.022	0.000*
	HRM	122	4.3361			

*$p < 0.001$

We verified that the p-value is < 0.001, meaning that the average "ECC Literacy" scores between the two groups are statistically different. Coaches have significantly higher "ECC Literacy" scores than HR professionals. So, after analyzing the two sub-hypotheses, H3 can be accepted, affirming that coaches are more literate than HR professionals in terms of EC conceptualization.

Hypothesis 4 affirms that individuals who have already participated in EC processes have lower scores on the GHQ (indicating a better level of MH). In this hypothesis, we considered HR individuals who answered "Coachee" to question number 64. We tested the variable "GHQ12 Total Score" against test value 18. The variable "GHQ12 Total Score" was operationalized through the sum of the 12 GHQ questions. The questions are evaluated from 0 to 3 and the sum of them is scored from 0 to 36 points, with the average reference value of 18, above which it is considered pathological. The average was found to 16.2237, indicating that HR professionals who have participated in a EC process as a coachee have lower GHQ scores, as shown in Table 9.8.

From the result of the test, we can see that the negative t-value (−27.279) indicates that the mean GHQ12 Total Score is statistically highly significant ($p < 0.001$), meaning that HR people who have already participated in EC processes as coachees have a better MH condition. With this result, we conclude that H4 is accepted.

Hypothesis 5—inductive, affirms that the perception of absenteeism and presenteeism is due to MH reasons and work-related stress factors. For this hypothesis, cross-tabulations were made between questions 66 and 69 and 67 and 70, as shown in Tables 9.9 and 9.10. Regarding absenteeism, we found that 64.3% of respondents believe it is due to MH reasons resulting from work-related stress factors. As for presenteeism, we found that 70.4% of respondents believe that it is conditioned by reasons of MH resulting from work-related stress factors.

Table 9.8 T-test of the GHQ-12 total score variable for Coachees HR professionals

Variable	n	Mean	*t*	*df*	*p* (bilateral)
GHQ12	76	16.2237	−27.279	75	0.000*

$^{*}p < 0.001$

Table 9.9 Perception of the reasons for absenteeism

		In your perception, is absenteeism, for MH reasons, a consequence of stress factors associated with work?		
		Yes%	No%	Other %
In your perception, does your organization record levels of absenteeism, due to MH reasons?	Yes%	64.3	14.3	21.4
	No%	53.0	24.2	22.7

Table 9.10 Perception about the reasons for Presenteeism

		In your perception, is presenteeism, for MH reasons, a consequence of stress factors associated with work?		
		Yes%	No%	Other%
In your perception, does your organization record levels of presenteeism, due to MH reasons?	Yes%	70.4	9.3	20.4
	No%	63.2	20.6	16.2

To conclude the analysis of the hypotheses, we also formulated an exploratory hypothesis, which affirms that there are significant differences in the answers to the GHQ-12 between coaches and HR professionals. For this hypothesis, we tested question 4—"Indicate your position when filling in this questionnaire" with the variable "GHQ 12 Total Score", as shown in Fig. 9.17.

By analyzing the Boxplot graph, which compares the GHQ 12 Total Score between the two groups of coaches and HR, we can see that both groups have similar medians. However, HRM professionals show a greater dispersion of scores, expressed by the number of outliers.

The variability within the group of HRM professionals is greater, indicating that there is a greater difference in GHQ-12 scores between individuals in this group. Coaches had less outliers and a more concentrated distribution around the median. HRM professionals show several outliers, suggesting the presence of individuals with more extreme MH states, both better and worse.

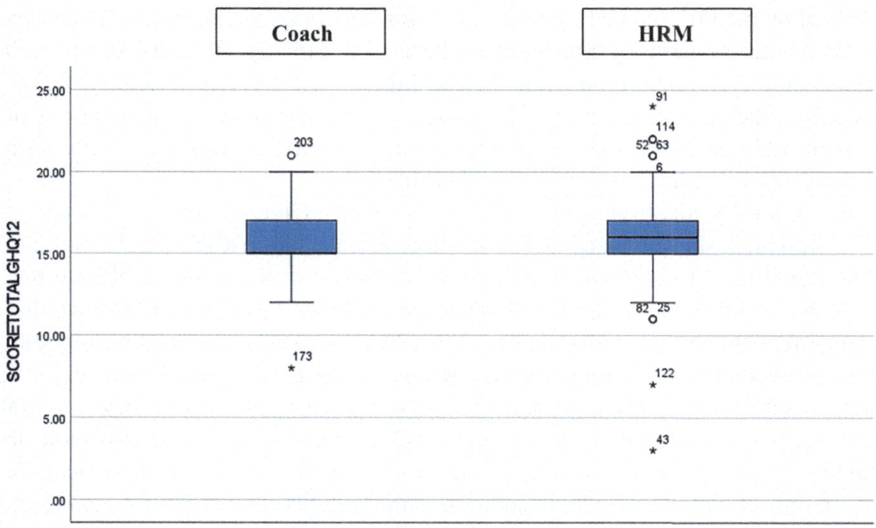

Fig. 9.17 Position of the respondent when filling out the GHQ-12

Table 9.11 T-test for "GHQ-12 Total Score" as a function of the "Position in which they are found to complete this Questionnaire

Position in which you find yourself when filling out this questionnaire						
Variable		n	Mean	*t*	*df*	*p* (bilateral)
GHQ12	Coach	134	15.4925	−1.847	254	0.066*
	HRM	122	16.0492			

*$p > 0.05$

We performed a T-test to compare the means between the GQH-12 Total Score of the two groups (Coache and HR), as shown in Table 9.11. We found that the p-value was 0.066, indicating that the means were similar and that there were no statistically significant differences between the means of the two groups. Therefore, the exploratory hypothesis is rejected (both have good MH).

9.5 Conclusion

The COVID-19 pandemic has caused a disruption in the functioning of our society, placing MH at the top of the WHO's priorities. Since then, leaders and managers have focused particularly closely on the mental and physical well-being of their employees as determining factors for organizational success, finding in EC an instrument to support personal and professional development.

Despite the ideological separation between the two contexts (EC and MH) imposed by the Professional Code of Ethics for Coaching, the intersection between the two is seen as mutually influential and beneficial (Boyatzis et al., 2023), although studies on the correlation between EC and MH are still limited (Geraghty, 2021). Therefore, the general purpose of this study was to investigate the perceptions of different EC stakeholders about its relationship with MH, as well as its impact on the prevention of prevalent psychosocial problems.

We began by identifying the participants' level of literacy on the concept of EC, whose data showed that coaches are more literate than HR professionals, highlighting the need to continue building a robust theoretical and practical base that facilitates a conceptually rigorous knowledge of the philosophy of EC. Nevertheless, participants in the study confirmed that EC has an effective relationship with and influence on MH, reinforcing the literature. This is because EC is a tool that facilitates the integral development of the person, opening the mind to new ways of thinking, being and acting (Boysen et al., 2021). In this sense, the results indicate that EC is a process with a socially positive representation in organizational culture and an impact on HRM, since, in addition to the benefits at an organizational level, reflected in improved performance and effectiveness (Athanasopoulou & Dopson, 2018; Ballesteros-Sánchez et al., 2019; Spence & Grant, 2007), it also has positive evidence in terms of psychological and socio-emotional development, which is substantiated in the literature. It makes it possible to manage stress, control anxiety and change attitudes towards the challenges of achieving work goals (Adi et al., 2022; Geraghty, 2021; Grover & Furnham, 2016; Nicolau & Constantin, 2021).

It is clear that these successful and intentional changes contribute significantly to developing leadership and management skills in coachees, which help them to be more effective (Boyatzis et al., 2022), finding a better sense of psychosocial well-being with an impact on MH (Greif & Palmer, 2022). This is suggested by the results obtained through the GHQ in HR who have already participated in EC processes as coachees, showing a better state of MH than those who have not participated in EC. Curiously, comparing the MH of the two sociodemographic groups using the GHQ, there are no statistically significant differences, suggesting that both have a balanced MH condition, although HR professionals indicate greater vulnerability with more extreme MH states, both better and worse.

It is important to note that the participants' perceptions of absenteeism and presenteeism are also associated with MH reasons and work-related stress factors, denoting an unbalance between personal and professional life (Bachkirova & Borrington, 2020; Toh et al., 2023), so this study offers yet another insight into how to minimize the problem in focus, which may involve the effective integration of EC. From this perspective, and in response to the starting question of the study—"How is the activity of Executive Coaching perceived in terms of its relationship and impact on improving MH?—we concluded that EC is perceived as a strategic HRM intervention tool that promotes MH. It is a social practice that predicts prosperity at work (Abid et al., 2020), including happiness (Romão et al., 2022), as it maximizes personal, social and organizational well-being, reflected in reciprocal gains, not only in people's satisfaction, but also in the profitability of companies, indirectly impacting on MH.

In conclusion, we recognize that EC is not just an organizational management trend, but a growing learning approach, the importance of valuing and democratizing which can benefit not only the financial health of the organization, but also promote the MH of human capital in organizations and, in the last instance, today's society in general. Both objectives are worthy, which is why there is an urgent need to invest in EC, since its role in improving the MH of workers, as well as in improving organizational results, seems to be unequivocal.

References

Abid, G., Ahmed, S., Qazi, T. F., & Sarwar, K. (2020). How managerial coaching enables thriving at work. A sequential mediation. *Journal of Entrepreneurship, Management & Innovation, 16*(2), 131–160. https://doi.org/10.7341/20201625

Adi, N.P., Odagami, K., & Nagata, T. (2022). Role of supervisor support in promoting mental health at the workplace. *The Indonesian Journal of Community and Occupational Medicine*. https://doi.org/10.53773/ijcom.v2i2.67.73-4

Agência Europeia para a Segurança no Trabalho. (2023). *Riscos psicossociais e saúde mental no trabalho*. AEST. https://osha.europa.eu/pt/themes/psychosocial-risks-and-mental-health

Albizu Gallastegui, E., Rekalde Abasolo, I., Landeta Rodríguez, J., & Fernández Ferrín, P. (2019). Analysis of executive coaching effectiveness: A study from the coachee perspective. *Cuadernos De Gestión, 19*(2), 33–52. https://doi.org/10.5295/cdg.170876ea

Ali, M., Raza, B., Waseer, W. A., & Imtiaz, N. (2020). Linking managerial coaching with employees' innovative work behaviors through affective supervisory commitment: Evidence from Pakistan. *International Review of Management and Marketing, 10*(4), 11–16. https://doi.org/10.32479/irmm.9715

American Management Association. (2019). *Coaching—a global study of successful practices.* AMA. https://www.amanet.org/assets/1/6/i4cp-coaching.pdf

AON (2022). *Aon health survey 2021 and beyond—the new era of personalized, virtual and inclusive care.* https://www.aon.com/2021-health-survey-results

Athanasopoulou, A., & Dopson, S. (2018). A systematic review of executive coaching outcomes: Is it the journey or the destination that matters the most? *The Leadership Quarterly, 29*(1), 70–88. https://doi.org/10.1016/j.leaqua.2017.11.004

Bachkirova, T. (2016). The self of the coach: Conceptualization, issues, and opportunities for practitioner development. *Consulting Psychology Journal: Practice and Research, 68*(2), 143–156. https://doi.org/10.1037/cpb0000055

Bachkirova, T., & Borrington, S. (2019). Old wine in new bottles: Exploring pragmatism as a philosophical framework for the discipline of coaching. *Academy of Management Learning & Education, 1*, 337–360. https://doi.org/10.5465/amle.2017.0268

Bachkirova, T., & Borrington, S. (2020). Beautiful ideas that can make us ill: Implications for coaching. *Philosophy of Coaching: An International Journal, 5*(1), 9–30. https://doi.org/10.22316/poc/05.1.03

Ballesteros-Sánchez, L., Ortiz-Marcos, I., & Rodríguez-Rivero, R. (2019). The impact of executive coaching on project managers' personal competencies. *Project Management Journal, 50* (3), 306–321. https://doi.org/10.1177/8756972819832191

Boyatzis, R. E., Hullinger, A., Ehasz, S. F., Harvey, J., Tassarotti, S., Gallotti, A., & Penafort, F. (2022). The grand challenge for research on the future of coaching. *The Journal of Applied Behavioral Science, 58*(2), 202–222. https://doi.org/10.1177/0021condic8863221079937

Boyatzis, R. E., Smith, M. L., Terblanche, N., & Passmore, J. (2023). The democratization of coaching. *Academy of Management Annual Meeting Proceedings.* https://doi.org/10.5465/AMPROC.2023.10515

Boysen, S. M., Terry Arya, T., & Page, L. (2021). Organizational and executive coaching: Creating a coaching culture in a non-profit. *International Journal of Evidence Based Coaching and Mentoring, 19*(2), 115–132. https://doi.org/10.24384/v95s-3g56

Brockbank, A., & McGill, I. (2012). *Faciltating reflective learning through mentoring and coaching* (2nd ed.). Kogan Page.

Brooks, P. J., Ripoll, P., Sánchez, C., & Torres, M. (2023). Coaching leaders towards favorable trajectories of burnout and engagement. *Frontiers in Psychology.* https://doi.org/10.3389/fpsyg.2023.1259672

Cavanagh, M., & Grant, A. M. (2004). Executive coaching in organisations: The personal is the professional. *International Journal of Coaching in Organisations, 2*(2), 6–15. https://www.researchgate.net/publication/228740106

Cavanagh, M. (2005). Mental-health issues and challenging clients in executive coaching. In M. Cavanagh, A. M. Grant, & T. Kemp. (Eds.), *Evidence-based coaching, theory, research and practice from the behavioural sciences* (vol. 1, pp. 21–36). Australian Academic Press.

Cohen, J. (1988). *Statistical power analysis for the behavioral sciences.* Erlbaum.

Damásio, B. F., de Machado, W. L., & da Silva, J. P. (2011). Estrutura Fatorial do Questionário de Saúde Geral (QSG-12) em uma Amostra de Professores. *Avaliação Psicológica, 10*(1), 99–105. http://pepsic.bvsalud.org/scielo.php?script=sci_arttext&pid=S1677-04712011000100011&lng=pt&tlng=pt

de Araújo, M. S. G. (2009). Preditores Individuais e Organizacionais de Bullying no Local de Trabalho, Tese de Doutoramento, Universidade do Minho. https://hdl.handle.net/1822/11041

De Haan, E. (2021). *What works in executive coaching—understanding coaching outcomes through quantitative research and practice-based evidence*. Routledge.

de Oliveira, C., Saka, M., Bone, L., & Jacobs, R. (2023). The role of mental health on workplace productivity: A critical review of the literature. *Applied Health Economics and Health Policy, 21*, 167–193. https://doi.org/10.1007/s40258-022-00761-w

Entidade Reguladora da Saúde (2023). *Acesso a serviços de saúde mental nos Cuidados de Saúde Primários*. ERS. https://www.ers.pt/media/slzpzdwk/estudo_saude_mental_02-2023.pdf

European Union (2021). *EU strategic framework on health and safety at work 2021–2027 Occupational safety and health in a changing world of work. Communication from the Commission to the European Parliament, the Council, the European Economic and Social Committee and the Committee of the Regions*. https://eur-lex.europa.eu/legal-content/EN/TXT/PDF/?uri=CELEX:52021DC0323

Gaspar, T., Jesus, S., Xavier, M., Dias, M. C., Machado, M. C., Correia, M. F., Pais- Ribeiro, J. L., Areosa, J., Canhão, H., Guedes, F. B., Cerqueira, A., & Matos, M. G. (2023). *Ecossistemas de Ambientes de Trabalho Saudáveis: Relatório 2022. Laboratório Português de Ambientes de Trabalho Saudáveis*. LAPATS. https://doi.org/10.15309/22psd230124

Geraghty, A. (2021). How do coaches work with clients showing signs of burnout? *International Journal of Evidence Based Coaching and Mentoring, S15*, 139–153. https://doi.org/10.24384/h73p-bn15

Goldberg, D. P. (1978). *Manual of the general health questionnaire*. NFER-NELSON Publishing.

Grant, A. M. (2007). A languishing—flourishing model of goal striving and mental health for coaching populations. *International Coaching Psychology Review, 2*(3), 250–264. Em J. Passmore & D. Tee (Eds.), *Coaching Researched: A Coaching Psychology Reader* (pp. 65–88). John Wiley & Sons Ltd. https://doi.org/10.1002/9781119656913

Grant, A. M. (2008). Past, present and future—the evolution of professional coaching and coaching psychology. In S. Palmer & A. Whybrow (Eds.), *Handbook of coaching psychology: A guide for practitioners* (pp. 23–39). Routledge/Taylor & Francis Group. https://psycnet.apa.org/record/2007-18824-002

Grant, A. M. (2020). An integrated model of goal-focused coaching: an evidence-based framework for teaching and practice. In J. Passmore, & D. Tee (Eds.), *Coaching researched: A coaching psychology reader*, (pp.115–139). John Wiley & Sons Ltd. https://doi.org/10.1002/9781119656913.ch7

Greif, S., & Palmer, S. (2022). Stress and stress management. Coaching. *International handbook of evidence-based coaching* (pp. 863–876). Publisher: Springer. https://doi.org/10.1007/978-3-030-81938-5_71

Grover, S., & Furnham, A. (2016). Coaching as a developmental intervention in organisations: A systematic review of its effectiveness and the mechanisms underlying it. *PLoS ONE, 11*(7), e0159137. https://doi.org/10.1371/journal.pone.0159137

International Coach Federation. (2020). *Global coaching study: Executive summary*. https://coachingfederation.org/app/uploads/2020/09/FINAL_ICF_GCS2020_ExecutiveSummary.pdf

International Commission on Occupational Health (2022). *ICOH survey on the first wave of COVID-19. Final Report*. https://www.icohweb.org/site/multimedia/pubblicazioni/icoh%20covid%202021.pdf

Jarosz, J. (2021). The impact of coaching on well-being and performance of managers and their teams during pandemic. *International Journal of Evidence Based Coaching and Mentoring, 19*(1), 4–27. https://doi.org/10.24384/n5ht-2722

Jordan, M., & Henderson, A. (2024). How do internal executive coaches make sense of organisational role boundaries? An interpretative phenomenological analysis study. *International Journal of Evidence Based Coaching and Mentoring, 22*(1), 234–249. https://doi.org/10.1111/joop.1211910.24384/90k3-5m55

Korotov, K. (2023). Accelerated development of organizational talent and executive coaching: a knowledge management perspective. *IN smart talent management: Managing people as knowledge assets* (pp. 67–83). Edward Elgar. https://doi.org/10.4337/9781802202717.00011

McCarthy, G., & Milner, J. (2020). Ability, motivation and opportunity: Managerial coaching in practice. *Asia Pacific Journal of Human Resources, 58*(1), 149–170.
Mental Health Europe. (2022). *Invitation for world mental health day 2022*. Mental Health Europe. https://www.mhe-sme.org/world-mental-health-day-2022-2/
Nicolau, A., Candel, O. S., Ticu Constantin, T., & Kleingeld, A. (2023). The effects of executive coaching on behaviors, attitudes, and personal characteristics: A meta-analysis of randomized control trial studies. *Frontiers in Psychology, 14*, 1089797. https://doi.org/10.3389/fpsyg.2023.1089797
Nicolau, A., & Constantin, T. (2021). How executive coaching helps managers increase their performance and deal with uncertainty: A systematic review. *Filodiritto Editore—Proceedings*, 99–105. https://www.researchgate.net/publication/353827883
OECD (2021). *COVID-19 and well-being: Life in the pandemic*. OECD Publishing. https://doi.org/10.1787/1e1ecb53-en
Oliveira-Silva, L. C., Werneck-Leite, C. D. S. N., Carvalho, P. S. F., Anjos, A. C., & Brandão, H. I. M. (2018). Desvendando o coaching: uma revisão sob a ótica da Psicologia. *Psicologia: Ciência e Profissão, 38*(2), 363–377. https://doi.org/10.1590/1982-3703000942017
Ordem dos Psicólogos Portugueses (2023). *Prosperidade e Sustentabilidade das Organizações – Relatório do Custo do Stresse e dos Problemas de Saúde Psicológica no Trabalho, em Portugal.*
Passarelli, A. M., Moore, S., & Van Oosten, E. B. (2022). How leaders and their coaches describe outcomes of coaching for intentional change. *Consulting Psychology Journal, 74*(4), 329–346. https://doi.org/10.1037/cpb0000240
Passmore, J., & Lai, Y.-L. (2020). Coaching psychology: Exploring definitions and research contribution to practice? In J. Passmore & D. Tee (Eds.), Coaching researched: *A coaching psychology reader for practitioners and researchers* (pp. 3–22). https://doi.org/10.1002/9781119656913.ch1
Peralta, H. C. (2023). Risco elevado de burnout em Portugal leva à criação do evento Happiness Camp. Revista *Forbes*. https://www.forbespt.com/risco-elevado-de-burnout-em-portugal-leva-a-criacao-do-evento-happiness-camp/
RAND Europe (2022–2023). *SPOTLIGTH 2022–2023*. https://www.rand.org/content/dam/rand/pubs/corporate_pubs/CPA1100/CPA1150-3/RAND_CPA1150-3.pdf#page=9
Romão, S., Ribeiro, N., Gomes, D. R., & Sharda, S. (2022). The impact of leaders' coaching skills on employees' happiness and turnover intention. *Administrative Sciences, 12*(84), 2–15. https://doi.org/10.3390/admsci12030084
Salvadorinho, J., & Teixeira, L. (2022). Leadership coaching framework tool-based to support worker engagement and retention in Industry 4.0. In *Proceedings of the5th European International Conference on Industrial Engineering and Operations Management Rome* (pp. 2733–2745).
Sarsur, A. M., & Parente, C. (2019). The coaching process seen from the daily (and controversial) perspective of experts and coaches. *Revista De Gestão, 26*(2), 126–142. https://doi.org/10.1108/REGE-07-2018-0082
Sasha Corporation. (2020). *The 2020 Sherpa executive coaching survey*. Sasha Corporation.
Spence, G. B., & Grant, A. M. (2007). Professional and peer life coaching and the enhancement of goal striving and well-being: An exploratory study. *The Journal of Positive Psychology, 2*(3), 185–194. https://doi.org/10.1080/17439760701228896
Theeboom, T., Beersma, B., & van Vianen, A. E. M. (2013). Does coaching work? A meta-analysis on the effects of coaching on individual level outcomes in an organizational context. *Journal of Positive Psychology, 9*(1), 1–18. https://doi.org/10.1080/17439760.2013.837499
Toh, S. H. Y., Lee, S. C., & Sündermann, O. (2023). Mobile behavioral health coaching as a preventive intervention for occupational public health: Retrospective longitudinal study. *JMIR Formative Research, 2023*(7), e45678. https://doi.org/10.2196/45678
Trevisan, I. P., Viegas, S. C., & Lopes, P. T. C. (2021). A Efetividade do uso do Coaching e do Mentoring na preparação para Provas e Concursos. *Revista Eletrônica em Gestão e Tecnologia.* https://www.researchgate.net/publication/355243365

UN (2015). *The UN's Sustainable Development Goals were put in place to transform the world by 2030.* https://www.un.org/sustainabledevelopment/sustainable-development-goals/

Vesoloski, S. P., & Garcia, L. N. (2022). Coaching: uma Ferramenta Percursora Capaz de (Re)criar o Ambiente Organizacional e Gerar Benefícios Significativos. *Revista Perspectiva, 46*(174), 131–141. https://ojs.uricer.edu.br/ojs/index.php/perspectiva/article/view/225

Wang, Q., Lai, Y.-L., Xu, X., & McDowall, A. (2022). The effectiveness of workplace coaching: A meta-analysis of contemporary psychologically informed coaching approaches. *Journal of Work-Applied Management, 14*(1), 77–101. https://doi.org/10.1108/JWAM-04-2021-0030

World Health Organization & International Labour Organization (2022). *Crucial changes needed to protect workers' health while teleworking.* https://www.who.int/news/item/02-02-2022-crucial-changes-needed-to-protect-workers-health-while-teleworking

World Health Organization (2018). *Mental health: Strengthening our response.* https://www.who.int/news-room/fact-sheets/detail/mental-health-strengthening-our-response

World Health Organization (2022a). *European framework for action on mental health 2021–2025.* https://www.who.int/europe/publications/i/item/9789289057813

World Health Organization (2022b). *World mental health report: Transforming mental health for all.* https://www.who.int/publications/i/item/9789240049338

World Health Organization (2022c). *Mental health.* https://www.who.int/news-room/fact-sheets/detail/mental-health-strengthening-our-response

World Health Organization (2022d). *WHO guidelines on mental health at work.* https://www.who.int/publications/i/item/9789240053052

Zhao, H., & Liu, W. (2020). Managerial coaching and subordinates' workplace well-being: A moderated mediation study. *Human Resource Management Journal, 30*(2), 293–311. https://doi.org/10.1111/1748-8583.12280

Chapter 10
Career Transitions: Insights from a Narrative Approach

Carla Malta, Cassiana Damo, and Viviana Meirinhos

Abstract Career transitions are complex processes that challenge individuals to redefine their professional identities, acquire new skills and adapt to constantly changing contexts. This study is based on a qualitative and narrative approach, using unstructured, in-depth narrative interviews to collect data and explore participants' career transition experiences. It employed content and narrative analysis methods, inspired by Bardin's (Edições 70, 2006) model. Narrative interviews were conducted with seven individuals with higher education, and the motivations, challenges, and impacts of career transitions on personal and professional development were analyzed. The study's results reveal the crucial role of support networks, reskilling and the redefinition of professional identity for successful transitions. Additionally, participants faced emotional, social and organizational challenges during the transition process but also identified opportunities for personal development, growth and professional fulfillment. It is concluded that these transitions are not merely moments of instability but significant opportunities for reinvention and development, both for individuals and organizations.

Keywords Career transition · Narrative approach · Human resources · Research methods

C. Malta (✉) · C. Damo · V. Meirinhos
CEOS.PP, ISCAP, Polytechnic of Porto, São Mamede de Infesta, Portugal
e-mail: carlasantosmalta@gmail.com

C. Damo
e-mail: cassiana@iscap.ipp.pt

V. Meirinhos
e-mail: vivianameirinhos@iscap.ipp.pt

C. Machado (ed.), *Human Resource Management in a Disrupted World*, Lecture Notes on Multidisciplinary Industrial Engineering,
https://doi.org/10.1007/978-3-031-89948-5_11

10.1 Introduction

Career transitions are multifaceted processes that challenge individuals to readjust their professional identities, develop new skills, and adapt to contexts that are often characterized by uncertainty and change (Akkermans et al., 2024; Roumell & Bolliger, 2024; Steindórsdóttir et al., 2023). These transitions are driven by a combination of factors, including labor market changes, personal or organizational needs, and shifts in motivations throughout the life cycle (Carstensen, 1995; Hobfoll et al., 2018; Spurk et al., 2019). In today's professional environment, characterized by volatility and diversification of trajectories, understanding these dynamics is essential to effectively support both individuals and organizations in managing change (Akkermans et al., 2024; Roumell & Bolliger, 2024).

An important concept in understanding career transitions is liminality, that refers to transition, period of uncertainty and disruption, characterized by the loss of professional identity (Roumell & Bolliger, 2024) but also by the potential for growth and reconfiguration. Liminality highlights the emotional and cognitive challenges faced by individuals, including ambiguity and disorientation (Roumell & Bolliger, 2024). Social support networks play a crucial role, providing a safe space for sharing experiences, emotional validation, and reinforcing self-efficacy (Bandura, 1977). Simultaneously, in this critical process, a reconfiguration process emerges that allows individuals to reconstruct their career narratives based on new opportunities, skills and values.

Career transitions are dynamic and iterative, unfolding in three main phases: before, during, and after (Akkermans et al., 2024). During the preparation phase, individuals mapping the environment, planning, set goals and engage in strategic actions (Akkermans et al., 2024). This initial phase is crucial for anticipating challenges and mobilizing resources, facilitating entry into new professional contexts. In the mediating phase, variables such as adaptability, resilience and networking are fundamental to facing the uncertainties and changes associated with transitions (Akkermans et al., 2024). The third phase focuses on adjusting into and learning a new role and variables such as career satisfaction and well-being (Akkermans et al., 2024).

The traditional model of linear careers, based on predictable and hierarchical progression, has been progressively replaced by approaches that value flexibility and horizontal mobility (Sullivan & Al Ariss, 2019). Boundary less careers promote adaptation to multiple contexts (Guan et al., 2019), while protean careers emphasize autonomy and continuous adjustment to change (Briscoe et al., 2021; Hall et al., 2018). These approaches reflect the contemporary reality, where horizontal transitions offer opportunities to explore new contexts, diversify skills, and achieve professional and personal fulfillment, even without hierarchical promotions (Steindórsdóttir et al., 2023). This paradigm highlights the importance of organizational policies that encourage continuous learning and internal mobility, enabling sustainable growth in dynamic labor markets (Bowen & Ostroff, 2004; Briscoe et al., 2021).

Psychological and social factors play a crucial role in career transitions. Social capital, represented by professional networks and interpersonal relationships, facilitates integration into new contexts and the reframing of previous skills, while psychological capital, comprising attributes such as resilience, self-efficacy, and confidence, provides the emotional resources needed to face the uncertainties associated with change (Teelken et al., 2024; Roumell & Bolliger, 2024). These dimensions interact synergistically, determining the success of transitions and underscoring the need for holistic interventions that combine individual and organizational support (Steindórsdóttir & Dysvik, 2023).

The Conservation of Resources Theory (COR) and the Socioemotional Selectivity Theory (SST) provide complementary theoretical frameworks for understanding transitions throughout the life cycle (Carstensen, 1995; Hobfoll et al., 2018). The COR emphasizes the accumulation and protection of resources a central strategy for achieving outcomes (Halbesleben et al., 2014), while the SST explores changing priorities over time (Carstensen, 1995), with younger individuals focusing on skills and exploration, and more experienced professionals prioritizing stability and meaningful connections (Steindórsdóttir et al., 2023). These models emphasize that the success of transitions depends on both intrinsic factors and external conditions, such as organizational support and available opportunities.

While challenging, career transitions offer opportunities for growth and reinvention. Integrating theories such as COR, SST, and dynamic models provides a solid foundation for understanding the dynamics of these processes. Effective strategies, such as support programs, internal mobility policies, and continuous training, can promote successful transitions, maximizing individual well-being and organizational value. In an increasingly diverse labor market, understanding and supporting these transitions is essential to creating meaningful and sustainable career trajectories.

10.2 Methods

To better understand these career transition experiences, a qualitative study was designed, including participants from different sectors and age groups. The following section presents the research methods, including participant selection, outlined objectives, and the tools used for data collection and analysis.

10.2.1 Objectives

This study aims to explore the career transitions of individuals with higher education through a narrative approach, in order to understand the subjective experiences and processes involved in career changes. By analyzing the participants' narratives, the study seeks to understand how career transitions are experienced, what factors influence these changes, the challenges faced and how requalification and the evolution of professional identity impact adaptation to the new job market.

To achieve this general objective, three specific objectives were structured, which are presented below. These objectives were defined to focus the analysis on the most significant aspects of career transitions, allowing for deeper and more targeted exploration:

- Analyze the motivations and factors that trigger career transitions: explore the reasons that lead participants to seek a career change, whether internal factors such as professional dissatisfaction or the search for self-fulfillment, or external factors, such as changes in the job market work or the need for requalification. The aim is to identify the factors that make the transition a choice or a necessity
- Examine the stages of the career transition process and the challenges faced: This objective focuses on the stages participants go through during their career transitions, main obstacles encountered, such as job insecurity or resistance to change, and how these difficulties are overcome throughout the process. The aim is to understand the dynamics of the transition process, critical moments and key decisions that shape the experiences of participants.
- Investigate the impact of requalification and the evolution of professional identity on career transitions. Explore how requalification and the acquisition of new skills influence the transition process and facilitate participants' adaptation to the new demands of the labor market. It also aims to understand how career transitions affect individuals' professional identity, self-esteem, self-confidence and sense of personal accomplishment. The objective is to understand how career changes can represent not only a professional change, but also a personal transformation, impacting participants' self-perception and confidence.

This chapter builds upon the research conducted in the master's dissertation, which also relied on the narrative approach. Drawing from the original study, which explored the motivations and challenges of career transitions, this chapter aims to deepen the analysis by incorporating new perspectives on how the career change process impacts professional identity and how reskilling plays a crucial role in adapting to the new job market.

10.2.2 Participants

The study included a carefully selected group of participants (P) to ensure representativeness and diversity in terms of career transition experiences. The participants were chosen based on heterogeneity criteria to capture a wide range of narratives and experiences.

Participants were selected according to specific criteria to ensure their relevance to the research objectives. To be eligible, participants had to:

Table 10.1 Participant (P) profile

P	Gender	Age	Previous career	Time in previous career	Current career	Time in current career
1	Feminine	30	Nutritionist	6 years	Product Manager	8 months
2	Feminine	32	Nurse	6 years and 6 months	Recruiter	1 year
3	Masculine	27	Researcher	1 year	Software Developer	6 months
4	Feminine	30	Laboratory technician	3 years	Software Engineer	3 years
5	Feminine	30	Nutritionist	1 year	Retail Manager	3 years
6	Feminine	51	Chemical engineer	15 years	Psychologist	12 years
7	Masculine	41	Archaeologist	5 years	Head of E-Commerce	12 years

- Hold a higher education degree in an initial field of study;
- Have worked in a professional role corresponding to their initial higher education qualification;
- Have returned to higher education to acquire a new academic qualification in a different field;
- Be working in a professional role in an area distinct from their initial field.

As we can see in Table 10.1, a total of seven individuals participated, including men and women aged between 27 and 51, from different professional sectors such as technology, education, healthcare, and engineering.

10.2.3 Study Design

This study adopted a qualitative approach, using the narrative method to explore the career transitions of individuals with higher education. The choice of the narrative method was based on its ability to capture the personal stories of the participants, allowing for an in-depth analysis of the subjective and contextual dimensions that shape professional changes. According to Muylaert et al. (2014) and Gomes (2003), narrative is not only a tool for sharing events but also a means of revealing underlying meanings and emotional, social, and cultural dynamics.

The narrative analysis applied in this study goes beyond the description of the events shared, also encompassing the elements that the narrators consider relevant. This approach provides a contextualized and systemic view of the participants' experiences, revealing their motivations, conflicts, and adaptation strategies, as well as the personal meanings they attribute to their transitions (Carvalho et al., 2021; Vilela et al., 2022).

The collected narratives underwent a two-stage analysis process, combining a thematic and narrative approach. Initially, a thematic analysis inspired by Bardin's (2006) model was conducted, which allowed for the identification of patterns and emerging themes in the narratives shared by the participants. This exploratory stage organized the data into general categories, reflecting the main aspects of career transitions. These emerging categories include motivation for career change, adaptation to new roles, personal growth, and organizational support during the transition.

Subsequently, a detailed narrative analysis was applied, based on the predefined narrative elements (Gancho, 2006; Kendal & Kendal, 2012), to explore how each participant structured their story and assigned meaning to their experiences. The collected narratives were analyzed based on predefined categories, grounded in the principles of narrative construction. (Gancho, 2006; Kendall & Kendall, 2012). These narrative elements allowed for the structured organization and interpretation of career transition narratives, capturing both the objective aspects and the subjective meanings of the experiences. Table 10.2 presents the narrative elements considered in the study.

The use of narrative elements allowed for a structured and in-depth exploration of career transitions, going beyond the superficial description of events. This approach enabled an understanding of transitions as multifaceted processes, in which emotional, social, and contextual factors interact in a dynamic manner.

Finally, the narrative elements served as an interpretive framework, guiding both the organization of the data and the analysis of the meanings attributed by the participants to their experiences of change.

The combination of these approaches provided a rich and integrated view, capturing both the general patterns and the particularities of each narrative.

10.2.4 Instruments and Procedures

The data were collected through narrative interviews, conducted in an unstructured and in-depth manner. This format allowed participants to share their career transition narratives spontaneously, without the limitations imposed by rigid questions. The use of narratives as a data collection tool proved to be suitable for capturing not only the objective events but also the subjective and emotional meanings attributed by the participants to their experiences. (Gomes, 2003; Muylaert et al., 2014). The Zoom platform was chosen for conducting the interviews due to its practicality, including video calls and the option to record sessions, ensuring the preservation of verbal and non-verbal expressions relevant for later analysis.

Before the interviews were conducted, participants received an informed consent statement. This document explained the purpose and objectives of the study, ensuring participants' right to confidentiality and anonymity of the collected data. It was guaranteed that participation was entirely voluntary and that they could withdraw from the study at any time, without any consequences. This procedure followed

Table 10.2 Identification and definition of narrative elements

Narrative elements	Definition
Point of view	The point of view refers to the perspective from which the narrative is told. It can be first-person (a character tells the story), third-person limited (an external narrator focused on a specific character), or third-person omniscient (a narrator who knows the thoughts and feelings of all characters) (Gancho, 2006; Twinkl, 2024)
Plot	The plot refers to the structure of the narrative, including the events that take place and how they relate to each other. It generally follows a linear progression, with an introduction, development, climax, and resolution. The plot is the backbone of the narrative, the sequence of events that keeps readers engaged (Gancho, 2006; Kendal & Kendal, 2012; Twinkl, 2024)
Characters	Characters are the individuals or entities that play roles in the narrative. They can be protagonists (heroes), antagonists (villains), or secondary characters. Character development involves building their personalities, motivations, and transformation arcs throughout the narrative (Gancho, 2006; Kendal & Kendal, 2012; Twinkl, 2024)
Settings	The setting is the physical and temporal environment where the narrative takes place. This includes descriptions of locations, time periods, cultures, and all elements that contribute to the narrative's context. The setting provides the backdrop for the events, influences the story's mood, and helps create an immersive atmosphere (Gancho, 2006; Twinkl, 2024)
Conflit	Conflict is the driving force of the story and represents the struggle or challenge the characters face throughout the narrative. It can exist between characters, between a character and society (external conflict), or even within the character themselves (internal conflict). Conflict creates tension and propels the narrative toward its climax and resolution (Gancho, 2006; Kendal & Kendal, 2012)
Theme	The theme refers to the central message or underlying idea of the narrative. It is the fundamental concept the story conveys about life, society, or the human condition, providing depth and meaning to the narrative (Gancho, 2006; Twinkl, 2024)
Tone and style	Tone refers to the emotional atmosphere of the narrative, while style pertains to the unique way the author expresses their ideas, including the use of language, rhythm, and stylistic choices (Steck, 2014)
Symbolism and motifs	Symbolism involves the use of symbols to represent broader ideas or concepts within the narrative. Motifs are recurring themes that help unify the work. Both provide additional depth and interpretive context (Ray, 2024)
Rising action	These are the events that build tension and conflict in the narrative, leading toward the climax (Bergman, 2017)
Climax	It is the point of greatest tension and interest in the narrative, often representing the crucial moment or the resolution of the conflict (Gancho, 2006)
Resolution	It refers to the conclusion of the narrative, where conflicts are resolved, and characters may experience some form of transformation or change (Gancho, 2006; Kendal & Kendal, 2012)

the ethical standards described by Godoy (1995) and Vilela et al. (2022), ensuring privacy and respect for the participants throughout the entire process.

The selection of participants was made through convenience sampling, based on the researcher's network of contacts. Individuals who had experienced significant career transitions and were willing to reflect on these experiences in depth were chosen. Each interview lasted approximately 45–60 min and was conducted individually. The sessions were designed to encourage participants to freely explore aspects of their career transitions, including motivations, challenges, overcoming obstacles, and learnings. The interviews were recorded with the participants' consent and fully transcribed to ensure accuracy and reliability in data analysis.

After the narratives were collected, the data were subjected to a thematic analysis, inspired by Bardin's (2006) content analysis model. This process consisted of three main stages. In the pre-analysis stage, the transcriptions were read several times for familiarization with the data and preliminary identification of significant units, such as sentences or excerpts that highlighted relevant aspects of the narratives. During the material exploration, these units were coded, grouped into categories, and subsequently organized into general themes. In the final stage, the results were processed to identify recurring patterns and singularities in the narratives, allowing an initial understanding of the reported experiences.

Subsequently, a narrative analysis was carried out, based on the predefined narrative elements, aiming to explore the structure of the narratives and the meanings attributed by participants to their career transition experiences, identify crucial moments, such as important decisions or turning points, and understand the central messages and underlying emotions in the narratives.

The analysis process also included coding the data in three levels. Open coding allowed the identification of emerging categories directly from the narratives, such as specific challenges or strategies used by the participants. In axial coding, connections were made between categories, such as the relationship between emotional barriers and support networks. Selective coding synthesized the data, identifying main patterns and relevant subcategories. The analysis also considered the context in which the narratives were told, exploring how external factors, such as changes in the labor market or economic conditions, influenced the narratives.

The final stage of the process involved a critical reflection by the researcher. This reflection included reviewing the interpretations made and considering potential biases or personal assumptions that might have influenced the analysis. Reflexivity was essential to ensure a rigorous and impartial approach, as recommended by Carvalho et al. (2021) and Vilela et al. (2022).

10.3 Results

The analysis of the participants' narratives reveals that career transition is a multifaceted process, where narrative elements play a crucial role in constructing the stories. These narratives, in turn, reveal not only the professional journey of the participants but also the meanings and emotions underlying their experience of change.

In the context of the study, four main emerging categories were identified: motivation for career change, adaptation to new roles, personal growth, and organizational support during the transition. Next, we will explore the narrative elements, linking them to these emerging categories, in order to illustrate how the narratives are constructed and how the narrative elements help to understand the complexity of the career transition.

10.3.1 Individual Analysis of Narrative Elements

The participants' narratives are constructed from various narrative elements, each playing a crucial role in organizing and giving meaning to the stories. These elements help to understand the emotional, psychological, and social dimensions of career transitions and reveal how participants interpret and communicate their experiences.

The predominant point of view in the narratives is the first person, allowing participants to share their experiences in an intimate and direct way. This perspective adds a subjective dimension to the narratives, placing the narrator's emotions and reflections at the center. Participants describe their fears, doubts, and motivations. For example, Participant 2 stated: "*(...) maybe I regret not having emigrated (...)*" and Participant 5 mentioned: "*I realized that I was also reaching the point where my basic education was no longer enough.*" This narrative choice reinforces the authenticity of the narratives, allowing for a deeper understanding of the transitions.

The plot of the narratives follows a linear structure, composed of well-defined moments: the introduction, which often begins with a description of dissatisfaction or discontent in the current career; the development, which includes the process of reflection, analyzing available options, and deciding to change; the climax, representing the moment of greatest tension, such as accepting a new job or overcoming a critical challenge; and finally, the resolution, where participants reflect on the change and achieve a sense of accomplishment. For example, Participant 4 described: "*In terms of the course, it was difficult, but I feel that I managed to finish also because of the help I had from him, because I think if it weren't for that, I don't know if I would have managed to finish successfully.* "

The plot organizes the events and offers cohesion to the narratives, highlighting key points of the transitions, as shown in Fig. 10.1.

In the career transition narratives collected, the central figure in each narrative is undoubtedly the participant. However, as the narratives unfold, it becomes clear that there are other characters playing significant roles in the career change trajectories, directly influencing the development of the storyline. Career transition is not an isolated process, and the impact of secondary figures, such as mentors, supervisors, family members, colleagues, or friends, can be just as, if not more, relevant than the protagonist's individual experience. These figures can assume roles of support, resistance, or motivation, and each of these characters has a profound impact on

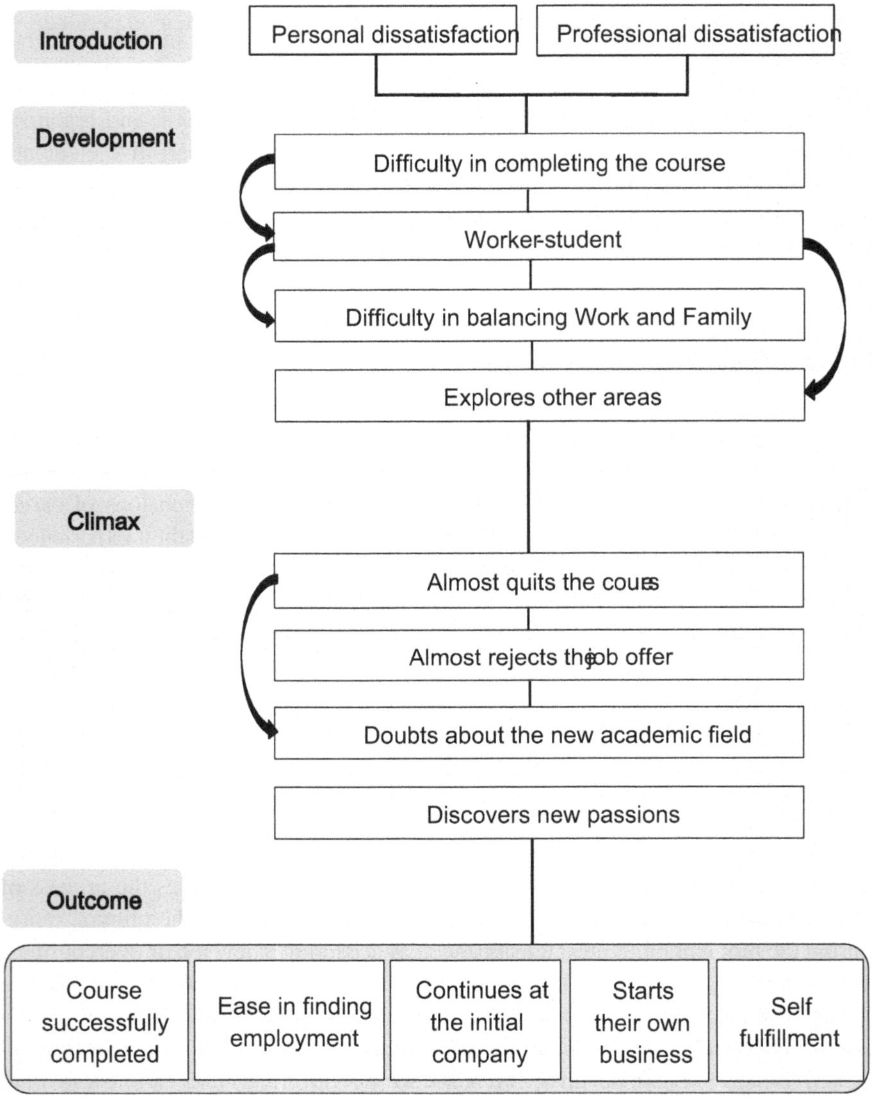

Fig. 10.1 Plot flowchart

how the participant perceives their career change journey. These characters emerge throughout the plot, shaping the narrative in a way that generates moments of tension, overcoming challenges, and learning.

When analyzing the collected narratives, the characters can be grouped into three main categories: heroes, villains, and secondary characters. Each of these categories contributes uniquely to understanding the challenges, motivations, and adaptation processes experienced by participants during their career transitions. These categories represent the different figures interacting with the protagonist and decisively influencing the unfolding of the plot.

The heroes in the participants' career transition narratives promote change and contribute to their progress, playing central roles by offering support, motivation, or guidance during key moments of the transition process. The heroic figures identified in the narratives are multifaceted, emerging from different contexts such as the workplace, education, family, and friends, reflecting the various sources of support that participants mobilize during their career transitions.

In the workplace context, the role of managers, leaders, and even the new company is highlighted. The presence of a new supervisor or a company that offers opportunities for professional development emerges as a determining factor for the success of the transition. The company and managers, by providing new opportunities and challenges, are seen as heroes due to the positive impact they have on the participant's achievement and adaptation to the new professional reality. The educational context also emerges as a significant source of support. Teachers, identified by participants as figures of guidance and inspiration, help build confidence and provide the necessary knowledge base for the transition, being described as mentors who play intellectual and motivational leadership roles.

In the family domain, family members such as a husband or sister are seen as pillars of emotional support and advice. These characters significantly contribute to the participant's emotional stability, helping them cope with challenges and make crucial decisions during the change process. Family support is a constant in the narratives, serving as an important reinforcement during moments of uncertainty.

Additionally, participants highlight friends, mentioning the importance of emotional support and the external perspectives that close friends offer, contributing to the analysis of career options and helping to reinforce the self-confidence needed to move forward with the transition. The narratives reveal figures outside the work context, such as the psychologist, who is presented as a valuable emotional support, offering the necessary guidance to deal with the psychological difficulties of the transition and helping the participant maintain focus on their goals.

Villains, or antagonists, are the figures that create obstacles and challenges faced by the protagonist throughout their transition, representing the difficulties that must be overcome for the participant to achieve the desired change. In the analyzed narratives, villains emerge prominently in the workplace context, social context, and ethical issues, each playing a specific role in creating tensions and conflicts. The workplace context is undoubtedly the category that most contributes to the identification of villains in the narratives. Figures such as the former supervisor, the former company, and the structural difficulties associated with Public Administration are mentioned as sources of resistance to change. The unfavorable work environment, lack of growth opportunities, and resistance from supervisors to new approaches or ideas are frequently cited as factors that create an environment of dissatisfaction, forcing the participant to seek new possibilities outside the organization. Furthermore, the social context also emerges as a source of antagonism. Social stereotypes and cultural resistance to career change are identified as significant obstacles, with participants referring to the difficulty of breaking away from established social expectations regarding their professional role. External pressures related to social conformity and public approval become sources of emotional tension for the participant, who feels compelled to challenge pre-existing norms.

Finally, the ethical issues addressed in the narratives of participants P3 and P4 appear as complicating factors in their trajectories. The difficulties associated with ethical pressures in advertising campaigns and ethical challenges in the professional field contribute to the construction of conflict and hinder the progress of career transitions.

Secondary characters play a crucial role by adding depth and context to the narrative. Although they are not directly responsible for driving the plot, they offer support, information, and additional perspectives that help shape the career transition experience. Secondary characters are mainly identified in the categories of education, family, friends, workplace context, and social context, and their contributions are essential for enriching the participants' narratives.

The education context is the category that most contributes to the emergence of secondary characters, with the presence of participants in lectures, teachers, trainers, and college leadership. These figures are fundamental in building knowledge and developing the skills necessary for the transition. The course director, trainers, and the marketing professor play key roles in providing the academic tools that allow participants to move forward to new challenges.

Family is also a category of secondary characters with significant impact on the narratives. Figures such as mother, father, grandmother, and other family members influence both the academic formation and professional development of the participants, providing emotional support and helping to resolve practical issues related to the transition.

Friends appear as sources of encouragement and opportunities, in addition to providing valuable advice that helps with professional decisions and overcoming difficulties during the transition.

Within the work context, the figure of the secretary emerges, playing a functional role in logistical support and in integrating the participant into the new work environment. Additionally, the medical information officer appears as a relevant figure outside the direct work context, providing essential information for some transitions to take place.

Society, as a secondary character, is identified as a source of resistance, representing the social expectations that conflict with the individual decision to change careers. Figure 10.2 illustrates the analysis of characters in career transition narratives reveals the complexity of social, professional, and personal dynamics that influence the choices and challenges faced by participants. Heroic figures provide support, guidance, and motivation, while villains represent the obstacles and difficulties participants must overcome. Secondary characters provide additional depth and interpretive of meaning and context, enriching the story and helping to shape each participant's journey. These interactions between characters are essential to understanding the career transition process and the various influences that shape each individual's professional journey.

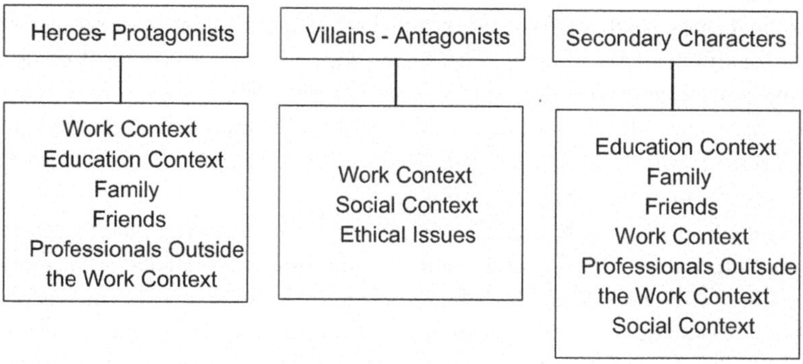

Fig. 10.2 Characters in career transition narratives

The narrative element of configurations plays a central role in understanding career transition narratives, as it defines the scenarios and conditions in which the participants' stories unfold. Configurations reveal the physical, emotional, and social contexts that shape the transition experiences, contributing to the construction of the plot and understanding the decisions and actions reported.

Based on the analyzed narratives, three main dimensions emerge that structure the Configurations element: Pre-Transition, Transition Process, and Effective Change. These dimensions reflect the progression of the participants' experience, from the initial conditions preceding the change to the consolidation in a new professional area.

Pre-Transition constitutes the context before the start of the career transition process, providing the backdrop for the subsequent decisions and actions. In this phase, four main categories emerge: Work Conditions, Work-Life Balance, Social and Economic Challenges, and Personal Issues. The narratives highlight unfavorable conditions as determining factors in shaping the emotional and professional climate. Issues such as low wages, excessive workload, lack of career progression, and toxic environments emerge as recurring elements. These factors create dissatisfaction and fuel the desire for change. For instance, P4 describes how monotony and the absence of professional goals contributed to their discontent, stating, "*It was always the same routine, with no goals… salary-wise, it was also lagging behind, which made me feel down.*"

The difficulty of balancing personal life with professional demands is another issue that permeates this phase. P1 reports that the workload and demanding shifts created an exhausting environment, stating, "*It ends up being very frustrating… the workload made it difficult to balance personal and professional life.*"

Economic and social contexts, such as job insecurity and labor market instability, appear as external barriers that amplify dissatisfaction. P3, for example, mentions that the economic crisis limited their chances of finding stability, stating, "*I tried to apply for some companies in the field, but due to the post-pandemic scenario, it wasn't possible.*"

Personal Issues, such as frustration and lack of fulfillment, emerge as internal catalysts for the need for change. These emotions create a scenario of discomfort, prompting participants to reflect on their trajectories. P5 describes how insecurity and the lack of scientific rigor in their initial field fueled their dissatisfaction, stating, "*I knew I didn't want to follow the path imposed by a brand… I had invested in my education and wanted to work according to it.*"

The Transition Process represents the phase in which participants begin to take action to change careers, facing emotional tensions and practical decisions that shape the unfolding of the narrative. One of the most significant categories in this phase is the Personal Context, where factors such as the internal need for change, frustration, lack of confidence, and the influence of personal events like marriage, the death of family members, and fertility treatments directly impact the decisions. P6, for example, describes how anxiety and nervousness led them to seek psychotherapy, stating, "*I started psychotherapy (…); (…), after two years of psychotherapy (…) I decided to study psychology, I enrolled in university.*"

In the Career domain, decision-making and reskilling stand out in this category, where participants invest in new training or explore opportunities to change fields. The organization of lectures, unemployment, and taking on family businesses are also mentioned as important milestones in the transition journey. P4, who decided to focus solely on the course, reflects on the impact of that choice, stating, "*I stopped working to dedicate myself to the course… it was a total investment, both financially and personally.*"

The Health category also plays a relevant role, highlighting how factors related to mental and physical health shape the transition experience. P2, while undergoing fertility treatments, mentions that the demanding hours of her initial field were incompatible with her personal goals, stating, "*I knew that nursing hours weren't compatible with what I wanted for my life.*"

Effective Change represents the consolidation of the transition, when participants embrace new careers and integrate into their chosen fields. This phase is marked by professional progression, adaptation, and personal fulfillment.

The narratives show that new professional opportunities are often achieved through networking or recent training. P3 describes how the postgraduate course facilitated direct access to her new field, stating, "*The postgraduate course provided direct access to the job market… it was a quick and very satisfying change.*" Adaptation to the new field is generally described as positive, with participants highlighting the impact of their acquired skills. P2 mentions that her previous experience helped her integrate quickly into her new role, stating, "*Since I already knew a bit about the field, it was easier… I adapted well to the work.*" Furthermore, the creation of personal businesses reflects a more independent approach to the transition. P6, who decided to establish her own space, states, "*I created my own space… I found an office for myself and started giving consultations.*" Academic networking also plays a crucial role in this phase. P7 mentions how her relationship with a professor resulted in a job opportunity, stating, "*I talked to the professor about what to do with my life… he arranged a meeting, and that day he gave me a job.*"

These dimensions illustrate the complexity of career transitions, showing how social, emotional, and professional contexts intertwine to influence individuals' choices and outcomes. Moreover, these phases vary in duration and intensity, reflecting participants' journeys from initial dissatisfaction to successful integration into a new professional area.

Pre-Transition is characterized by a period of dissatisfaction and reflection, in which participants experience adverse conditions and begin to question their career paths. This period can be prolonged, as in the case of P6, who remained 15 years in her initial field, or shorter, as in the case of P3, who stayed less than a year before deciding to change. Participants frequently report difficult working conditions, such as low wages, lack of career progression, and toxic environments. P4, for example, describes how the monotonous routine and the absence of professional goals contributed to her demotivation, stating, "*It was a very monotonous routine... salary-wise it was lagging behind, which made me demotivated.*"

The duration of the Transition Process varies between immediate changes and longer processes. In the case of P5, the transition was nearly immediate, while P6 took five years to complete her change. In several narratives, participants highlight the need to overcome frustrations and find strength to move forward. P1, for example, states that although she spent a year reflecting on the change, only at the end of that period did she begin to take action, stating, "*I was frustrated and thinking about it for a year, but only in the last few months did I take concrete steps.*"

In Effective Change, many participants mention that academic training was decisive in facilitating integration into the job market. P3, for example, states that the postgraduate course provided direct access to the new professional field, mentioning *that* "*the postgraduate course provided direct access to the job market... it was a quick and very satisfying change.*" Adaptation to the new field is often described as positive. P2 reports that she adapted quickly to the new environment, stating, "*Since I already knew a bit about the field, it was easier... I adapted well to the work.*" P6, who decided to open her own business, describes how she managed to establish her own space, saying, "*I created my own space... I found an office for myself and started giving consultations.*"

The analysis of the temporal perspective of the narrative configurations allows us to understand how career transitions develop over time, revealing the influence of context and actions at each stage. Pre-transition highlights the factors that generate dissatisfaction and the desire for change; the Transition Process reflects the dynamism of decisions and the difficulties faced; and effective change demonstrates the consolidation and success in the new career. These phases, although unique for each participant, follow common patterns that highlight how time plays an essential role in shaping narratives, molding the experiences and outcomes of career transitions.

Conflict, as a narrative element, plays a central role in structuring career transition stories, acting as a driver that propels the narratives and creates tension, leading participants to crucial reflections and decision-making. In the analyzed narratives, conflicts emerge in two main categories: external conflicts, related to forces and circumstances beyond the participants' control that impact their trajectory, and internal conflicts, reflecting psychological, emotional, and moral dilemmas experienced during the transition process.

External conflicts arise from circumstances outside the participants' direct control and have a significant influence on their choices. Among the identified external conflicts, problematic relationships in the work context, challenges imposed by society, gender-related issues, and difficulties faced with professors and higher education institutions stand out.

In the work context, participants often face challenges related to the employer or the conditions of the profession. P2 describes the lack of career progression opportunities in nursing, reflecting the frustration experienced: "*In nursing, studying doesn't help at all... in the public sector, it's very difficult to advance in the career.*" Meanwhile, P4, who worked in nutrition, mentions the impact of comments and prejudices in the field, stating, "*There are many comments around what nutrition is... I was aware that this was not the path I wanted to take.*"

Challenges imposed by society also shape the narratives, reflecting stereotypes and external pressures. P6, for example, mentions the conflicts generated by working in a predominantly male environment, highlighting: "*I worked in a metalworking factory, in an extremely male-dominated environment... I always had a girlish appearance, which made everything harder.*" These conflicts illustrate the cultural and social barriers that influence the participants' trajectories.

With regard to gender, P6 describes the difficulties faced not only in the work environment but also in the academic context, stating "*There were professors who didn't find it amusing because I discussed issues with them... maybe because I was older and didn't accept the first thing they put in our heads.*" These experiences demonstrate how external conflict can arise from hierarchical relationships or established social norms, contributing to the tension in the narratives.

Internal conflicts are also prevalent in the narratives, reflecting the emotional and psychological dilemmas faced by participants during their transitions. These conflicts include academic and professional indecisiveness, feelings of inferiority, and challenges associated with returning to studying or leaving work. Academic indecision frequently appears at the beginning of the narratives, marking the uncertainty about which path to take. P7, while recounting their first course, admits "*I did the course and right in the first year, I thought about quitting... I started to realize that maybe it wasn't really for me.*" This feeling of doubt and uncertainty is recurrent and extends to the choice of new areas of study, as in the case of P3, who faced the dilemma of choosing between a master's in education or a postgraduate course in IT, stating "*I had to weigh two options... a two-year master's or a postgraduate with direct access to the job market.*"

Feelings of inferiority are mentioned by participants who, even after achieving change, still question their abilities. P4, for example, says "*Three and a half years later, I still feel that sensation of not being on the same level as others.*" These reflections highlight the emotional challenges that accompany career transitions.

Another significant internal conflict is the challenge of leaving work to return to studying, mentioned by participants such as P3, who describes the tension of giving up financial stability to invest in education, stating: "*It's a huge challenge to stop working and receiving a salary to dedicate myself to studying.*" Similarly, P6 recounts the doubts that arose during the transition: "*Should I have gotten involved in this? Maybe it's not for me.*"

Additionally, the lack of professional fulfillment, even after achieving previously set goals, reflects the complexity of the emotions experienced during career transitions. P2 describes this feeling of emptiness "*We achieved our goal, but we don't feel completely fulfilled... it always seems like something is missing.*"

External and internal conflicts do not operate in isolation but interact to shape the narratives, creating a climate of tension that drives the participants' actions and decisions. For example, P7 recounts a moment of deep reflection while working in the family business, saying "*I was sitting at the computer, in silence, and I thought: what am I doing here? I don't know if this is for me.*" This moment reflects the overlap of internal conflicts (doubts about the choice) and external conflicts (pressures from the work environment).

Similarly, P1 mentions that accumulated frustration led them to take action, stating "*I had been thinking about this for a year, but only in the last few months did I decide to do something concrete.*" This account illustrates how internal conflict transforms into a driver of change, influenced by external pressures.

External conflicts, such as challenges in the work environment, social pressures, and gender issues, reflect the structural and cultural barriers faced by the participants. Internal conflicts, on the other hand, highlight the emotional and psychological dilemmas accompanying transitions, ranging from indecision to feelings of inferiority. When these dimensions of conflict interact, they create rich and complex stories, revealing the depth of the challenges faced by the participants. This narrative element emphasizes the importance of understanding both external forces and internal reflections in the study of career transitions, providing a more comprehensive perspective on the professional change process.

The themes identified in the analyzed narratives reflect three major categories: Work Context, Personal Context and Professional Development Context. These categories organize the reflections and experiences of the participants, offering a broad perspective on career change trajectories and the underlying meanings of their narratives.

In the Work Context, the themes focus on issues related to the work environment, professional recognition and the challenges faced by the participants. These themes reveal the tensions between individual expectations and organizational realities, highlighting factors such as insensitivity from leadership, lack of progression opportunities and the need to balance personal and professional life. P2 reflects on the difficulties in the public sector, mentioning the lack of recognition and demanding working conditions, stating "*There's a lack of people... but there's also no understanding from leadership to help us in this regard.*" On the other hand, P4 addresses the need for courage to change and find a more rewarding work environment, emphasizing that "*the search for a place where I felt heard and respected was essential for my decision to change.*" P3's narrative also illustrates the tension between the desire for stability and the limitations imposed by the professional context, stating "*I didn't see myself there... without any kind of stability.*" These themes highlight how the work environment can deeply influence transition decisions, acting both as an obstacle and a catalyst for career change.

In the Personal Context, the themes reflect the emotional and psychological capabilities of the participants to deal with life challenges and career changes. This context is characterized by narratives of resilience, overcoming, courage and self-discovery, which shape the participants' decisions and influence their paths. P1 highlights the importance of resilience by recounting how they faced adversities throughout their journey, stating "*I thought many times about quitting, but I didn't... I made it to the end, took the exam, passed, and earned my diploma.*" Similarly, P4 reflects on the courage required to change and persist, mentioning that "*flexibility and the ability to adapt were key to overcoming challenges.*" P6's narrative explores deeper themes of self-discovery and purpose, emphasizing that "*having an awareness of the impact it had on me and wanting to do the same for others... a greater purpose*" was central to their career transition. These themes reveal how personal and emotional challenges shape participants' decisions, revealing internal insights that drive change.

In the Context of Professional Development, the themes address the strategies used by participants to enhance their skills and deal with career transitions from both a psychological and practical perspective. The narratives highlight the importance of openness to new opportunities, adaptability to change and the search for a better future. P5 reflects on the flexibility required to explore different career paths, stating "*I entered the retail world without really knowing what I was getting into... but I realized that nutrition wasn't where I wanted to continue.*" On the other hand, P7 emphasizes openness to unexpected opportunities, mentioning "*It was a series of coincidences, but I was open to those coincidences.*" Adaptability is a recurring theme, as shown by P4, who highlights "*They realized that I could easily adapt to new changes, which was very important in this area.*" These narratives demonstrate how career transitions are often guided by development and learning strategies, which allow participants to face new challenges and explore different possibilities.

The emerging themes in the narratives interconnect to form a broader framework of career transitions, reflecting the multiple contexts that influence the participants' stories. For example, P3's narrative, which seeks professional stability while dealing with limitations imposed by the work environment, integrates the Work Context with the Personal Context, showing how professional and emotional experiences interact to shape decisions. On the other hand, P6, by exploring self-discovery and purpose, links the Personal Context with the Professional Development Context, presenting how internal motivations influence the strategies used to overcome challenges and achieve personal and professional fulfillment. These thematic relationships highlight the complexity of career transitions and illustrate how participants go through multiple contexts and experiences to construct their narratives.

Tone and Style play a central role in the analysis of career transition narratives, influencing both the interpretation of content and the way in which the stories are presented by the participants. The tone, which sets the emotional climate of the narrative, is divided into two main categories: Joy and Introspection. These categories reflect the participants' subjective perception of their transition processes and shape how these experiences are reported.

In narratives with a more joyful tone, participants express optimism, adventure, and even humor. P4 describes their transition with confidence and enthusiasm, stating "*I was lucky to find another path... I knew it would offer opportunities and open many doors.*" Meanwhile, P7 adopts an adventurous tone when recounting their experiences in the field of archaeology, saying "*I went to excavate a monastery... I felt like I was truly interacting with History.*" This positive tone is also manifested in the humor of narratives such as P1, who reflects on the challenges of balancing work and studies with humor "*I would get home from work, study, and think: my God, what am I doing with my life?*".

On the other hand, narratives with an introspective tone reveal deep reflections, criticism and sometimes melancholy. P2 describes the lack of opportunities in nursing with a discouraged tone "*In nursing, studying doesn't help... it's very difficult to progress in the career.*" P3, in turn, reflects on the emotional impact of their choices, asking "*There was a time when I stopped and thought: in 20 years, do I still want to be doing this?*" These introspective narratives show the internal dilemmas that shape participants' decisions and highlight the emotional and psychological challenges faced during career transitions.

The style of the narratives is also diverse, reflecting the expressive choices of the participants. The reflective style is characterized by deep and introspective analyzes of the chosen paths. P1, for example, questions herself about her professional trajectory "*Should I have gotten into this?*". Similarly, P2 uses a reflective style to consider the limitations of the nursing sector, stating "*If the leadership in nursing in Portugal were different, maybe the story would be another.*" Another style identified in the narratives is the conversational style, which adopts a more accessible and direct approach. P3 narrates their choices in a colloquial manner, saying "*It was all very random... well, I ended up going into biology.*" In some cases, participants enrich their stories with descriptive details, such as P7, who recalls childhood episodes "*My grandmother would make me study during the holidays, while my friends went to the beach... so I would set the clock ahead to avoid studying for an entire hour.*" Irony also appears as a narrative style, giving a critical and humorous tone to the stories. P6 exemplifies this style when reflecting on difficulties in their academic journey "*Theoretically, I had a supervisor... but I was the one who wrote the reports for her to submit to the board.*" This use of irony reveals the ability to transform frustrations into narratives that reflect resilience and a critical view of the transition process.

The interaction between tone and style provides valuable insights into how participants interpret and communicate their experiences. Narratives with an introspective tone and a reflective style, like P3's, highlight the internal dilemmas faced during the transition. In contrast, narratives with a joyful tone and a conversational style, like P7's, reflect the ability to find enthusiasm even in the face of challenges. Moreover, the use of irony, as demonstrated by P6, highlights participants' creativity and resilience in dealing with difficulties using humor to critique or subvert negative experiences. These narratives not only recount events but also reveal the emotional and expressive strategies participants adopt to navigate their career transitions.

Tone and Style, as narrative elements, are fundamental to understanding how participants experience and articulate their career transitions.These elements not only shape the content of the narratives but also offer a unique perspective on the subjectivities involved. The analysis of these aspects highlights the richness and diversity of the narratives, providing a deeper understanding of the motivations, challenges, and achievements that characterize career changes.

Throughout the career transition journey, participants' narratives reveal symbols that reflect milestones and transformations. In this process, each symbol relates to unique and contextual experiences that lead to change. The Door, for example, appears as a symbol of transition, representing the passage to new opportunities. P1 associates the Door with the positive impact of their education, stating "*This post-graduate course opened doors for the role I'm currently in.*" This symbolism reveals the perception that education not only expands career options but also symbolizes an entry into more favorable professional environments.

Similarly, the Mentor emerges as a catalyst for change, representing figures of support and inspiration throughout the transition. P7 describes the influence of the teacher as a significant turning point in their career: "*I went to talk to LR, and that day he gave me a job… and that's how my life changed.*" This symbolism reinforces the role of support networks and inspiring figures in facilitating successful professional transitions.

Other symbols, such as the Journey, represent the continuous process of exploration and personal development. P3's narrative reflects this search, stating "*I stopped my master's degree to go work in Croatia… after finishing, I realized it wasn't what I wanted.*" This symbol highlights the exploratory nature of career transitions, often accompanied by self-discovery and re-evaluation of goals and objectives.

The Voice emerges as a symbol reflecting the participants' difficulty in being heard and recognized in the professional context. This symbol is often associated with a sense of powerlessness and a lack of autonomy in the workplace. P2 describes this experience "*You don't have a voice, no one listens to you.*" This metaphor highlights the participants' struggle for recognition and value, which are crucial for satisfaction and motivation in the professional environment. Symbols like the Network and the Boat underline the importance of security and stability in the transition process. P2 uses the Boat to describe the financial support from her husband, essential for enabling the career change "*My husband had financial stability to keep the boat afloat, in case something went wrong.*" These symbols highlight the relevance of financial conditions as prerequisites for successful transitions, emphasizing the role of the economic and social context.

The Motives identified in the narratives fall into five main categories: Intrinsic, Extrinsic, Rational, Emotional, and Short-Term. These categories help analyze the forces driving the decisions to change, reflecting both internal needs and external pressures.

Related to personal desires and the search for fulfillment, Intrinsic motives are central in several narratives. P1, for example, mentions frustration with her professional life as a decisive factor: "*I felt frustration related to my professional life.*" These motives reflect dissatisfaction with the current state and the search for alignment between personal values and professional identity. Extrinsic motives include external factors, such as compensation and career advancement. P4 describes the decision to change due to insufficient financial conditions "*I liked what I did, but the pay wasn't enough.*" These motives highlight how dissatisfaction with external conditions can trigger transition, often accompanied by a reassessment of priorities. Rational motives reflect decisions based on logic and the analysis of opportunities. P3 exemplifies this approach, stating "*I couldn't stay in that role because it didn't offer stability.*" These narratives underline how the search for more favorable conditions and economic security shapes career choices. Emotional motives, on the other hand, are linked to subjective perceptions of fulfillment and happiness. P4 describes the monotony of the work as a demotivating factor: "*I felt like every day was the same.*" These reasons reflect the importance of finding work that meets our individual values and aspirations. Short-term motives are associated with immediate objectives, such as solving an urgent situation or securing a job. P7 describes how a meeting with a teacher resulted in a quick change: "*That day, he gave me a job, and my life changed.*" These types of reasons demonstrate how decisive moments and unexpected opportunities can be catalysts for professional change and mark the beginning of new professional paths.

The Symbolisms and Motives do not operate in isolation but interact to shape career transition narratives. For example, the symbolism of the Door, associated with the expansion of opportunities, frequently reflects both rational and emotional motives. Participants who mention this symbol highlight both the logical decision to acquire new skills and the emotional desire to overcome previous limitations. Similarly, the figure of the Mentor interacts with intrinsic and extrinsic motives, functioning as a source of inspiration and a means of access to new professional opportunities. P7's narrative, describing the impact of the professor, highlights how the presence of a supportive figure can motivate both emotionally and rationally during transitions. This analysis reveals that Symbolisms and Motives are essential elements in constructing narratives, helping interpret the internal and external forces shaping decisions and experiences of change. These elements highlight the challenges faced and the meanings attributed to these experiences, providing an enriching and detailed perspective on the dynamics of career transitions.

In the analyzed narratives, the Rising Action is driven by a series of internal and external tensions faced by the participants, fueling a plot of professional challenges, personal doubts and moments of crisis. In most narratives, it is frustration with the current professional situation that appears as the starting point for upward action. Dissatisfaction with the profession, difficulties in achieving goals and lack of personal fulfillment signal the beginning of the transition process. Insecurity caused by the lack of job satisfaction is one of the main forces that drive participants to look for alternatives, and frustration with work results in the decision to change careers. P2 also expresses the difficulty of making an important decision when she had to leave a stable position to accept an uncertain offer. She describes that moment as one of the hardest in her transition: "*That was the hardest moment in my transition. Leaving something certain for something uncertain. The contract was for 1 year.*" This decision-making conflict, between the security of the known and the uncertainty of the future, constitutes a significant turning point in the career transition narrative.

The need to balance work with education to change fields also emerges as a crucial point of tension. The challenge of continuing in the current job while investing in new qualifications can generate great insecurity. P3 describes the dilemma of choosing a new area of study, which would require more years of study, or a shorter course, but with immediate access to the job market: "*I had to weigh two options, one in education, which I really liked, but it was a 2-year master's, or getting a postgraduate degree in IT with direct access to the job market.*" This type of decision is a classic example of how career transitions involve not only choosing a new field but also confronting uncertainties about the professional future and the need to weigh the risks and benefits of each alternative.

In some narratives, participants also face the insecurity of leaving their job to study full-time, which creates fear of being unable to complete the education or failing. P6, for example, expresses the internal pressure of having to invest in her education, while also feeling the difficulty of aligning her new aspirations with external expectations: "*Because if I had invested in my education, I had to have the freedom to work on my education and not under what a brand said I should implement.*" The insecurity about the ability to meet academic goals reflects the emotional and professional tensions that characterize this phase of the career transition.

Career transition narratives are often shaped by unforeseen events that act as catalysts for change. The loss of a family member, for example, can be a moment of reflection that forces a reevaluation of life and career choices. P7 describes the death of her stepfather as a decisive moment that reoriented her professional trajectory "*I'm in the Algarve, I come to Porto for my grandparents' 50th wedding anniversary, and my stepfather dies of a heart attack.*" This tragic event forced the participant to rethink her priorities and make significant changes, accelerating the career transition process. Moreover, P5, by accepting a job offer without having a clear idea of what to expect, also demonstrates how impulsive decisions or those made in moments of uncertainty can be decisive in one's professional trajectory. P5 says "*I accepted a job offer without knowing exactly what I was getting into.*" This type of change, motivated by the desire for something new, exemplifies how career transitions can be influenced by decisions that, at first glance, seem uncertain but end up providing new growth opportunities.

The Rising Action in career transition narratives is full of tension, where participants face emotional, professional, and personal challenges. Dissatisfaction with work, insecurity about new education and unforeseen events like personal losses contribute to an intensification of the plot. These moments of increasing tension prepare the ground for the effective change, reflecting the complexity and unpredictability of career transitions. The rising action in the narratives shows how participants deal with adversity and the difficult choices they have to make and their uncertainties, highlighting the difficulties overcome and the continuous process of adaptation that lead to the desired professional transformation.

The Climax is one of the most significant elements in career transition narratives, as it represents the point of greatest intensity, where participants face the most significant challenges and make crucial decisions that will define the direction of their professional journey. This culminating moment does not resolve all the issues but marks the peak of the previously presented conflicts and shapes the change. The climax emerges as the turning point, where doubts and tensions reach their peak, resulting in decisions that, in some way, profoundly alter the lives of the participants. In the analysis of the narratives of the participants in this study, three climax moments can be identified: the Moment of Education, the Professional Moment, and the Moment of Transformation. Each of these dimensions is closely linked to different aspects of the transition experience, from investment in new qualifications to making significant professional decisions that result in substantial career change.

The Moment of Education is one of the most important categories in the climax of the narratives, especially because it is associated with significant academic challenges and moments of overcoming. The completion of a postgraduate degree is often described as the high point of this phase, where resolving internal conflicts and achieving an academic goal become milestones in the career transition. Upon completing their postgraduate degree, P1 describes the emotional impact of this victory, which for them symbolizes the ability to overcome personal and professional obstacles "*I think it emphasized my desire to overcome challenges, and this was also clearly a personal victory in that sense. Because so many times I thought about quitting, and I didn't, and I got to the end.*" This academic moment is not only a formal conclusion but also a psychological and emotional milestone that reinforces the decision to continue pursuing a new career. The climax of the Moment of Education is also visible in P3's narrative, where they face uncertainty when choosing a new area of study without prior experience. The decision to invest in IT, an unfamiliar field, creates deep internal conflict, as they express: "*It was never an area I had experienced, never tried it, so I could... go there and... have thrown the work away, quit, not have applied for the AECs, and gone there, applied, paid for the postgraduate and not liked it. It was a risk, and it scared me in the beginning.*" This climax moment, where the risk of failure is present, reflects the tension felt by participants when making decisions that radically change their professional paths.

The Professional Moment is another fundamental category of the climax, where decisions about career changes or new professional opportunities become the turning point in the narratives. P2, upon receiving a job offer in a completely different field, expresses the dilemma of accepting the offer, facing uncertainty about the future "*I thought a lot, I thought a lot, and I almost said no, honestly; but I also thought, if not now, it won't be later, and even if it goes wrong... there are other opportunities, I'll continue with the master's, and I can go a year or two without working.*" This moment reveals the tension that arises when it is necessary to leave a stable situation to embark on an uncertain change, with this decision being an important milestone in the participant's professional transition. Another example arises when P5 describes their true vocation as being in the retail and management world. This recognition arises gradually, reflecting a deep alignment with their professional values and interests "*The involvement in the retail world, combining people management with commercial management, was what sparked my greatest interest.*" This moment of discovery represents a clear assertion of their new professional identity, marking a decisive turning point in their career journey.

The Moment of Transformation is a category of the Climax that involves a professional change and a profound change in the individual's personal identity. This moment is characterized by self-discovery and a reconfiguration of the person's values and goals, influencing both their professional and personal life. P6 exemplifies this transformation by describing the impact of psychotherapy, which played a central role not only in their mental health but also in their career transition journey. The participant describes this moment as a true change of purpose, going beyond just a professional choice: "*What led me to seek psychology was realizing the impact it had on me and wanting to do the same for others. A bit of a higher purpose. In Maslow's hierarchy, we're at self-actualization.*" This climax moment not only redefines P6's professional path but also results in an internal transformation, reflecting a shift in their consciousness about their role and purpose in life.

The climax reveals the complexity of career transitions, where not only the choice of a new profession but also the internal transformation process are essential for the individual's success and fulfillment. This turning point is fundamental in understanding the journey of each participant, as it marks the intensification of tensions and the preparation for an effective change in their future professional life.

Conclusion is not limited to problem solving, but also involves the integration of lived experiences and adaptation to new contexts, both personal and professional. It reflects the final balance of the internal and external transformations that characterized the participants' trajectory throughout the career transition process.

In the analyzed narratives, the conclusions can be classified into two main categories: Personal Context and Professional Context. The personal context involves the fulfillment of individual expectations and the well-being resulting from the change, while the professional context refers to adapting to the new job market and career progression, accompanied by professional achievement.

In the Personal Context, the conclusion of the narratives reflects a sense of personal fulfillment, overcoming challenges, and well-being resulting from the career transition. The participants express a feeling of self-realization as a consequence of their inner transformation, which manifests through the professional change. P1, after facing various obstacles during her transition, reports that her journey culminates in a sense of personal accomplishment "*So, obviously, this was also a stage of personal fulfillment and overcoming.*" This feeling of overcoming, linked to the career change, reflects an experience that goes beyond the acquisition of new professional skills, involving a profound change in the perception of oneself and one's personal life. P2 reports his personal conclusion about the career change, highlighting the positive impact on his well-being and self-esteem. The participant emphasizes how the transition led to a greater appreciation of her skills and the ability to be heard and respected in the new work environment "*Now I feel that I am more heard, that I have an opinion within the company, that they listen to what I say, that they respect me, that they want to know my opinion.*" This conclusion emphasizes how personal appreciation and the confidence emerging from professional adaptation are closely linked to well-being and self-esteem strengthening.

The Professional Context, in turn, highlights the participants' adaptation to the new job market, alignment with new areas of activity, and recognition of their skills in the new professional context. This adaptation can be challenging, as it includes obstacles in the integration process and the search for stability, however, it also provides a new sense of professional fulfillment. P5's example illustrates this aspect well. After her career transition, she reports a sense of appreciation, mentioning how the new role in retail and management makes her feel fulfilled "*Now I am a Retail Manager, in another area, and I really enjoy it, and I hope I can also change between other areas.*" This conclusion reveals a significant shift in that the participant not only successfully adapted to a new field of work but also found a position aligned with her interests, which contributes to her growth and professional satisfaction.

Another significant example comes from P2, who, after changing sectors, feels that appreciation and professional recognition were key factors in her adaptation. Reflecting on her journey, she highlights how she now feels more respected and valued in the new professional context: "*Now I feel that I am more heard, that I have an opinion within the company, that they listen to what I say, that they respect me, that they want to know my opinion.*" This conclusion shows that the career change not only enabled the realization of new professional goals but also brought about the recognition she was previously denied, resulting in the appreciation of her skills and experience.

Finally, P3 illustrates how adapting to the new job market is one of the biggest achievements after the transition. Changing fields was challenging as she had concerns about her ability to adapt, but in the end, she found satisfaction in the process "*I thought a lot, I thought a lot, and I almost said no, honestly; but I also thought, if not now, then never, and if it goes wrong… there are other opportunities, I'll continue with the master's, and I can stay a year or two without working.*" This example highlights the relevance of flexibility and a long-term vision for participants who, despite uncertainty, chose to invest in the transition as a commitment to the future.

The conclusion of career transition narratives goes beyond resolving the conflicts experienced, also encompassing the integration of change processes. The participants' narratives show that adapting to the new job market and aligning with a new professional identity are not isolated events. On the contrary, these processes are deeply related to personal changes, such as strengthening self-esteem and the desire for self-realization. Furthermore, the conclusions reflect a combination of overcoming, well-being and future vision, indicating that the success of the career transition is not only professional, but also a continuous process of personal growth and development.

Career transition, as evidenced by the participants' narratives, is a complex and dynamic process, where professional fulfillment and personal well-being are related. Each career transition story is unique, but they all reveal a search for purpose, adaptation and self-affirmation, culminating in a conclusion that is not just an end, but the beginning of a new chapter in the participants' professional and personal lives.

10.3.2 Relationship Between Narrative Elements and Emerging Categories

Narrative analysis provides a robust framework for understanding career transitions, allowing an investigation into how individuals experience, interpret, and manage significant changes in their professional trajectories. The narrative elements reveal underlying processes that help explain the emerging categories of this study: motivation for career change, adaptation to new roles, personal growth, and organizational support during the transition. These elements not only organize the participants' narratives but also elucidate the complexity of their experiences and strategies.

Motivation for career change is often triggered by experiences of dissatisfaction, frustration and lack of professional fulfillment. P4 describes, the impact of monotony and lack of professional recognition "*It was a very repetitive routine, always the same. There were no objectives, no goals to achieve, so it ended up being a bit demotivating, and since the salary didn't match either, it left me feeling a bit down.*" This perspective highlights the introspective and critical nature of the analyzed narratives, where the internal conflict intensifies until reaching a climax, culminating in the decision-making to change career.

Adapting to new roles demonstrates how participants face challenges related to learning and integrating into unfamiliar professional environments. Narrative settings, which include both physical and social environments, have a significant influence on this process. P5 describes "*I entered the retail world without really knowing what I was getting into, but I realized that managing people was what interested me most.*" This narrative demonstrates how the adaptation process includes overcoming internal conflicts and discovering new passions, often symbolized by moments of progress and acceptance of new responsibilities.

Personal growth emerges as a central dimension in career transitions, fueled by themes of resilience, overcoming adversity and self-discovery. P6's narrative demonstrates the transformative impact of psychotherapy on both her professional and personal journey "*I discovered the impact that it has on life, and the desire to do the same for other's. A bit of a greater purpose. In Maslows hierarchy, we are in self-actualization.*" This narrative emphasizes how the search for a greater purpose and self-fulfillment become fundamental for profound and significant changes, revealing that personal development transcends the professional sphere and is related to emotional satisfaction and general well-being.

Organizational support during the transition plays an essential role, shaping the course of the narratives and influencing their outcomes. Teachers and leaders appear as key figures in creating opportunities and validating participants' choices. P7 narrates the impact of a teacher "*Well, I made the appointment, T made the meeting, I went to talk to L, and on that day he gave me a job. And that's how my life went from archaeologist to company S.*" On the other hand, the absence of organizational support can be a factor that intensifies conflict, as P2 states "*You have no voice, no one listens to you, you can't give an opinion on anything because everything is already decided*." These examples illustrate how power and recognition dynamics influence the experience of transition.

Narrative elements are organized effectively to expose the emotional, social, and practical dimensions of career transitions and changes. This perspective offers an intimate look at personal experiences, while the plot structures events around pivotal moments. The characters symbolize external influences that enrich or hinder the journey, and the scenarios place the actions in a physical and temporal context. Conflict and theme highlight challenges faced and core values, while symbolism and motifs add depth and cohesion to stories.

The outcomes of the narratives reflect the integration of personal and professional dimensions, showing that career transitions are multifaceted processes. P1's narrative represents this sense of fulfillment "*Therefore, obviously, this was also a stage of personal fulfillment and overcoming*." This account highlights how professional changes not only meet external needs but also provide a deeper sense of purpose and well-being.

By relating the narrative elements to the emerging categories, this study underscores the complexity of career transitions and provides a theoretical framework that illuminates the underlying processes. This framework not only enriches the understanding of the dynamics of career transitions but also offers practical perspectives for professionals and organizations seeking to support individuals in times of change.

10.4 Conclusions

Career transitions challenge individuals to redesign their professional identities and adapt to continuously changing contexts. This study, grounded in narrative analysis, revealed the depth and complexity of these career transitions, supplying important insights into the individual and organizational factors that shape the experience of change. By integrating narrative elements such as point of view, plot, conflicts and symbolism, with emerging categories, motivation for career change, adaptation to new roles, personal growth and organizational support, this study delivered an innovative perspective to understanding the dynamics that support these trajectories.

The results indicate that the motivation to change career is often associated with internal and external conflicts, reflecting tensions between personal aspirations and the limitations of their context. Adaptation to new roles appears as a dimension mediated by the ability to re-signify skills and integrate new practices in challenging environments, in line with the concepts of proactivity and resilience. Deeply embedded in themes of overcoming adversity and self-discovery, personal growth highlights the relevance of narrative symbols as indicators of personal change. Organizational support proves crucial in mitigating periods of liminality offering structure and validation to the trajectories of individuals in transition.

This study reinforces the contributions of the Conservation of Resources Theory (COR) and the Socio-Emotional Selectivity Theory (SST). It was possible to demonstrate that individual narratives not only reveal the psychological complexities of changes, but also reveal how the organizational and social context influences the success of these transitions. The analysis of the narratives allowed us to understand how individuals construct meanings throughout the process and how they align their personal objectives with external demands, in a dynamic of negotiation throughout the entire career transition process.

With important practical implications, the results indicate that effective transition support strategies must take into account the subjective dimension of individual experiences. Organizational policies that encourage autonomy, flexibility and access to support networks that can increase individuals' self-confidence, and mentoring and coaching programs can promote personal development and adaptation. This study highlights that career transitions, rather than just periods of instability, constitute opportunities for reinvention and development, both for individuals and for the organizations that embrace them.

References

Akkermans, J., da Motta Veiga, S. P., Hirschi, A., & Marciniak, J. (2024). Career transitions across the lifespan: A review and research agenda. *Journal of Vocational Behavior, 148*, 103957. https://doi.org/10.1016/j.jvb.2023.103957

Bandura, A. (1977). Self-efficacy: Toward a unifying theory of behavioral change. *Psychological Review, 84*(2), 191–215. https://doi.org/10.1037/0033-295X.84.2.191

Bardin, L. (com Reto, L. A., & Pinheiro, A.). (2006). Análise de conteúdo. *Edições* 70.

Bergman, B. (2017). *Rising action* [LitCharts LLC]. https://www.litcharts.com/

Bowen, D. E., & Ostroff, C. (2004). Understanding HRM-firm performance linkages: The role of the "strength" of the HRM system. *The Academy of Management Review, 29*(2), 203–221. https://doi.org/10.2307/20159029

Briscoe, J. P., Kaˇse, R., Dries, N., Dysvik, A., Unite, J. A., Adeleye, I., Andresen, M., Apospori, E., Babalola, O., & Bagdadli, S. (2021). Here, there, & everywhere: Development and validation of a cross-culturally representative measure of subjective career success. *Journal of Vocational Behavior*, *130*, 103612. https://doi.org/10.1016/j.jvb.2021.103612

Carstensen, L. L. (1995). Evidence for a life-span theory of socioemotional selectivity. *Current Directions in Psychological Science*, *4*(5), 151–156. http://www.jstor.org/stable/20182356

Carvalho, T. M. D., Silva, C. R. D., & Bianchi, E. M. P. G. (2021). Análise Crítica da Pesquisa Narrativa. *Research, Society and Development*, *10*(8), Artigo 8. https://doi.org/10.33448/rsd-v10i8.17743

Gancho, C. V. (2006). *Como Analisar Narrativas* (9ª edição). Editora Ática.

Godoy, A. S. (1995). Introdução à pesquisa qualitativa e suas possibilidades. *Revista de Administração de Empresas*, *35*(2), Artigo 2. https://doi.org/10.1590/S0034-75901995000200008

Gomes, A. (2003). *A Narrativa enquanto Instrumento de Investigação e de Autoconhecimento. Um estudo da narrativa pessoal de uma professora de educação física acerca da sua experiência no projeto "Férias em Português em Timor Lorosa'e"*. Dissertação de Mestrado em Ciência do Desporto, na Área de Especialização de Desporto de Crianças e Jovens. FCDEF - UP, Porto.

Guan, Y., Arthur, M. B., Khapova, S. N., Hall, R. J., & Lord, R. G. (2019). Career boundarylessness and career success: A review, integration and guide to future research. *Journal of Vocational Behavior, 110*, 390–402. https://doi.org/10.1016/j.jvb.2018.05.013

Halbesleben, J. R., Neveu, J.-P., Paustian-Underdahl, S. C., & Westman, M. (2014). Getting to the "COR" understanding the role of resources in conservation of resources theory. *Journal of Management, 40*(5), 1334–1364. https://doi.org/10.1177/0149206314527130

Hall, D. T., Yip, J., & Ron, K. (2018). Protean careers at work: Self-direction and values orientation in psychological success. *Annual Review of Organizational Psychology and Organizational Behavior*, 5(1), 6.1–6.28. https://doi.org/10.1146/annurev-orgpsych-032117-104631

Hobfoll, S. E., Halbesleben, J., Neveu, J. P., & Westman, M. (2018). Conservation of resources in the organizational context: The reality of resources and their consequences. *Annual Review of Organizational Psychology and Organizational Behavior, 5*, 103–128. https://doi.org/10.1146/annurev-orgpsych-032117-104640

Kendall, J. E., & Kendall, K. E. (2012). Storytelling as a qualitative method for IS research: Heralding the heroic and echoing the mythic. *Australasian Journal of Information Systems, 17*(2), Artigo 2. https://doi.org/10.3127/ajis.v17i2.697

Muylaert, C. J., Sarubbi Jr, V., Gallo, P. R., Neto, M. L. R., & Reis, A. O. A. (2014). Narrative interviews: An important resource in qualitative research. *Revista da Escola de Enfermagem da USP, 48*(spe2), Artigo spe2. https://doi.org/10.1590/S0080-623420140000800027

Ray, R. (2024). *Explore tema, símbolo e motivo com exemplos gratuitos.* storyboardthat. https://www.storyboardthat.com/pt/articles/e/temas-s%C3%ADmbolos-motivos

Roumell, E. A., & Bolliger, D. U. (2024). Transition strategies for career change: A narrative approach. *Career Development Quarterly, 72*(1), 45–63. https://doi.org/10.1002/cdq.12345

Spurk, D., Hirschi, A., & Dries, N. (2019). Antecedents and outcomes of objective versus subjective career success: Competing perspectives and future directions. *Journal of Management, 45*(1), 35–69. https://doi.org/10.1177/0149206318786563

Steck, E. E. (2014). *How to Be a Writer: Adopting style and tone of a publication.* https://blog.quiet.ly/community/adopting-style-tone-of-a-publication/

Steindórsdóttir, B., & Dysvik, A. (2023). Career success through horizontal career transitions: An example from a Norwegian organization. In *Understanding careers around the globe* (pp. 231–239). Edward Elgar Publishing. https://doi.org/10.4337/9781035308415.00036

Steindórsdóttir, B. D., Sanders, K., Arnulf, J. K., & Dysvik, A. (2023). Career transitions and career success from a lifespan developmental perspective: A 15-year longitudinal study. *Journal of Vocational Behavior, 140*, 103809. https://doi.org/10.1016/j.jvb.2022.103809

Sullivan, S. E., & Al Ariss, A. (2019). Making sense of different perspectives on career transitions: A review and agenda for future research. *Human Resource Management Review, 31*(1), Article 100727. https://doi.org/10.1016/j.hrmr.2019.100727

Teelken, C., van der Weijden, I., & Heusinkveld, S. (2024). Reimagining doctorate holders' motivations to make career transitions: Exploring post-PhD career prospects within and outside academia. In M. Thunnissen & P. Boselie (Eds.), *Talent management in higher education* (pp. 99–118). Emerald Publishing. https://doi.org/10.1108/978-1-80262-685-820241006

Twinkl. (2024). *How to write a good story checklist.* Twinkl. https://www.twinkl.pt/resource/au-l-195-how-to-write-a-good-story-checklist

Vilela, E. G., Lopes Borrego, C., & Barroso De Azevedo, A. (2022). Pesquisa Narrativa: Uma proposta metodológica a partir da experiência. *Revista de Estudos Aplicados em Educação, 6*(12), Artigo 12. https://doi.org/10.13037/rea-e.vol6n12.8129

Part VI
Thematic Area: Change Management and Organizational Resilience

Chapter 11
Telework and Well-Being: Investigating the Role of Motivation in Shaping Remote Work Experiences

Rita Couto, Sílvia Lopes, Vânia Sofia Carvalho, and Maria José Chambel

Abstract Teleworking has emerged as a significant paradigm in the organisation of contemporary work worldwide. This work format has gained prominence in Portugal, with 20.2% of the active population engaged in telework by the second half of 2024. Grounded in the Self-Determination Theory (SDT), this study investigates how autonomous and controlled motivation relates to individuals' well-being, both in the workplace (e.g., engagement, burnout) and beyond (e.g., general life satisfaction). A sample of 262 employees participated, with 200 (76.3%) adopting a hybrid telework model and 62 (23.7%) working exclusively remotely. The findings highlight that autonomous motivation positively correlates with work engagement and life satisfaction, whereas controlled motivation is linked to higher levels of burnout, specifically exhaustion and cynicism. These results underscore the relevance of SDT principles in understanding the dynamics of well-being in teleworking environments.

Keywords Telework motivation · Telework well-being · Work engagement · Burnout

11.1 Introduction

Before the COVID-19 pandemic, teleworking was often viewed as a privilege, offered to high-performing employees as a reward or made available to those who voluntarily opted for it, typically under policies promoting well-being and family support (Wang et al., 2023). However, the pandemic abruptly transformed teleworking into a necessity, removing the element of choice and voluntariness. Employees were

R. Couto (✉) · S. Lopes · V. S. Carvalho · M. J. Chambel
CICPSI, Faculty of Psychology, University of Lisbon, Lisboa, Portugal
e-mail: ritacouto@edu.ulisboa.pt

S. Lopes
Faculty of Psychology and Educational Sciences, University of Coimbra, Coimbra, Portugal

C. Machado (ed.), *Human Resource Management in a Disrupted World*, Lecture Notes on Multidisciplinary Industrial Engineering,
https://doi.org/10.1007/978-3-031-89948-5_12

compelled to adapt rapidly to this shift, which redefined their work dynamics and led to new perspectives on productivity and work-life balance (Bhat et al., 2023; Bin et al., 2021; Geldart, 2022). In the post-COVID-19 era, as the nature of (tele)working continues to evolve, it is crucial to examine the underlying dynamics that have shaped new perspectives on teleworking. These perspectives are essential for addressing emerging challenges and seizing opportunities that scientific literature and organisations must recognise. Motivation is a key factor in this transformation, and it plays a pivotal role in shaping teleworkers' experiences and outcomes. However, the study of motivation in teleworking remains underexplored despite its critical impact on job satisfaction, well-being, and fulfilling individuals' basic psychological needs. This gap highlights an urgent need to understand how motivation influences the teleworking experience, particularly as technological advancements redefine the workplace (Gagné et al., 2022).

Given the continuous growth in teleworking, its understanding and impact are essential, both from the worker's point of view and from the organisation's point of view (Faustino et al., 2023). Even in "post-pandemic" times, the prospect of teleworking in the future continues to be a modality that is gaining more and more followers under the condition of the possibility of deciding on its intensity (Aczel et al., 2021; Bérastégui, 2021; Biasi et al., 2022); In Portugal, the growing relevance of teleworking has been verified by the progressive adoption where, in the second half of year 2023, National Statistics Institute (INE), indicated that 18.4% of people belonging to the active population were teleworking. According to the same source, in the second half of 2024, the percentage is 20.2%.

Teleworking is a flexible work modality that balances physical distance from the traditional workplace (often at home) with a portion of typical working hours, ranging from a few hours per week to nearly full-time. This arrangement heavily relies on Information and Communication Technologies (ICT) to facilitate work tasks and colleague interactions. Furthermore, the nature of teleworking is shaped by its intensity and the specific locations where it is carried out (Allen et al., 2015; Beckel et al., 2023).

Scientific literature has focused on the impact of teleworking, covering aspects such as workers' well-being and mental health (Blahopoulou et al., 2022), work performance (Liu et al., 2021), work-family relationships (Lonska et al., 2021) and perceived work demands (Bhat et al., 2023). However, the effects of teleworking on well-being are inconsistent. On the one hand, studies show that those who prefer telework experience less psychological distress and positive well-being and work-life balance than those who work in person (Erro-Garcés et al., 2022; Otsuka et al., 2021). On the other hand, it still seems inconclusive how teleworking affects the well-being of individuals, as research shows us that this modality can both be a challenge (i.e., increasing the performance and motivation of teleworkers), and that it can be seen as a stressor (i.e., one that increases tension levels), creating an environment that is difficult to manage, stimulating the emergence of adverse consequences (e.g., lower levels of well-being at work) (Dias et al., 2022; Perry et al., 2018).

Teleworking necessitates physical and cognitive adjustments in how individuals perform their job functions, and these changes can significantly influence workplace well-being, particularly in terms of burnout and engagement (Sardeshmukh et al., 2012). Beyond the workplace, when teleworking aligns with employees' motivations and fulfils their needs, it is likely to yield more favourable outcomes, including enhanced life satisfaction (Thompson et al., 2022). Despite the challenges associated with teleworking, such as blurred boundaries between personal and professional life and the perception of extended work hours, research has shown that teleworking is often linked to increased engagement (Demerouti et al., 2014; Gerards et al., 2018; Henke et al., 2016; Tavares, 2017; Brummelhuis et al., 2012). However, the impact of teleworking extends beyond work dynamics alone. Understanding its broader implications, particularly on life satisfaction, is crucial, as there is limited scientific evidence exploring the factors and mechanisms of teleworking that influence work-life satisfaction (Tan et al., 2024).

The reasons individuals choose to telework establish the context and conditions for specific behaviours, influencing whether they remain in or leave this work modality (Thompson et al., 2022). Consequently, understanding the motivations for teleworking and its potential consequences in the workplace is essential (Deci et al., 2017; Ren et al., 2021; Valero et al., 2015). The motivations play a pivotal role in shaping teleworkers' behaviours (Ajzen, 1991; Ton et al., 2022), and the ability to predict employees' actions is closely tied to understanding their work motivations (Ren et al., 2021; Zhang, 2019). In alignment with this, recent studies (e.g., Beckel & Fisher, 2022; Nakrošienė et al., 2019; Ngamkroeckjoti et al., 2022; Thompson et al., 2022; Urien, 2023) have underscored the importance of examining the motivations underlying individuals' adoption of teleworking. Further investigation into the factors influencing teleworkers' motivation levels as work culture evolves remains critical (Bhattacharya & Mittal, 2020).

In this context, Self-Determination Theory (SDT) stands out as a robust theoretical framework for understanding human motivation, offering high applicability in organisational settings for predicting individuals' well-being, attitudes, and behaviours (Deci et al., 2017). SDT explains motivation through activation and intention processes, which account for the energy, direction, persistence, and equifinality with which individuals engage in specific actions (Lopes & Chambel, 2017; Ryan & Deci, 2000). The reasons individuals initiate and sustain particular behaviours can vary in the degree to which their motivations are autonomous or controlled, with the quality of these motivations significantly influencing individual outcomes (Deci & Ryan, 2000).

Importantly, SDT provides a framework for understanding the consequences resulting from motivations for teleworking. Evidence suggests a positive relationship between autonomous motivations and well-being (Chambel et al., 2015), including enhanced engagement (Li et al., 2015; Lopes & Chambel, 2017). Conversely, controlled motivations are linked to less favourable well-being outcomes, such as higher levels of burnout (Gagné et al., 2015). These findings reflect the nature of autonomous motivation, which involves engaging in intrinsically rewarding activities aligned with personal characteristics and needs. In contrast, controlled motivation is driven by external pressures or obligations (Deci et al., 2017).

Analysing teleworking through the lens of SDT offers valuable insights and predictions for future research on flexible work arrangements. The context in which work is performed significantly influences individual motivation and outcomes, making SDT a crucial tool for understanding teleworking dynamics (Gillet et al., 2016; McAnally & Hagger, 2024; Ren et al., 2021).

Building on SDT, this study aims to examine the relationship between autonomous motivation, controlled motivation, and individuals' well-being both at and outside work. Workplace well-being will be assessed through engagement (Schaufeli et al., 2002) and burnout (Maslach et al., 1996), while well-being beyond work will be evaluated through general life satisfaction (Diener et al., 1985). This study seeks to fill a gap in the scientific literature by being the first to empirically investigate the motivations for teleworking and their relationships with well-being inside and outside work. Furthermore, the findings are expected to yield practical implications for organisations, enabling them to develop policies that foster autonomous motivation among teleworkers and enhance their overall well-being.

11.1.1 Telework

Teleworking, although dependent on ICT, has been mentioned throughout history, even in times where ICT was non-existent (Allen et al., 2015; Wang et al., 2024). An example of this was in the 1960s and 1970s when the idea of teleworking emerged as a possibility to reduce travel or even as an energy conservation strategy (Kraemer, 1982; Kraemer & King, 1982; Owen, 1962; Wang et al., 2024). More recently, the Covid-19 pandemic has boosted the adoption of teleworking so much that it has reached levels of adherence never before recorded in the entire history of work (Wang et al., 2024). Furthermore, there is evidence that teleworking will be the preferred work arrangement in the future, as its usefulness and practicality in business and economic continuity have been demonstrated in recent years (Dimian et al., 2023). Since teleworking was no longer required, many employees decided to leave this modality, but others have remained teleworking to this day (Kalter et al., 2021; Wang et al., 2024).

Teleworking is a type of work carried out remotely from the organisation's facilities, predominantly from the worker's home, using ICT (Allen et al., 2015; Beckel et al., 2023; Carvalho et al., 2022; Nakrošienė et al., 2019; Sinclair et al., 2020). This modality depends on some characteristics that must be taken into consideration when studying it, namely: (flexibility of) schedule; the extent of teleworking (the length of time or regularity of teleworking); the experience; the possibility of choice (voluntarism or not); the existence of a designated place for it (when this is carried out at home); the presence of people other than your co-workers (during the period of performing their duties), and, finally, whether there is a formalisation of the same, in the organisation/employment contract (Allen et al., 2015; Beckel & Fisher, 2022; Madsen, 2003; Nakrošienė & Butkevičienė, 2016; Nakrošienė et al., 2019).

Additionally, when considering the option for teleworking, a range of individual (e.g., openness to experience, self-discipline), family (e.g., work-family boundaries) and organisational (e.g., work characteristics such as autonomy) characteristics must be taken into account in order to determine the benefits felt in this type of work (Allen et al., 2015; Anderson et al., 2015; Perry et al., 2018; Wang et al., 2021).

11.1.2 Self-determination Theory

Self-Determination Theory (SDT) provides a comprehensive framework for understanding how different types of motivation influence functioning, behaviour, and outcomes (Ryan & Deci, 2000). It posits that individual reasons for initiating and sustaining specific actions vary along a continuum of autonomous to controlled motivations. The quality of these motivations significantly affects attitudes, behaviours, and well-being, particularly in the workplace (Deci & Ryan, 2000). Autonomous motivation, characterised by voluntarism, reflects a sense of autonomy and self-determination in pursuing activities. In contrast, controlled motivation, marked by involuntariness, stems from external pressures or contingencies (Bernhard-Oettel et al., 2013; Dias et al., 2022; Lopes & Chambel, 2014). These distinct motivations are associated with qualitatively different outcomes, including reasons for teleworking (Dias et al., 2022; Ryan & Deci, 2019). According to SDT, the quality of motivation directly influences work outcomes such as well-being, mediated by contextual and individual variables, including perceptions of technology mastery and the nature of the teleworking environment (Dias et al., 2022; Ryan & Deci, 2019).

Motivation quality and the persistence of behaviour are also shaped by the extent to which basic psychological needs (i.e., autonomy, competence, and relatedness) are satisfied (McAnally & Hagger, 2024). These universal needs are essential for cognitive, social, and functional balance. Environments that fail to meet these needs can undermine well-being and overall functioning (Batista & Chambel, 2016; Deci & Ryan, 2000). Autonomy represents the need for individuals to act in alignment with their values and sense of identity (McAnally & Hagger, 2024). Competence involves feeling effective and capable in performing tasks (McAnally & Hagger, 2024; Olafsen et al., 2015). Relatedness reflects the need for social support, acceptance, and belonging in one's environment (McAnally & Hagger, 2024). Satisfaction of these needs is linked to persistence in adaptive behaviours and positive outcomes, including well-being and life satisfaction (Deci & Ryan, 2000; McAnally & Hagger, 2024; Tang et al., 2021).

SDT describes the motivation continuum from the least self-determined (extrinsic motivation) to the most self-determined (intrinsic motivation). At the lowest end is a motivation, characterised by a lack of intention, devaluation of activities, feelings of incompetence, and a sense that actions yield no meaningful results (Ryan & Deci, 2000). Among extrinsic motivations of a controlled nature, external motivation involves engaging in activities to obtain rewards (e.g., salary) or avoid punishment (e.g., unemployment). Introjected motivation reflects internalised external regulations, where actions are driven by pride or avoidance of guilt and shame (Ryan & Deci, 2000). Moving toward more autonomous forms, identified motivation entails recognising and valuing the importance of specific actions (e.g., career advancement). In contrast, integrated motivation aligns external regulations with personal values and life goals (Ryan & Deci, 2000). At the highest end of the continuum, intrinsic motivation arises from genuine interest and enjoyment, leading to engagement and satisfaction in the activity (Ryan & Deci, 2000).

While SDT primarily addresses motivations, it highlights other factors, such as ego involvement and intrinsic interests, that can drive behaviour (Gagné & Hewett, 2024; Ryan & Deci, 2019; Ryan et al., 2022). However, over-reliance on methods that emphasise external regulation can undermine more adaptive forms of motivation, leading to less favourable outcomes (Van den Broeck et al., 2021). Therefore, understanding the dynamics of motivation and the satisfaction of psychological needs is essential for fostering well-being and sustained engagement, especially in contexts like teleworking.

11.1.2.1 Motivations as Reasons for Teleworking

Choosing to telework is an individual decision that reflects a preference for this work mode, driven by a positive attitude. This choice transcends the constraints imposed by the COVID-19 pandemic. Indeed, even before the pandemic, studies had established a link between positive attitudes toward teleworking and a preference for this arrangement (Peters et al., 2004; Urien, 2023). Flexible work practices like teleworking are crucial in applying SDT (McAnally & Hagger, 2024). For instance, working from home can foster greater autonomy and overall satisfaction than working on-site at a traditional workplace (McAnally & Hagger, 2024). Consequently, the work context significantly influences individuals' perceptions of psychological need satisfaction and subsequent motivation quality (McAnally & Hagger, 2024).

Several studies have applied SDT in work contexts, demonstrating that practices fostering autonomous motivation are associated with greater engagement and job satisfaction, as well as lower levels of stress and burnout (Cerasoli et al., 2014; Fernet et al., 2004; McAnally & Hagger, 2024; Raza et al., 2015). Workers who cited autonomous reasons for performing specific tasks and actions have been shown to exhibit more communication and orientation-focused behaviours (Park & Rainey, 2012) and report higher levels of job satisfaction (Omansky et al., 2016). Conversely, workers driven by controlled motivations, such as verbal or monetary rewards, reported reduced engagement and job satisfaction (Li et al., 2015; Roche & Haar, 2020). Additionally, controlled motivations have been linked to increased exhaustion in workers (Lopes et al., 2022; Ryan & Deci, 2000), as well as heightened levels of burnout and psychological tension (Chambel et al., 2015; Fernet et al., 2015).

The anticipated benefits of teleworking serve as antecedents to the motivations for adopting this work arrangement and influence the prevalence of its adoption (Wang et al., 2024). These motivations determine several key aspects of teleworking.

First, they establish whether individuals voluntarily choose this modality, reflecting a sense of voluntarism (i.e., whether they actively request teleworking). Second, they shape the extent or intensity of the desire to telework (i.e., the preference level for this option). Finally, they influence the frequency of teleworking (Hjorthol, 2006; Thompson et al., 2022). Teleworking characteristics, such as increased time for family, greater flexibility, and enhanced autonomy in terms of location and schedule, play a crucial role in shaping the reasons for adopting this work arrangement (Dimian et al., 2023; Wang et al., 2024). These characteristics align with efforts to meet workers' needs, further reinforcing the appeal of teleworking.

It is essential to contextualise motivations within teleworking, as the satisfaction of psychological needs in the workplace is closely linked to autonomous motivation, which in turn leads to adaptive outcomes such as greater engagement, improved work performance, and higher job satisfaction (Slemp et al., 2018; Tafvelin & Sterling, 2018). Individuals may have various motivations for working remotely (Allen et al., 2015; Bailey & Kurland, 2002). Examples of autonomous motivations include the opportunity to define and work during their most productive hours and minimise distractions caused by coworkers (Golden & Veiga, 2008; Martínez-Sánchez et al., 2008; Tremblay & Genin, 2007). Additional motivations involve personal and family demands, such as achieving a better balance between professional and personal life, reducing conflicts between the two domains, increasing productivity, and avoiding the time and stress associated with commuting to a physical workplace (Wang et al., 2024).

Interestingly, research findings (Dimian et al., 2023) indicate that potential challenges such as computer security risks, interruptions, and the necessity of online meetings did not significantly deter participants from choosing telework.

On the other hand, controlled motivations are influenced by various contextual and organisational factors. Examples include external circumstances such as the COVID-19 pandemic, inadequate technological or ergonomic resources, and organisational constraints such as a lack of choice in work modality, inflexible schedules, or insufficiently supportive leadership (McAnally & Hagger, 2024; Park & Cho, 2022; Vleeshouwers et al., 2022; Wang et al., 2023). These factors can undermine the autonomy of teleworkers, driving controlled motivation instead of autonomous motivation.

11.1.3 Work Engagement

Enthusiasm and energy are the sensations that best characterise employees with high levels of engagement (Bakker & Demerouti, 2008). Engagement represents a positive and rewarding mental state toward work, described by three core dimensions: vigour, dedication, and absorption (Schaufeli et al., 2002). Specifically, engagement is marked by higher energy levels and resilience while working (vigour), intense involvement in work with a sense of meaning, enthusiasm, and challenge (dedication), and complete concentration and happiness in work itself (absorption) (Parent-Lamarche, 2022). Regardless of the work setting, engagement is a critical factor for organisations, as it underpins both organisational efficiency and employee well-being (Brummelhuis et al., 2012; Huo et al., 2024). Consequently, human resource strategies should prioritise fostering engagement through talent management initiatives to acquire and retain high-performing employees (Carasco-Saul et al., 2015; Kim & Hyun, 2017; Parent-Lamarche, 2022; Wollard & Shuck, 2011).

However, some aspects of teleworking can influence employee engagement, both positively and negatively, making it essential for organisations to mitigate adverse effects (Wang et al., 2023). The impact of teleworking on engagement remains a topic of active discussion (Huo et al., 2024). On the one hand, teleworking can enhance engagement by providing flexibility, autonomy, and better work-life balance, which foster motivation and sustain commitment to work. On the other hand, it may reduce engagement by increasing work-related pressures and creating negative emotions due to distractions from domestic and social activities, the absence of supervision and support, limited feedback, and higher role ambiguity (Bakker, 2010; Barber & Santuzzi, 2015; Brummelhuis et al., 2012; Chênevert et al., 2022; Huo et al., 2024; Naqshbandi et al., 2024; Peters et al., 2014; Sardeshmukh et al., 2012; Wang et al., 2023).

The flexibility offered by teleworking allows employees to manage their time better, engage in leisure or pleasurable activities, and build resilience to stress (Guimaraes & Dallow, 1999; Hobfoll, 1989; Mann & Holdsworth, 2003; Sardeshmukh et al., 2012). Flexibility, autonomy, control, and work-family balance are key benefits of teleworking that promote motivation and sustain commitment and enthusiasm for work (Huo et al., 2024; Kurland & Bailey, 2000; Parent-Lamarche, 2022; Strauss & Parker, 2014). Intrinsic motivation, in particular, plays a crucial role in engagement as it highlights the inherent value and meaning of the work itself (Huo et al., 2024; Jia & Zhang, 2018; Schaufeli et al., 2002).

When personal interests align with motivations to engage in teleworking, individuals experience congruence between their values and work, enabling them to participate fully and enthusiastically in their activities and yield positive outcomes (Ren et al., 2021). Given that autonomous motivation is one of the most favourable states in the workplace (Mesri et al., 2024), it is expected to be positively related to the engagement of teleworkers.

11.1.4 Burnout

Since Freudenberger introduced the concept of burnout in 1974, its definition and understanding have undergone continuous refinement (Bakker et al., 2023; Demerouti, 2024). Maslach and Jackson (1981) made significant contributions by systematically conceptualising burnout as a syndrome characterised by three dimensions: emotional exhaustion (feelings of being overwhelmed and drained due to work), depersonalisation (a negative, detached response toward others), and personal accomplishment (feelings of incompetence and lack of fulfilment both personally and professionally) (Demerouti, 2024).

Years later, Maslach and colleagues (1996) revised the depersonalisation component, replacing it with cynicism, which they described as a detached attitude toward work in general rather than specifically toward other people (Demerouti, 2024). More recently, in 2019, the World Health Organization (WHO) recognised burnout as an occupational phenomenon, validating its significance as a global workplace issue. The WHO's conceptualisation aligns with the theoretical framework proposed by Maslach et al. (2001) (Demerouti, 2024). Notably, burnout has garnered particular attention in Portugal, one of the few European Union countries to officially recognise it as an occupational disease (Demerouti, 2024; Lastovkova et al., 2018). Burnout is classified as the result of prolonged exposure to emotional and interpersonal stressors in the workplace, characterised by exhaustion, cynicism, and ineffectiveness (Demerouti, 2024; Maslach et al., 2001).

Working remotely, away from company facilities and regular interpersonal interactions, has several disadvantages. These include an increased risk of social and professional isolation (Bentley et al., 2016; Crawford et al., 2011; Faustino et al., 2023; Tavares, 2017), fewer career advancement opportunities (Buomprisco et al., 2021; Crawford et al., 2011; Tavares, 2017), perceptions of overworking (Crawford et al., 2011; Kelliher & Anderson, 2010; Tavares, 2017), and higher levels of presenteeism (Steidelmüller et al., 2020; Tavares, 2017). These factors can create a greater likelihood of negative emotions and feelings detrimental to the well-being of teleworkers. Supporting this, research has shown that individuals who teleworked more than they desired reported worse well-being than those who teleworked less than they preferred (Heiden et al., 2023).

If teleworkers lack a clear preference for their work arrangement, feel forced to telework, or fail to identify reasons for their choice, their experience is likely driven by controlled motivation. This form of motivation has been widely linked to adverse work outcomes, such as turnover intentions, work anxiety, and burnout (Gagné et al., 2015; Gillet et al., 2013, 2016; Ren et al., 2021). Symptoms of burnout—such as physical and emotional exhaustion, diminished cognitive and emotional capacities, and detachment—are rooted in frustration and a perceived inability to continue, leading to a loss of will and energy (Demerouti, 2024; Schaufeli & Taris, 2005; Schaufeli et al., 2020; Shirom & Melamed, 2006).

Because controlled motivation is associated with adverse outcomes, teleworkers who feel pressured to think, act, or behave in ways misaligned with their values and interests may experience burnout. This misalignment leads to dissatisfaction and emotional strain. Consequently, burnout is expected to be positively related to controlled motivation (Ren et al., 2021).

11.1.5 Life Satisfaction

Life satisfaction is a cognitive process through which individuals evaluate their perception of satisfaction with their current quality of life (Diener et al., 1985). Achieving general well-being requires consideration of multiple dimensions that collectively contribute to an integrated state of satisfaction and fulfilment. These dimensions include emotional, occupational, physical, social, intellectual, and spiritual aspects, supporting a healthy and balanced life (Diener et al., 1985; Hettler, 1986; Miller, 2005). Work is an integral part of human life, and its influence on life satisfaction is unavoidable, as even minor disruptions in the work domain can significantly affect subjective well-being and mental health (Guest, 2017). For instance, flexible work arrangements have considerably impacted workers' well-being (Ray & Pana-Cryan, 2021).

Research indicates that teleworking is associated with enhanced life satisfaction. This is mainly due to the flexibility it offers, allowing individuals to choose where and when they work, thereby increasing autonomy and personal control (Anderson et al., 2015; Carillo et al., 2021; Kortsch et al., 2022; Tan et al., 2024). Teleworking also meets individuals' needs, such as balancing professional and family life, by enabling them to organise work around their priorities. This flexibility improves quality of life and overall well-being (Macêdo et al., 2020; Petcu et al., 2023; Ríos Villacorta et al., 2024). However, life satisfaction may decrease in cases where teleworking extends beyond established business hours or becomes excessive (Tan et al., 2024; Yang et al., 2023).

Empirical evidence further supports the notion that workers with higher levels of engagement report greater life satisfaction (Diener et al., 1985), while those experiencing burnout report reduced life satisfaction (Demerouti et al., 2001).

Concerning the motivations for teleworking, individuals who perceive complete freedom of choice and can structure their work to achieve self-determined goals experience greater pleasure and satisfaction at work (Rietveld et al., 2021). Therefore, it is reasonable to predict that life satisfaction will positively correlate with autonomous motivation and engagement while negatively correlating with controlled motivation and burnout.

11.1.6 The Role of Organisations in Promoting Motivations and Its Consequences in Terms of Engagement, Burnout and General Satisfaction with Life

To feel engaged and fulfilled, individuals must perceive their work as meaningful. According to SDT principles, organisations should strive to create meaningful jobs to meet employees' basic psychological needs (Gagné & Hewett, 2024; Martela et al., 2021). The attribution of meaning to work can be influenced by social, individual, organisational, and leadership factors, all of which contribute to satisfying these needs

(Gagné & Hewett, 2024; Lysova et al., 2019; Martela et al., 2021). These factors, in turn, shape the nature and quality of motivations. Consequently, organisations offering teleworking opportunities must understand the motivations and needs of their employees to act more effectively (Beckel et al., 2023).

Aligned with SDT, creating environments that support psychological needs increases the likelihood of fostering desirable behaviours and achieving adaptive outcomes (McAnally & Hagger, 2024). In teleworking, critical factors include the availability of organisational resources, the quality of professional relationships, the efficiency of digital infrastructure, work-life balance, and the potential for professional isolation. These factors, which are constantly evolving, directly influence the satisfaction of employees' basic needs and their ability to achieve work goals, ultimately enhancing workplace well-being (Gagné et al., 2022; Stoian et al., 2022). Beckel et al. (2023) emphasise that organisations should carefully consider how employees choose to telework, highlighting the importance of analysing motivations to design effective strategies tailored to workers' needs.

SDT identifies autonomy support as a key tool for promoting motivation and positive behaviours. Organisations and leaders are encouraged to build motivational and social environments that meet employees' psychological needs (McAnally & Hagger, 2024). Autonomy, a central characteristic of teleworking, should be supported through proportional organisational actions, such as allowing employees to choose how they engage in tasks, providing clear explanations of work responsibilities, and encouraging employees to express opinions about how tasks are performed (McAnally & Hagger, 2024; Reeve, 2006).

Research suggests that employees' perception of autonomy support from social agents—such as leaders and colleagues—leads to positive outcomes, including increased autonomous motivation and reduced controlled motivation (Dahling & Lauricella, 2017; Gillet et al., 2013; van Hooff & van Hooff, 2017). To achieve mutually beneficial outcomes, leaders should identify motivational elements that best align with strategies to promote employee engagement, reduce disinterest, and enhance job satisfaction (McAnally & Hagger, 2024). Employees motivated to adopt specific behaviours or strategies are more likely to experience greater engagement, resulting in improved performance and job satisfaction (Galletta et al., 2019; McAnally & Hagger, 2024; Moon et al., 2019; Slemp et al., 2021). Conversely, organisations that neglect the motivations for teleworking may inadvertently foster controlled motivation, leading to unmet needs and adverse outcomes such as burnout.

Additionally, well-being encompasses multiple life domains that interact with one another. Considering well-being outside of work is equally important, mainly as teleworking offers flexibility to align work with personal preferences, significantly influencing life satisfaction (Tan et al., 2024). Practices aimed at optimising teleworking should also address this dimension, recognising teleworking's potential to deliver positive outcomes by leveraging its inherent advantages (Tan et al., 2024).

11.2 Materials and Methods

11.2.1 Study Design

The present study adopts an empirical quantitative approach, utilising descriptive and correlational analyses to examine the data. Data collection was conducted through an online survey administered in August 2024. Consequently, the sample was selected using a convenience sampling method, making it non-probabilistic (Montero & León, 2007).

11.2.2 Participants

The sample comprised 262 individuals, including participants with hybrid work (76.3%) and participants who worked 100% in remote work (23.7%). The majority of participants are women (65.5%), aged between 21 and 70 years ($M = 36.40$, $SD = 9.71$), and possess a master's degree (64.8%). Full-time work was the most reported (96.6%), and most participants had an effective contract (71%). Additionally, most participants have worked in the present organisation between 3 and 5 years (28.6%), followed by 1–2 years (22.9%).

11.2.3 Variables and Instruments

The research protocol included the following instruments: (1) a sociodemographic questionnaire; (2) a scale of motivations for teleworking, developed based on Lopes and Chambel (2014) and Gagné et al. (2015); (3) the Ultra-Short Measure for Work Engagement developed by Schaufeli et al. (2019); (4) the Maslach Burnout Inventory de Maslach al. (1996); and (5) the Satisfaction With Life Scale developed by Diener et al. (1985). More information on the instruments used is outlined below.

11.2.3.1 Telework Motivations

This scale, headed by the phrase "Why do you telework?", is composed of 12 items, grouped into four factors that measure intrinsic motivation ("*Because I like teleworking.*"); integrated regulation ("*Because teleworking is what best suits my personal needs and goals.*"); introjected regulation ("*Because I need to prove to myself that I can telework.*"); and external regulation ("*Teleworking is a way to avoid control and surveillance by supervisor(s)*".). This study was based on the motivational continuum, from autonomous motivation (intrinsic motivation and integrated regulation) to controlled motivation (introjected motivation and external regulation).

The response format was a Likert scale from 1 to 7, with one corresponding to "not at all" and seven corresponding to "completely". High scores for one type of motivation indicate that this factor better reflects the worker's reasons for teleworking.

11.2.3.2 Engagement

To measure engagement, we used a 3-item version of the Utrecht Work Engagement Scale: (1) "*At my work, I feel bursting with energy*" (corresponds to vigour); (2) "*I am enthusiastic about my job*" (dedication); (3) "*I am immersed in my work*" (absorption). The participants answered the items using a seven-point Likert scale, ranging from 1 ("never") to 7 ("always, every day").

11.2.3.3 Burnout

Burnout was measured using the Maslach Burnout Inventory (Maslach et al., 1996) used in previous studies with Portuguese samples (e.g., Lopes et al., 2022; Lopes & Chambel, 2017). The measure included two dimensions: exhaustion (5 items, e.g., "*I feel emotionally drained by my work*") and cynicism (5 items, e.g., "*I question the significance of my work*"). Respondents answered the items of both subscales on a seven-point scale that ranged from "never" (1) to "every day" (7).

11.2.3.4 Life Satisfaction

Satisfaction with life was assessed with the 5-item scale of Diener et al. (1985) (e.g., "*I am satisfied with my life*"). Items were scored on a seven-point rating scale that ranged from "totally disagree" (1) to "totally agree" (7).

11.2.4 Procedure

The questionnaire was hosted on the Google Forms platform to gather responses from participants. On average, completing the form took approximately 12 min. Participants provided their consent by agreeing to an informed consent form. Their anonymity was guaranteed, and they were offered the opportunity to receive feedback. No monetary or otherwise incentives were provided for participation in this study.

11.2.5 Data Analysis

Using the SPSS 29.0 software, we conducted an exploratory factor analysis for each measure used. Subsequently, we computed descriptive statistics (means and standard deviations) and conducted a reliability analysis to assess the internal consistency of the study measures using Cronbach's alpha. Intercorrelations among the study variables were also calculated using SPSS 29.0. To test the hypotheses, we applied model 4 of the PROCESS macro developed by Hayes (2013), which employs the bootstrapping method to estimate mediation effects. Specifically, we used a bootstrap approach with 5,000 samples to calculate 95% bias-corrected confidence intervals (CIs) for the standardised indirect effects.

11.3 Results

11.3.1 Exploratory Factor Analysis

As mentioned earlier, we performed an exploratory factor analysis for each variable under study. The item responses associated with each variable were analysed using principal components exploratory factor analysis with varimax rotation. To determine the number of factors to retain, we applied multiple criteria (Ford et al., 1986; Kim & Mueller, 1978; Stevens, 1992), including eigenvalues exceeding 1.0 and explained variance greater than 60%. Additionally, we aimed to retain only items with loadings of 0.5 or higher on the intended factor and below 0.3 on other factors.

Regarding the measurement of telework motivations, the exploratory factor analysis revealed a three-factor solution comprising 12 items (Table 11.1). The eigenvalues for the factors were as follows: factor 1 (4.34), factor 2 (2.26), and factor 3 (1.05), collectively accounting for 63.85% of the variance. The first factor represented autonomous motivation, the second reflected introjected regulation, and the third corresponded to external regulation. Given that the exploratory factor analysis indicated a three-factor solution, a confirmatory factor analysis was conducted using AMOS 29.0 to test a two-factor model. In this model, autonomous motivation was represented as a first-order latent variable, while introjected regulation and external regulation were combined into a single first-order latent variable labelled controlled motivation. The two-factor model tested was based on the motivational continuum, from autonomous motivation (intrinsic motivation and integrated regulation) to controlled motivation (introjected motivation and external regulation) as predicted in the self-determination theory (Deci & Ryan, 2000). The two-factor model demonstrated a good fit to the data [χ^2 (50) = 89.00, $p < 0.01$, SRMR = 0.05; CFI = 0.97; IFI = 0.97; RMSEA = 0.06].

The principal components analysis for the work engagement measure identified a single-factor solution, accounting for 68.74% of the variance, with factor loadings ranging from 0.71 to 0.89.

Table 11.1 Exploratory factorial analysis with varimax rotation

	Factor 1	Factor 2	Factor 3
Autonomous motivation			
1. Because teleworking is what best suits my personal needs and goals	0.785	0.011	0.022
2. Because teleworking gives me a better quality of life	0.872	0.008	−0.112
3. It allows me more flexibility in balancing other areas of my life (for example, balancing work and family, domestic tasks, and studies)	0.747	−0.012	−0.240
4. Because teleworking gives me more well-being and satisfaction	0.888	0.023	−0.036
5. Because I like teleworking	0.878	−0.079	−0.044
6. Because I prefer to telework	0.869	−0.050	0.084
Introjected regulation			
7. Because I need to prove to myself that I can telework	0.031	0.524	0.363
8. Because if I could not telework, I would feel ashamed	−0.009	0.863	−0.047
9. I will feel bad about myself if I cannot telework	−0.072	0.779	0.267
External regulation			
10. Teleworking is a way of pleasing other people (for example, my superiors, colleagues, family, clients…)	−0.117	0.110	0.652
11. Teleworking is a way to avoid control and surveillance by supervisor(s)	−0.032	0.010	0.707
12. Teleworking allows me to avoid being criticised by others (e.g., my superiors, colleagues, my family, clients…)	−0.005	0.284	0.673
% Variance explained	36.19%	18.84%	8.82%

Concerning the burnout measure, the principal components analysis yielded a two-factor solution, explaining 68.07% of the variance. The first factor represented exhaustion, while the second captured cynicism, with factor loadings between 0.53 and 0.83.

Similarly, the principal components analysis for the life satisfaction measure revealed a single-factor solution, accounting for 68.48% of the variance, with factor loadings ranging from 0.74 to 0.91.

The Table 11.1 demonstrates the exploratory factor analysis. The twelve items used to measure telework motivations were distributed across three factors. The first factor represents autonomous motivation, the second introjected regulation, and the third corresponds to external regulation.

Table 11.2 Descriptive statistics and reliability analysis

Variables	Range	Mean	Standard Deviation	Cronbach Alpha
AM	1–7	6.01	1.15	0.91
CM	1–7	1.64	0.75	0.66
Engagement	1–7	4.34	1.05	0.76
Exhaustion	1–7	2.38	1.47	0.91
Cynicism	1–7	1.64	1.16	0.77
LS	1–7	4.98	1.09	0.87

Note AM = Autonomous Motivation; CM = Controlled Motivation; LS = Life Satisfaction

11.3.2 Reliability Analysis and Descriptive Statistics

Table 11.2 displays the studied variables' means, standard deviations, and reliability analysis. The reliability analysis showed that the variables of this study had an acceptable Cronbach's alpha.

On a seven-point Likert scale, higher mean scores were observed for autonomous motivation (M = 6.01; SD = 1.15), engagement (M = 4.34; SD = 1.05), and life satisfaction (M = 4.98; SD = 1.09). These findings suggest that autonomous motivation is the most salient motivation for teleworking rather than controlled motivation, accompanied by moderate engagement and life satisfaction.

In contrast, scores for controlled motivation (M = 1.64; SD = 0.75), exhaustion (M = 2.38; SD = 1.47), and cynicism (M = 1.64; SD = 1.16) suggest that employees experience some degree of controlled motivation and perceive moderate levels of exhaustion and cynicism in their work environment.

The Table 11.2 describes descriptive statistics and reliability analysis. The table shows an acceptable Cronbach's alpha for each variable. Autonomous motivation scored higher than controlled motivation, and the levels of engagement, life satisfaction and burnout were moderate.

11.3.3 Correlations

Observing Table 11.3, the results showed a positive correlation between sex with autonomous motivation ($r = 0.141$, $p < 0.05$), and with exhaustion (r = 0.196, $p <$ 0.01). Also, the findings showed a negative correlation between sex with controlled motivation ($r = -0.133$, $p < 0.05$) and engagement ($r = -0.188$, $p < 0.01$). Overall, these findings suggest that women present higher autonomous motivation than men. In contrast, men feel more controlled motivation than women. The results also indicate that men feel more engaged at work, and women feel more exhausted.

Concerning age, results demonstrate a negative relationship with exhaustion ($r = -0.146$, $p < 0.05$), cynicism ($r = -0.150$, $p < 0.05$), and a positive relationship

Table 11.3 Correlation analysis

	1	2	3	4	5	6	7	8	9	10	11
1. SEX											
2. AGE	−0.199**										
3. EDU	0.086	−0.097									
4. WF	0.045	−0.077	−0.034								
5. WL	0.011	−0.108	−0.059	0.006							
6. TR	−0.159*	0.491**	−0.044	−0.147*	0.009						
7. AM	0.141*	−0.043	−0.061	0.097	0.042	−0.040					
8. CM	−0.133*	−0.057	−0.028	0.004	−0.070	0.005	−0.103				
9. ENG	−0.188**	0.147*	0.082	0.112	−0.017	0.008	0.129*	−0.071			
10. EX	0.196**	−0.146*	−0.093	−0.140*	−0.064	0.108	−0.029	0.183**	−0.442**		
11. CYN	0.055	−0.150*	−0.033	−0.085	−0.025	0.033	−0.124*	0.279**	−0.530**	0.638**	
12. LS	0.020	0.145*	0.216**	0.113	0.020	0.002	0.132*	−0.158*	0.385**	−0.326**	−0.367**

Note **$p < 0.01$; *$p < 0.05$.; EDU = Education; WF = Work Format; WL = Workload; TR = Tenure; AM = Autonomous Motivation; CM = Controlled Motivation; ENG = Engagement; EX = Exhaustion; CYN = Cynicism; LS = Life Satisfaction with life satisfaction ($r = 0.145$, $p < 0.05$). Those results indicate that younger participants feel more exhausted and cynical, unlike older participants, who feel more life satisfaction.

Additionally, people with more qualifications may be more satisfied with their lives than those with fewer academic qualifications ($r = 0.216, p < 0.01$). Concerning the work format, this variable was negatively related to exhaustion ($r = -0.140$ $p < 0.01$), indicating that workers in the hybrid format report feeling more exhausted than those who exclusively perform telework.

As anticipated, the relationship between autonomous motivation and controlled motivation was not statistically significant ($r = -0.103$, $p > 0.05$). Autonomous motivation showed a positive correlation with engagement ($r = 0.129, p < 0.05$) and life satisfaction ($r = 0.132$, $p < 0.05$), suggesting that higher levels of autonomous motivation are associated with greater well-being both at work (e.g., engagement) and outside of work (e.g., life satisfaction). Additionally, autonomous motivation was negatively correlated with cynicism ($r = -0.124$, $p < 0.05$), indicating that higher autonomous motivation is linked to lower levels of cynicism.

On the other hand, controlled motivation was positively correlated with exhaustion ($r = 0.183$, $p < 0.05$) and cynicism ($r = 0.279$, $p < 0.05$), and negatively correlated with life satisfaction ($r = -0.158$, $p < 0.05$). These findings imply that higher levels of controlled motivation are associated with lower well-being at work (evidenced by increased burnout indicators) and reduced well-being outside of work (e.g., diminished life satisfaction).

The Table 11.3 presents a correlation analysis. Autonomous motivation was associated with more positive outcomes and controlled motivation was associated with negative outcomes. Sociodemographic variables were also presented.

11.3.4 Regression Analysis with Testing Hypotheses

Following the descriptive analysis and correlation results, we obtained a broad idea of the pattern of the relationships among all the dimensions. Using the PROCESS macro in SPSS IBM Statistics 29.0 software with model 4, we completed a regression analysis and tested the hypotheses.

The results show that the more autonomous motivation for teleworking, the more engagement ($\beta = 0.13, p < 0.05$), the more satisfaction with life ($\beta = 0.13, p < 0.05$) and less cynicism in the work (β = -0.12, $p < 0.05$). Engagement and cynicism at work also increase and decrease life satisfaction, respectively ($\beta = 0.24, p < 0.01$; β = -0.15, $p < 0.05$.)

The results of autonomous motivation are illustrated. The Fig. 11.1 shows that increased autonomous motivation is related to increased engagement and life satisfaction. It also demonstrates how life satisfaction is positively related to engagement and negatively related to cynicism.

On the other hand, the more controlled motivation for teleworking, the more exhaustion ($\beta = 0.18$, $p < 0.01$), the more cynicism ($\beta = 0.13$, $p < 0.05$) and less

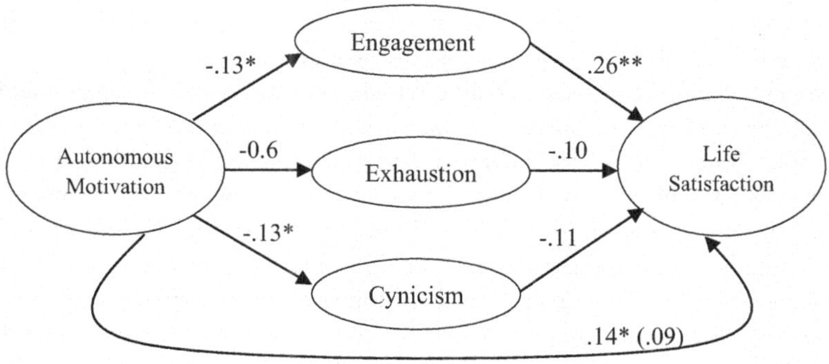

Fig. 11.1 Autonomous motivation results

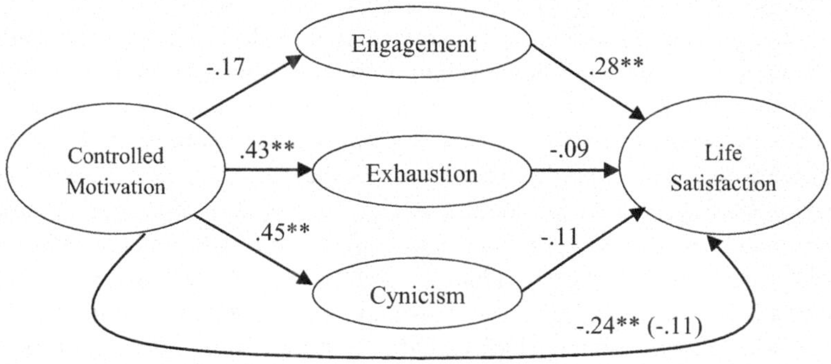

Fig. 11.2 Controlled motivation results

satisfaction with life ($\beta = 0.28, p < 0.01$). Additionally, work engagement increases life satisfaction ($\beta = 0.26, p < 0.01$).

Controlled motivation results are illustrated. The Fig. 11.2 shows that increased controlled motivation is related to increased exhaustion and cynicism. It also demonstrates that work engagement increases life satisfaction.

11.4 Discussion

This study investigated teleworking workers' motivations and well-being based on the assumptions of the Self-Determination Theory (Deci & Ryan, 2000). The results reinforce the relevance of this approach to understanding the different forms of motivation and their impacts on teleworkers' well-being, particularly in indicators such as engagement, burnout, and life satisfaction.

In general, the aforementioned results reinforce the applicability of the fundamental principles of SDT in the context of teleworking. These results show that different types of motivation (autonomous vs. controlled) are linked to different impacts on the well-being of individuals who work remotely.

The positive association between autonomous motivation, work engagement, and life satisfaction and the negative association with cynicism (indicating that higher autonomous motivation is linked to lower levels of cynicism) strengthens the central principle of SDT, which suggests that the feeling of autonomy and self-determination promoted by autonomous motivation contributes to well-being (Chambel et al., 2015; Lopes & Chambel, 2014). Once individuals behave with a full sense of volition, freedom and choice (i.e. higher levels of autonomous motivation; higher voluntariness), they are more likely to experience more positive states and show better outcomes (i.e., work engagement, life satisfaction) than individuals who behave with a feeling of pressure and control that makes them feel distant from their work (i.e. cynicism) (Lopes et al., 2022; Ryan & Deci, 2019). As such, our first hypothesis was partially confirmed (e.g., autonomous motivation is positively related to engagement and life satisfaction and negatively related to both dimensions of burnout). The fact that we did not find significant relationships between autonomous motivation and exhaustion can be explained by the perceived moderate levels of exhaustion in their work environment (M = 2.38; SD = 1.47), which generally does not present a negative result for the sample in question. Furthermore, future studies related to teleworking should consider the context in which teleworking is carried out because the adjustment between an individual and his/her work environment affects their well-being and job satisfaction (Andela & van Der Doef, 2019; Gander et al., 2020; Heiden et al., 2023). However, despite this non-significant relationship observed, it should be noted that the findings indicated that autonomous motivation was the most salient motivation for teleworking rather than controlled motivation (M = 6.01; SD = 1.15; M = 1.64, SD = 0.75, respectively). Therefore, we can assume that teleworking is beneficial for workers, which is in line with other studies (e.g., Allen et al., 2015; Charalampous et al., 2019; Hallman et al., 2021; Heiden et al., 2021, 2023; Widar et al., 2021).

On the other hand, the positive relationship between controlled motivation and burnout dimensions (exhaustion and cynicism) supports the idea defended by the SDT that externally regulated forms of motivation can result in negative psychological consequences, which are reflected in well-being inside and outside the workplace (i.e., less work engagement and life satisfaction) (Fernet et al., 2015; Gagné et al., 2018; Lopes et al., 2022; Ryan & Deci, 2000). This result is in line with our expectations, and we can say that our second hypothesis was supported (i.e., controlled motivation is positively related to exhaustion and cynicism and negatively related to engagement and life satisfaction). Overall, the current findings demonstrate the importance of this study, suggesting that the type of motivation for teleworking provides an important background for understanding the different results in the context of teleworking.

The results indicate that older participants are more engaged at work and less exhausted and cynical. This data coincides with previous studies that indicate that work engagement increases with age (Mori et al., 2024). This pattern may be associated with accumulated experience and a more remarkable ability to manage the challenges of remote work. In addition, the present study also revealed that participants with higher levels of academic qualifications have greater life satisfaction. This result suggests that access to knowledge and development opportunities contributes to a more positive view of quality of life (An et al., 2020; Salinas-Jiménez et al., 2011).

In a recent study by Gaspar et al. (2024), for professionals in hybrid work, age, marital status, psychosocial environment, professional engagement, and stress management skills best explained psychosocial risks related to mental health. The results of the present study also allow us to add a new variable to this list of factors, i.e., sex. More precisely, the current study's findings indicated that women, compared to men, presented higher autonomous motivation and exhaustion, and lower controlled motivation and engagement. Gender is a factor that influences teleworking choices, indicating that woman seems to be more autonomously motivated to telework (Aboelmaged & Elamin, 2009; Wang et al., 2024). This result may be related to the dual role played by many women, who juggle professional and family responsibilities, resulting in higher levels of burnout (Allen et al., 2015). Furthermore, those gender-related differences are not something new. Previous studies have demonstrated that telework can provide enhanced gender inequalities (Asgari et al., 2014, 2023; Cannito & Scavarda, 2020; Neculaesei & Tocar, 2023; Zhang et al., 2020).

Regarding the work format, the present study showed that individuals in a hybrid work regime reported higher levels of exhaustion than individuals who exclusively perform telework. This finding is inconsistent with previous studies that found hybrid work is the best combination of the two types of work (Urien, 2023) and can have a burnout-reducing effect (Höcker et al., 2024). This finding can be explained since most of the participants in the current study were women (65.5%), and the results demonstrated that they reported exhaustion levels more. This data also suggests that the elimination of frequent travel and the possibility of fully managing your work environment at home can contribute to reducing emotional exhaustion (Uru et al., 2022). On the other hand, it is also not known whether these participants had flexible working hours. This could be another explanation, as greater flexibility is associated with fewer burnout symptoms (Kaduk et al., 2019). Therefore, future studies should consider this variable.

11.5 Limitations and Future Research

This study has limitations that should be mentioned. First, its cross-sectional design limits the ability to draw causal inferences. Longitudinal studies, which assess variables at multiple time points, would provide a more comprehensive understanding of teleworking and the dynamic relationships between the variables examined in this study (Taris & Kompier, 2014). Second, the reliance on self-report questionnaires introduces the potential for common method bias and social desirability effects.

which could influence the validity of the results (Podsakoff et al., 2003). Furthermore, the imbalance in sample sizes between hybrid workers ($n = 200$) and fully remote workers ($n = 62$) constitutes another limitation, as it may affect the robustness and generalizability of the comparisons between work modalities. Finally, the scale used to measure motivations for teleworking requires further psychometric validation to ensure reliability and validity across diverse populations. More specifically, the study was conducted in Portugal, a European country with distinct cultural and organisational contexts, which could limit the extent to which the results can be applied to other regions. Future research could expand on this by exploring teleworking practices in a variety of cultural and economic settings to provide a broader understanding of its impact on employees. Addressing these limitations in future research will enhance the understanding of teleworking's implications for employees and organisations.

11.6 Practical Implications

Based on the results obtained, several practical recommendations can be proposed to help organisations implement policies and practices that promote motivation and well-being in teleworking, considering the differences in gender, age, qualifications, and work format.

Organisations should consider offering full teleworking arrangements for functions that do not require regular physical presence, ensuring employees have adequate conditions to work remotely, such as quality equipment and access to adequate information technologies (Golden & Veiga, 2008; Tremblay & Genin, 2007). Therefore, organisations should adopt policies that promote a balance between work and personal life, such as flexible working hours and parental support initiatives, paying particular attention to women. Furthermore, it is crucial to develop targeted mental health programs to reduce the impact of burnout on women, recognising their contributions and creating an inclusive and equitable environment (Purvanova & Muros, 2010).

On the other hand, organisations should value more experienced workers, providing them with opportunities to mentor younger colleagues, thus promoting the exchange of knowledge and the construction of positive interpersonal relationships in the virtual environment (Martínez-Sánchez et al., 2008; Schaufeliet al., 2002). In the same way, organisations must invest in ongoing training and personalised career plans, allowing employees to develop skills aligned with their interests and objectives (Tremblay & Genin, 2007).

Additionally, when teleworking, it is essential to communicate clear and realistic goals that allow employees to realise the positive impact of their work. This practice reinforces engagement by linking individual tasks to broader organisational and social objectives (Golden & Veiga, 2008). From a practical point of view, providing continuous training, even remotely, can increase the perception of competence and promote autonomous motivation, resulting in higher levels of engagement (Tremblay & Genin, 2007).

To decrease controlled motivation, overly intrusive remote supervision practices should be avoided, and goals adjusted to the reality of teleworking should be established (Demerouti, 2024). Furthermore, taking regular breaks, setting clear boundaries between work and personal life, and offering online psychological support can help mitigate teleworking exhaustion (Parent-Lamarche, 2022). To prevent cynicism about work, it is suggested that opportunities for virtual social interactions, such as informal meetings or collaborative projects, be created (Tremblay & Genin, 2007).

11.7 Conclusion

This study contributes to deepening knowledge about the motivational dynamics in teleworking, offering relevant insights for implementing organisational practices that enhance well-being and productivity in remote working. The results reinforce the relevance of SDT to understanding different forms of motivation and their impacts on the well-being of teleworkers, particularly in indicators such as engagement, burnout, and life satisfaction. In short, autonomous motivation is associated with more favourable outcomes, and controlled motivation is associated with more negative outcomes. Therefore, organisations must implement policies that promote the autonomous motivation of teleworkers.

References

Aboelmaged, M. G., & Elamin, A. M. (2009). Teleworking in United Arab Emirates (UAE): An empirical study of influencing factors, facilitators, and inhibitors. *International Journal of Business Science and Applied Management, 4*(1), 36–49. https://doi.org/10.69864/ijbsam.4-1.36

Aczel, B., Kovacs, M., Van Der Lippe, T., & Szaszi, B. (2021). Researchers working from home: Benefits and challenges. *PLoS ONE, 16*(3), e0249127. https://doi.org/10.1371/journal.pone.0249127

Ajzen, I. (1991). The theory of planned behavior. *Organizational Behavior and Human Decision Processes*. https://doi.org/10.1016/0749-5978(91)90020-T

Allen, T. D., Golden, T. D., & Shockley, K. M. (2015). How effective is telecommuting? Assessing the status of our scientific findings. *Psychological Science in the Public Interest, 16*(2), 40–68. https://doi.org/10.1177/1529100615593273

An, H. Y., Chen, W., Wang, C. W., Yang, H. F., Huang, W. T., & Fan, S. Y. (2020). The relationships between physical activity and life satisfaction and happiness among young, middle-aged, and older adults. *International Journal of Environmental Research and Public Health, 17*(13), 4817. https://doi.org/10.3390/ijerph17134817

Andela, M., & van Der Doef, M. (2019). A comprehensive assessment of the person–environment fit dimensions and their relationships with work-related outcomes. *Journal of Career Development, 46*(5), 567–582. https://doi.org/10.1177/0894845318789512

Anderson, A. J., Kaplan, S. A., & Vega, R. P. (2015). The impact of telework on emotional experience: When, and for whom, does telework improve daily affective well-being? *European Journal of Work and Organizational Psychology, 24*(6), 882–897. https://doi.org/10.1080/1359432X.2014.966086

Asgari, H., Jin, X., & Mohseni, A. (2014). Choice, frequency, and engagement: Framework for telecommuting behavior analysis and modeling. *Transportation Research Record, 2413*(1), 101–109. https://doi.org/10.3141/2413-11

Asgari, H., Gupta, R., & Jin, X. (2023). Impacts of COVID-19 on future preferences toward telework. *Transportation Research Record, 2677*(4), 611–628. https://doi.org/10.1177/03611981221115078

Bailey, D. E., & Kurland, N. B. (2002). A review of telework research: Findings, new directions, and lessons for the study of modern work. *Journal of Organizational Behavior: THe International Journal of Industrial, Occupational and Organizational Psychology and Behavior, 23*(4), 383–400. https://doi.org/10.1002/job.144

Bakker, A. B. (2010). *Engagement and "job crafting": Engaged employees create their own great place to work*. Edward Elgar Publishing. https://doi.org/10.4337/9781849806374.00027

Bakker, A. B., & Demerouti, E. (2008). Towards a model of work engagement. *Career Development International, 13*(3), 209–223. https://doi.org/10.1108/13620430810870476

Bakker, A. B., de Vries, J. D., & Dollard, M. F. (2023). Burnout: A review of theory and measurement. *International Journal of Environmental Research and Public Health, 20*(5), 3881. https://doi.org/10.3390/ijerph20053881

Barber, L. K., & Santuzzi, A. M. (2015). Please respond ASAP: Workplace telepressure and employee recovery. *Journal of Occupational Health Psychology, 20*(2), 172. https://doi.org/10.1037/a0038278

Batista, F., & Chambel, M. J. (2016). Work engagement: Individual needs and resources as drivers for workplace well-being. *Psicologia: Reflexão e Crítica, 29*(1), Article 47. https://doi.org/10.1186/s41155-016-0047-3

Beckel, J. L., & Fisher, G. G. (2022). Telework and worker health and well-being: A review and recommendations for research and practice. *International Journal of Environmental Research and Public Health, 19*(7), 3879. https://doi.org/10.3390/ijerph19073879

Beckel, J. L., Kunz, J. J., Prasad, J. J., Finch, H. M., & Kaldahl, K. N. (2023). The impact of telework on conflict between work and family: A meta-analytic investigation. *Occupational Health Science*. https://doi.org/10.1007/s41542-023-00158-8

Bentley, T. A., Teo, S. T. T., McLeod, L., Tan, F., Bosua, R., & Gloet, M. (2016). The role of organisational support in teleworker wellbeing: A socio-technical systems approach. *Applied Ergonomics, 52*, 207–215. https://doi.org/10.1016/j.apergo.2015.07.019

Bérastégui, P. (2021). Teleworking in the aftermath of the Covid-19 pandemic: Enabling conditions for a successful transition. *ETUI Research Paper-Policy Brief*.

Bernhard-Oettel, C., Rigotti, T., Clinton, M., & de Jong, J. (2013). Job insecurity and well-being in the temporary workforce: Testing volition and contract expectations as boundary conditions. *European Journal of Work and Organizational Psychology, 22*(2), 203–217. https://doi.org/10.1080/1359432X.2011.647409

Bhat, Z. H., Yousuf, U., & Saba, N. (2023). Revolutionizing work-life balance: Unleashing the power of telecommuting on work engagement and exhaustion levels. *Cogent Business & Management, 10*(2). https://doi.org/10.1080/23311975.2023.2242160

Bhattacharya, S., & Mittal, P. (2020). The impact of individual needs on employee performance while teleworking. *Australasian Accounting, Business and Finance Journal, 14*(5), 5–22. https://doi.org/10.14453/aabfj.v14i5.2

Biasi, P., Checchi, D., & De Paola, M. (2022). Remote working during COVID-19 outbreak: Workers' well-being and productivity. *Politica Economica*. https://doi.org/10.1429/104443

Bin, W., Liu, Y., Qian, J., & Parker, S. K. (2021). Achieving effective remote working during the COVID-19 pandemic: A work design perspective. *Applied Psychology, 70*, 16–59. https://doi.org/10.1111/apps.12290

Blahopoulou, J., Ortiz-Bonnin, S., Montañez-Juan, M., Espinosa, G. T., & García-Buades, M. E. (2022). Telework satisfaction, wellbeing and performance in the digital era: Lessons learned during COVID-19 lockdown in Spain. *Current Psychology*. https://doi.org/10.1007/s12144-022-02873-x

Brummelhuis, L. L. T., Hetland, J., Keulemans, L., & Bakker, A. B. (2012). Do new ways of working foster work engagement?. *Psicothema*.

Buomprisco, G., Ricci, S., Perri, R., & De Sio, S. (2021). Health and telework: New challenges after COVID-19 pandemic. *European Journal of Environment and Public Health, 5*(2), em0073. https://doi.org/10.21601/ejeph/9705

Cannito, M., & Scavarda, A. (2020). Childcare and remote work during the COVID-19 pandemic. Ideal worker model, parenthood and gender inequalities in Italy. *Italian Sociological Review, 10*(3S), 801–801. https://doi.org/10.13136/isr.v10i3S.399

Carasco-Saul, M., Kim, W., & Kim, T. (2015). Leadership and employee engagement: Proposing research agendas through a review of literature. *Human Resource Development Review, 14*(1), 38–63. https://doi.org/10.1177/1534484314560406

Carillo, K., Cachat-Rosset, G., Marsan, J., Saba, T., & Klarsfeld, A. (2021). Adjusting to epidemic-induced telework: Empirical insights from teleworkers in France. *European Journal of Information Systems, 30*(1), 69–88. https://doi.org/10.1080/0960085X.2020.1829512

Carvalho, V. S., Imam, H., Chambel, M. J., & Santos, M. (2022). Family supportive supervisor behavior and work-family boundary control in teleworkers during a lockdown: Portugal and Pakistan comparison. *Frontiers in Psychology*. https://doi.org/10.3389/fpsyg.2022.1008992

Cerasoli, C. P., Nicklin, J. M., & Ford, M. T. (2014). Intrinsic motivation and extrinsic incentives jointly predict performance: A 40-year meta-analysis. *Psychological Bulletin, 140*(4), 980. https://doi.org/10.1037/a0035661

Chambel, M. J., Castanheira, F., Oliveira-Cruz, F., & Lopes, S. (2015). Work context support and Portuguese soldiers' well-being: The mediating role of autonomous motivation. *Military Psychology, 27*(5), 297–310. https://doi.org/10.1037/mil0000087

Charalampous, M., Grant, C. A., Tramontano, C., & Michailidis, E. (2019). Systematically reviewing remote e-workers' well-being at work: A multidimensional approach. *European Journal of Work and Organizational Psychology, 28*(1), 51–73. https://doi.org/10.1080/1359432X.2018.1541886

Chênevert, D., Fournier, P. L., Menvielle, L., Bruneau, J., Jutras-Aswad, D., & Bissonnette, A. (2022). Lockdowns and telework: Psychological and work-related consequences. *Relations industrielles/Industrial Relations, 77*(1). https://doi.org/10.7202/1088552ar

Crawford, J. O., MacCalman, L., & Jackson, C. A. (2011). The health and well-being of remote and mobile workers. *Occupational Medicine, 61*(6), 385–394. https://doi.org/10.1093/occmed/kqr071

Dahling, J. J., & Lauricella, T. K. (2017). Linking job design to subjective career success: A test of self-determination theory. *Journal of Career Assessment, 25*(3), 371–388. https://doi.org/10.1177/1069072716639689

de Macêdo, T. A. M., Cabral, E. L. D. S., Silva Castro, W. R., de Souza Junior, C. C., da Costa Junior, J. F., Pedrosa, F. M., & Másculo, F. S. (2020). Ergonomics and telework: A systematic review. *Work, 66*(4), 777–788. https://doi.org/10.3233/WOR-203224

Deci, E. L., & Ryan, R. M. (2000). The "what" and "why" of goal pursuits: Human needs and the self-determination of behavior. *Psychological Inquiry, 11*(4), 227–268. https://doi.org/10.1207/S15327965PLI1104_01

Deci, E. L., Olafsen, A. H., & Ryan, R. M. (2017). Self-determination theory in work organizations: The state of a science. *Annual Review of Organizational Psychology and Organizational Behavior, 4*(1), 19–43. https://doi.org/10.1146/annurev-orgpsych-032516-113108

Demerouti, E. (2024). Burnout: A comprehensive review. *Zeitschrift Für Arbeitswissenschaft, 78*, 1–13. https://doi.org/10.1007/s41449-024-00452-3

Demerouti, E., Bakker, A. B., Nachreiner, F., & Schaufeli, W. B. (2001). The job demands-resources model of burnout. *Journal of Applied Psychology, 86*(3), 499–512. https://doi.org/10.1037/0021-9010.86.3.499

Demerouti, E., Derks, D., Ten Brummelhuis, L. L., & Bakker, A. B. (2014). New ways of working: Impact on working conditions, work–family balance, and well-being. *The Impact of ICT on Quality of Working Life*. https://doi.org/10.1007/978-94-017-8854-0_8

Dias, P., Lopes, S., & Peixoto, R. (2022). Mastering new technologies: Does it relate to teleworkers' (in)voluntariness and well-being? *Journal of Knowledge Management, 26*(10). https://doi.org/10.1108/JKM-01-2021-0003

Diener, E. D., Emmons, R. A., Larsen, R. J., & Griffin, S. (1985). The satisfaction with life scale. *Journal of personality assessment, 49*(1), 71–75. https://doi.org/10.1207/s15327752jpa4901_13

Dimian, G. C., Gheorghe, M., Boldeanu, D. M., & Maftei, M. (2023). How digitalization, work-family balance, and work efficiency can influence employees' preferences for teleworking in the future. *Engineering Economics, 34*(2), 139–157. https://doi.org/10.5755/j01.ee.34.2.30090

Erro-Garcés, A., Urien, B., Čyras, G., & Janušauskienė, V. M. (2022). Telework in Baltic countries during the pandemic: Effects on wellbeing, job satisfaction, and work-life balance. *Sustainability, 14*(10), 5778. https://doi.org/10.3390/su14105778

Faustino, D., Sousa, M., Gonçalves, M. M., Soares, I., & Oliveira, J. T. (2023). Perceção do Impacto do Teletrabalho na Carreira: Validação da Escala do Impacto do Teletrabalho na Carreira para a População Portuguesa. *Revista iberoamericana de diagnóstico y evaluación psicológica, 1*(67), 5–17. https://doi.org/10.21865/RIDEP67.1.01

Fernet, C., Guay, F., & Senécal, C. (2004). Adjusting to job demands: The role of work self-determination and job control in predicting burnout. *Journal of Vocational Behavior, 65*(1), 39–56. https://doi.org/10.1016/S0001-8791(03)00098-8

Fernet, C., Trépanier, S. G., Austin, S., Gagné, M., & Forest, J. (2015). Transformational leadership and optimal functioning at work: On the mediating role of employees' perceived job characteristics and motivation. *Work & Stress, 29*(1), 11–31. https://doi.org/10.1080/02678373.2014.1003998

Ford, J. K., MacCallum, R. C., & Tait, M. (1986). The application of exploratory factor analysis in applied psychology: A critical review and analysis. *Personnel Psychology, 39*, 291–314. https://doi.org/10.1111/j.1744-6570.1986.tb00583.x

Gagné, M., & Hewett, R. (2024). Assumptions about human motivation have consequences for practice. *Journal of Management Studies*. https://doi.org/10.1111/joms.13092

Gagné, M., Forest, J., Vansteenkiste, M., Crevier-Braud, L., Van den Broeck, A., Aspeli, A. K., & Westbye, C. (2015). The multidimensional work motivation scale: Validation evidence in seven languages and nine countries. *European Journal of Work and Organizational Psychology, 24*(2), 178–196. https://doi.org/10.1080/1359432X.2013.877892

Gagné, M., Parker, S. K., Griffin, M. A., Dunlop, P. D., Knight, C., Klonek, F. E., & Parent-Rocheleau, X. (2022). Understanding and shaping the future of work with self-determination theory. *Nature Reviews Psychology, 1*(7), 378–392. https://doi.org/10.1038/s44159-022-00056-w

Gagné, M., Deci, E. L., & Ryan, R. M. (2018). Self-determination theory applied to work motivation and organizational behaviour. In D. S. Ones, N. Anderson, C. Viswesvaran, & H. K. Sinangil (Eds.), *The SAGE handbook of industrial, work and organizational psychology: Organizational psychology* (pp. 97–121). Sage Reference.

Galletta, M., Vandenberghe, C., Portoghese, I., Allegrini, E., Saiani, L., & Battistelli, A. (2019). A cross-lagged analysis of the relationships among workgroup commitment, motivation and proactive work behaviour in nurses. *Journal of Nursing Management, 27*(6), 1148–1158. https://doi.org/10.1111/jonm.12786

Gander, F., Hofmann, J., & Ruch, W. (2020). Character strengths: Person–environment fit and relationships with job and life satisfaction. *Frontiers in Psychology, 11*, 1582. https://doi.org/10.3389/fpsyg.2020.01582

Gaspar, T., Jesus, S., Farias, A. R., & Matos, M. G. (2024). Healthy work environment ecosystems for teleworking and hybrid working. *Procedia Computer Science, 239*, 1132–1140. https://doi.org/10.1016/j.procs.2024.06.279

Geldart, S. (2022). Remote work in a changing world: A nod to personal space, self-regulation and other health and wellness strategies. *International Journal of Environmental Research and Public Health, 19*, 4873. https://doi.org/10.3390/ijerph19084873

Gerards, R., de Grip, A., & Baudewijns, C. (2018). Do new ways of working increase work engagement? *Personnel Review, 47*(2), 517–534. https://doi.org/10.1108/PR-02-2017-0050

Gillet, N., Gagné, M., Sauvagère, S., & Fouquereau, E. (2013). The role of supervisor autonomy support, organizational support, and autonomous and controlled motivation in predicting employees' satisfaction and turnover intentions. *European Journal of Work and Organizational Psychology, 22*(4), 450–460. https://doi.org/10.1080/1359432X.2012.665228

Gillet, N., Fouquereau, E., Lafrenière, M. A. K., & Huyghebaert, T. (2016). Examining the roles of work autonomous and controlled motivations on satisfaction and anxiety as a function of role ambiguity. *The Journal of Psychology, 150*(5), 644–665. https://doi.org/10.1080/00223980.2016.1154811

Golden, T. D., & Veiga, J. F. (2008). The impact of superior–subordinate relationships on the commitment, job satisfaction, and performance of virtual workers. *The Leadership Quarterly, 19*(1), 77–88. https://doi.org/10.1016/j.leaqua.2007.12.009

Guest, D. E. (2017). Human resource management and employee well-being: Towards a new analytic framework. *Human Resource Management Journal, 27*(1), 22–38. https://doi.org/10.1111/1748-8583.12139

Guimaraes, T., & Dallow, P. (1999). Empirically testing the benefits, problems, and success factors for telecommuting programmes. *European Journal of Information Systems, 8*(1), 40–54. https://doi.org/10.1057/palgrave.ejis.3000317

Hallman, D. M., Januario, L. B., Mathiassen, S. E., Heiden, M., Svensson, S., & Bergström, G. (2021). Working from home during the COVID-19 outbreak in Sweden: Effects on 24-h time-use in office workers. *BMC Public Health, 21*, 1–10. https://doi.org/10.1186/s12889-021-10582-6

Hayes, A. F. (2013). *Introduction to mediation, moderation, and conditional process*. Guilford Publications.

Heiden, M., Widar, L., Wiitavaara, B., & Boman, E. (2021). Telework in academia: Associations with health and well-being among staff. *Higher Education, 81*, 707–722. https://doi.org/10.1007/s10734-020-00569-4

Heiden, M., Hallman, D. M., Svensson, M., Mathiassen, S. E., Svensson, S., & Bergström, G. (2023). Mismatch between actual and preferred extent of telework: Cross-sectional and prospective associations with well-being and burnout. *BMC Public Health, 23*(1), 1736. https://doi.org/10.1186/s12889-023-16683-8

Henke, R. M., Benevent, R., Schulte, P., Rinehart, C., Crighton, K. A., & Corcoran, M. (2016). The effects of telecommuting intensity on employee health. *American Journal of Health Promotion, 30*(8), 604–612. https://doi.org/10.4278/ajhp.141027-QUAN-544

Hettler, B. (1986). Strategies for wellness and recreation program development. *New Directions for Student Services*. https://doi.org/10.1002/ss.37119863404

Hjorthol, R. J. (2006). Teleworking in some Norwegian urban areas. *Urban Geography, 27*, 610–627. https://doi.org/10.2747/0272-3638.27.7.610

Hobfoll, S. E. (1989). Conservation of resources: A new attempt at conceptualizing stress. *American Psychologist, 44*(3), 513–524. https://doi.org/10.1037/0003-066X.44.3.513

Höcker, M. C., Bachtal, Y., Voll, K., & Pfnür, A. (2024). Healthy, healthier, hybrid work: The burnout-reducing potential of remote work and the mediating effect of work autonomy. *International Journal of Workplace Health Management, 17*(4), 319–334. https://doi.org/10.1108/IJWHM-02-2024-0036

Huo, W., Liang, B., Liang, Y., Song, M., & Wang, Y. (2024). Optimizing telework with exercise: An exploration of the curvilinear relationship between telework intensity and work engagement. *Behavioral Sciences, 14*(8), 674. https://doi.org/10.3390/bs14080674

Instituto Nacional de Estatística. (2023, agosto). Módulo ad hoc do Inquérito ao Emprego "Trabalho a partir de casa". Instituto Nacional de Estatística. https://www.ine.pt/xportal/xmainxpid=INE&xpgid=ine_destaques&DESTAQUESdest_boui=593942616&DESTAQUESmodo=2. Consultado a 10 de outubro de 2023.

Jia, F., & Zhang, Y. (2018). The impact of positive affect on entrepreneurial motivational outcomes-a self-regulatory perspective. *International Review of Entrepreneurship, 16*(4).

Kaduk, A., Genadek, K., Kelly, E. L., & Moen, P. (2019). Involuntary vs. voluntary flexible work: Insights for scholars and stakeholders. *Community, Work and Family, 22*(4), 412–442. https://doi.org/10.1080/13668803.2019.1616532

Kalter, M. J. O., Geurs, K. T., & Wismans, L. (2021). Post COVID-19 teleworking and car use intentions. Evidence from large scale GPS-tracking and survey data in the Netherlands. *Transportation Research Interdisciplinary Perspectives 12*, 100498. https://doi.org/10.1016/j.trip.2021.100498

Kelliher, C., & Anderson, D. (2010). Doing more with less? Flexible working practices and the intensification of work. *Human Relations, 63*(1), 83–106. https://doi.org/10.1177/0018726709349199

Kim, W., & Hyun, Y. S. (2017). The impact of personal resources on turnover intention: The mediating effects of work engagement. *European Journal of Training and Development, 41*(8), 705–721. https://doi.org/10.1108/EJTD-05-2017-0048

Kim, J. O., & Mueller, C. W. (1978). *Factor analysis: Statistical methods and practical issues.* Sage.

Kortsch, T., Rehwaldt, R., Schwake, M. E., & Licari, C. (2022). Does remote work make people happy? Effects of flexibilization of work location and working hours on happiness at work and affective commitment in the German banking sector. *International Journal of Environmental Research and Public Health, 19*(15), 9117. https://doi.org/10.3390/ijerph19159117

Kraemer, K. L., & King, J. L., (1982). Telecommunications/transportation substitution and energy conservation. Part 2. *Telecommunication Policy, 6*(2), 87–99. https://doi.org/10.1016/0308-5961(82)90027-1

Kraemer, K.L., (1982). Telecommunications/transportation substitution and energy conservation. Part 1. *Telecommunication Policy 6*(1), 39–59. https://doi.org/10.1016/0308-5961(82)90004-0

Kurland, N. B., & Bailey, D. E. (2000). Telework: The advantages and challenges of working here, there, anywhere, and anytime. *IEEE Engineering Management Review, 28*(2), 49–60.

Lastovkova, A., Carder, M., Rasmussen, H. M., Sjoberg, L., De Groene, G. J., Sauni, R., & Pelclova, D. (2018). Burnout syndrome as an occupational disease in the European Union: An exploratory study. *Industrial Health, 56*(2), 160–165. https://doi.org/10.2486/indhealth.2017-0132

Li, M., Wang, Z., You, X., & Gao, J. (2015). Value congruence and teachers' work engagement: The mediating role of autonomous and controlled motivation. *Personality and Individual Differences, 80*, 113–118. https://doi.org/10.1016/j.paid.2015.02.021

Liu, L., Wan, W., & Fan, Q. (2021). How and when telework improves job performance during COVID-19? Job crafting as mediator and performance goal orientation as moderator. *Psychology Research and Behavior Management, 14*, 2181. https://doi.org/10.2147/PRBM.S340322

Lonska, J., Mietule, I., Litavniece, L., Arbidane, I., Vanadzins, I., Matisane, L., & Paegle, L. (2021). Work–life balance of the employed population during the emergency situation of COVID-19 in Latvia. *Frontiers in Psychology, 12*, 682459. https://doi.org/10.3389/fpsyg.2021.682459

Lopes, S., & Chambel, M. J. (2014). Motives for being temporary agency worker: Validity study of one measure according to the self-determination theory. *Social Indicators Research, 116*(1), 137–152. https://doi.org/10.1007/s11205-013-0273-3

Lopes, S., & Chambel, M. J. (2017). Temporary agency workers' motivations and well-being at work: A two-wave study. *International Journal of Stress Management, 24*(4), 321–346. https://doi.org/10.1037/str0000041

Lopes, S., Dias, P. C., Sabino, A., Cesário, F., & Peixoto, R. (2022). Employees' fit to telework and work well-being:(in) voluntariness in telework as a mediating variable? *Employee Relations: THe International Journal, 45*(1), 257–274. https://doi.org/10.1108/ER-10-2021-0441
Lysova, E. I., Allan, B. A., Dik, B. J., Duffy, R. D., & Steger, M. F. (2019). Fostering meaningful work in organizations: A multi-level review and integration. *Journal of Vocational Behavior, 110*, 374–389. https://doi.org/10.1016/j.jvb.2018.07.004
Madsen, S. R. (2003). The effects of home-based teleworking on work-family conflict. *Human Resource Development Quarterly, 14*(1), 35–58. https://doi.org/10.1002/hrdq.1049
Mann, S., & Holdsworth, L. (2003). The psychological impact of teleworking: Stress, emotions and health. *New Technology, Work and Employment, 18*(3), 196–211. https://doi.org/10.1111/1468-005X.00121
Martela, F., Gómez, M., Unanue, W., Araya, S., Bravo, D., & Espejo, A. (2021). What makes work meaningful? Longitudinal evidence for the importance of autonomy and beneficence for meaningful work. *Journal of Vocational Behavior, 131*, 103675. https://doi.org/10.1016/j.jvb 2021.103675
Martínez-Sánchez, A., Pérez-Pérez, M., José Vela-Jiménez, M., & de-Luis-Carnicer, P. (2008). Telework adoption, change management, and firm performance. *Journal of Organizational Change Management, 21*(1), 7–31. https://doi.org/10.1108/09534810810847011
Maslach, C., & Jackson, S. E. (1981). The measurement of experienced burnout. *Journal of Organizational Behavior, 2*(2), 99–113. https://doi.org/10.1002/job.4030020205
Maslach, C., Jackson, S. E., & Leiter, M. P. (1996). *Maslach burnout inventory manual* (3rd ed.) Consulting Psychologists Press.
Maslach, C., Schaufeli, W. B., & Leiter, M. P. (2001). Job burnout. *Annual Review of Psychology, 52*(1), 397–422. https://doi.org/10.1146/annurev.psych.52.1.397
McAnally, K., & Hagger, M. S. (2024). Self-determination theory and workplace outcomes: A conceptual review and future research directions. *Behavioral Sciences, 14*(6), 428. https://doi.org/10.3390/bs14060428
Mesri, G., Sameri, M., & Keyhan, J. (2024). The relationship between active work behaviors, work values, transformational and interactive leadership style, and work engagement: The mediating role of innovation culture and autonomous motivation. *Current Psychology*. https://doi.org/10.1007/s12144-024-06088-0
Miller, J. W. (2005). *Wellness: The history and development of a concept* (pp. 84–106).
Montero, I., & León, O. G. (2007). A guide for naming research studies in psychology. *International Journal of Clinical and Health Psychology, 7*(3), 847–862.
Moon, T. W., Hur, W. M., & Hyun, S. S. (2019). How service employees' work motivations lead to job performance: The role of service employees' job creativity and customer orientation. *Current Psychology, 38*, 517–532. https://doi.org/10.1007/s12144-017-9630-8
Mori, K., Odagami, K., Inagaki, M., Moriya, K., Fujiwara, H., & Eguchi, H. (2024). Work engagement among older workers: a systematic review. *Journal of Occupational Health, 66*(1), uiad008. https://doi.org/10.1093/joccuh/uiad008
Nakrošienė, A., & Butkevičienė, E. (2016). Nuotolinis darbas Lietuvoje: Samprata, privalumai ir iššūkiai darbuotojams. *Filosofija. Sociologija, 27*(4), 364–372.
Nakrošienė, A., Bučiūnienė, I., & Goštautaitė, B. (2019). Working from home: Characteristics and outcomes of telework. *International Journal of Manpower, 40*(1), 87–101. https://doi.org/10.1108/IJM-07-2017-0172
Naqshbandi, M. M., Kabir, I., Ishak, N. A., & Islam, M. Z. (2024). The future of work: Work engagement and job performance in the hybrid workplace. *The Learning Organization, 31*(1), 5–26. https://doi.org/10.1108/TLO-08-2022-0097
Neculaesei, A. N., & Tocar, S. (2023). Determinants of perceived performance during telework: Evidence from Romania. *Sustainability, 15*(8), 6334. https://doi.org/10.3390/su15086334
Ngamkroeckjoti, C., Klongthong, W., & Thavorn, J. (2022). Motivational factors influencing telework during the COVID-19 pandemic. *Emerging Science Journal, 6*(0), 229–246. https://doi.org/10.28991/esj-2022-SPER-016

Olafsen, A. H., Halvari, H., Forest, J., & Deci, E. L. (2015). Show them the money? The role of pay, managerial need support, and justice in a self-determination theory model of intrinsic work motivation. *Scandinavian Journal of Psychology, 56*(4), 447–457. https://doi.org/10.1111/sjop.12211

Omansky, R., Eatough, E. M., & Fila, M. J. (2016). Illegitimate tasks as an impediment to job satisfaction and intrinsic motivation: Moderated mediation effects of gender and effort-reward imbalance. *Frontiers in Psychology, 7*, 1818. https://doi.org/10.3389/fpsyg.2016.01818

Otsuka, S., Ishimaru, T., Nagata, M., Tateishi, S., Eguchi, H., Tsuji, M., Ogami, A., Matsuda, S., & Fujino, Y. (2021). A cross-sectional study of the mismatch between telecommuting preference and frequency associated with psychological distress among Japanese workers in the COVID-19 pandemic. *Journal of Occupational and Environmental Medicine, 63*(9), e636–e640. https://doi.org/10.1097/JOM.0000000000002318

Owen, W. (1962). Transportation and technology. *The American Economic Review, 52*(2), 405–413.

Parent-Lamarche, A. (2022). Teleworking, work engagement, and intention to quit during the COVID-19 pandemic: Same storm, different boats? *International Journal of Environmental Research and Public Health, 19*(3), 1267. https://doi.org/10.3390/ijerph19031267

Park, S., & Cho, Y. J. (2022). Does telework status affect the behavior and perception of supervisors? Examining task behavior and perception in the telework context. *The International Journal of Human Resource Management, 33*(7), 1326–1351. https://doi.org/10.1080/09585192.2020.1777183

Park, S. M., & Rainey, H. G. (2012). Work motivation and social communication among public managers. *The International Journal of Human Resource Management, 23*(13), 2630–2660. https://doi.org/10.1080/09585192.2011.637060

Perry, S. J., Rubino, C., & Hunter, E. M. (2018). Stress in remote work: Two studies testing the demand control-person model. *European Journal of Work and Organizational Psychology, 27*(5), 577–593. https://doi.org/10.1080/1359432X.2018.1487402

Petcu, M. A., Sobolevschi-David, M. I., Crețu, R. F., Curea, S. C., Hristea, A. M., Oancea-Negescu, M. D., & Tutui, D. (2023). Telework: A social and emotional perspective of the impact on employees' wellbeing in the COVID-19 pandemic. *International Journal of Environmental Research and Public Health, 20*(3), 1811. https://doi.org/10.3390/ijerph20031811

Peters, P., Tijdens, K. G., & Wetzels, C. (2004). Employees' opportunities, preferences, and practices in telecommuting adoption. *Information & Management, 41*(4), 469–482. https://doi.org/10.1016/S0378-7206(03)00085-5

Peters, P., Poutsma, E., Van der Heijden, B. I., Bakker, A. B., & Bruijn, T. D. (2014). Enjoying new ways to work: An HRM-process approach to study flow. *Human Resource Management, 53*(2), 271–290. https://doi.org/10.1002/hrm.21588

Podsakoff, P. M., MacKenzie, S. B., Lee, J.-Y., & Podsakoff, N. P. (2003). Common method biases in behavioral research: A critical review of the literature and recommended remedies. *Journal of Applied Psychology, 88*(5), 879–903. https://doi.org/10.1037/0021-9010.88.5.879

Purvanova, R. K., & Muros, J. P. (2010). Gender differences in burnout: A meta-analysis. *Journal of Vocational Behavior, 77*(2), 168–185. https://doi.org/10.1016/j.jvb.2010.04.006

Ray, T. K., & Pana-Cryan, R. (2021). Work flexibility and work-related well-being. *International Journal of Environmental Research and Public Health, 18*(6), 3254. https://doi.org/10.3390/ijerph18063254

Raza, M. Y., Akhtar, M. W., Husnain, M., & Akhtar, M. S. (2015). The impact of intrinsic motivation on employee's job satisfaction. *Management and Organizational Studies, 2*(3), 80–88. https://doi.org/10.5430/mos.v2n3p80

Reeve, J. (2006). Teachers as facilitators: What autonomy-supportive teachers do and why their students benefit. *The Elementary School Journal, 106*(3), 225–236. https://doi.org/10.1086/501484

Ren, F., Zhang, Q., & Wei, X. (2021). Work autonomous and controlled motivation on Chinese employees' work performance and innovative work behaviour: The moderating role of financial stress. *Frontiers in Psychology, 12*, 676063. https://doi.org/10.3389/fpsyg.2021.676063

Rietveld, J. R., Hiemstra, D., Brouwer, A. E., & Waalkens, J. (2021). Motivation and productivity of employees in higher education during the first lockdown. *Administrative Sciences, 12*(1), 1. https://doi.org/10.3390/admsci12010001

Ríos Villacorta, M. A., Ramos Farroñán, E. V., Arbulú Ballesteros, M. A., Otiniano León, M. Y., Bravo Jaico, J. L., Suysuy Chambergo, E. J., & Alarcón García, R. E. (2024). Human-centric telework and sustainable well-being: Evidence from Peru's public sector. *Sustainability, 16*(22), 9713. https://doi.org/10.3390/su16229713

Roche, M., & Haar, J. (2020). Motivations, work–family enrichment and job satisfaction: An indirect effects model. *Personnel Review, 49*(3), 903–920. https://doi.org/10.1108/PR-06-2019-0289

Ryan, R. M., & Deci, E. L. (2000). Self-determination theory and the facilitation of intrinsic motivation, social development, and well-being. *American Psychologist, 55*(1), 68–78. https://doi.org/10.1037/0003-066X.55.1.68

Ryan, R. M., Duineveld, J. J., Di Domenico, S. I., Ryan, W. S., Steward, B. A., & Bradshaw, E. L. (2022). We know this much is (meta-analytically) true: A meta-review of meta-analytic findings evaluating self-determination theory. *Psychological Bulletin, 148*(11–12), 813. https://doi.org/10.1037/bul0000385

Ryan, R. M., & Deci, E. L. (2019). Brick by brick: The origins, development, and future of self-determination theory. In *Advances in motivation science* (vol. 6, pp. 111–156). Elsevier. https://doi.org/10.1016/bs.adms.2019.01.001

Salinas-Jiménez, M. D. M., Artés, J., & Salinas-Jiménez, J. (2011). Education as a positional good: A life satisfaction approach. *Social Indicators Research, 103*, 409–426. https://doi.org/10.1007/s11205-010-9709-1

Sardeshmukh, S. R., Sharma, D., & Golden, T. D. (2012). Impact of telework on exhaustion and job engagement: A job demands and job resources model. *New Technology, Work and Employment, 27*(3), 193–207. https://doi.org/10.1111/j.1468-005X.2012.00284.x

Schaufeli, W. B., & Taris, T. W. (2005). The conceptualization and measurement of burnout: Common ground and worlds apart. *Work & Stress, 19*(3), 256–262. https://doi.org/10.1080/02678370500385913

Schaufeli, W. B., Salanova, M., González-Romá, V., & Bakker, A. B. (2002). The measurement of engagement and burnout: A two sample confirmatory factor analytic approach. *Journal of Happiness Studies, 3*, 71–92. https://doi.org/10.1023/A:1015630930326

Schaufeli, W. B., Shimazu, A., Hakanen, J., Salanova, M., & De Witte, H. (2019). An ultra-short measure for work engagement: The UWES-3 validation across five countries. *European Journal of Psychological Assessment, 35*(4), 577–591. https://doi.org/10.1027/1015-5759/a000430

Schaufeli, W. B., Desart, S., & De Witte, H. (2020). Burnout Assessment Tool (BAT)—development, validity, and reliability. *International Journal of Environmental Research and Public Health, 17*(24), 9495. https://doi.org/10.3390/ijerph17249495

Shirom, A., & Melamed, S. (2006). A comparison of the construct validity of two burnout measures in two groups of professionals. *International Journal of Stress Management, 13*(2), 176. https://doi.org/10.1037/1072-5245.13.2.176

Sinclair, R. R., Allen, T., Barber, L., Bergman, M., Britt, T., Butler, A., Ford, M., Hammer, L., Kath, L., Probst, T., & Yuan, Z. (2020). Occupational Health Science in the Time of COVID-19: Now more than Ever. *Occupational Health Science, 4*(1–2). https://doi.org/10.1007/s41542-020-00064-3

Slemp, G. R., Kern, M. L., Patrick, K. J., & Ryan, R. M. (2018). Leader autonomy support in the workplace: A meta-analytic review. *Motivation and Emotion, 42*(5), 706–724. https://doi.org/10.1007/s11031-018-9698-y

Slemp, G. R., Zhao, Y., Hou, H., & Vallerand, R. J. (2021). Job crafting, leader autonomy support, and passion for work: Testing a model in Australia and China. *Motivation and Emotion, 45*(1), 60–74. https://doi.org/10.1007/s11031-020-09850-6

Steidelmüller, C., Meyer, S.-C., & Müller, G. (2020). Home-based telework and presenteeism across Europe. *Journal of Occupational & Environmental Medicine, 62*(12), 998–1005. https://doi.org/10.1097/JOM.0000000000001992

Stevens, J. (1992). *Applied multivariate statistics for the social sciences*. Erlbaum.

Stoian, C. A., Caraiani, C., Anica-Popa, I. F., Dascălu, C., & Lungu, C. I. (2022). Telework systematic model design for the future of work. *Sustainability, 14*(12), 7146. https://doi.org/10.3390/su14127146

Strauss, K., & Parker, S. K. (2014). Effective and sustained proactivity in the workplace: A self-determination theory perspective. *The Oxford handbook of work engagement, motivation, and self-determination theory*, 50–71. https://doi.org/10.1093/oxfordhb/9780190276997.013.3

Tafvelin, S., & Stenling, A. (2018). Development and initial validation of the need satisfaction and need support at work scales: A validity-focused approach. *Scandinavian Journal of Work and Organizational Psychology, 3*(1). https://doi.org/10.16993/sjwop.30

Tan, J., Lim-Soh, J., & Tan, P. L. (2024). The impact of teleworking on women's work-life balance and life satisfaction: A longitudinal study from Singapore. *Applied Research in Quality of Life*. https://doi.org/10.1007/s11482-024-10340-x

Tang, M., Wang, D., & Guerrien, A. (2021). The contribution of basic psychological need satisfaction to psychological well-being via autonomous motivation among older adults: A cross-cultural study in China and France. *Frontiers in Psychology, 12*, 734461. https://doi.org/10.3389/fpsyg.2021.734461

Taris, T. W., & Kompier, M. A. J. (2014). Cause and effect: Optimizing the designs of longitudinal studies in occupational health psychology. *Work & Stress, 28*(1), 1–8. https://doi.org/10.1080/02678373.2014.878494

Tavares, A. I. (2017). Telework and health effects review. *International Journal of Healthcare, 3*(2), 30–36. https://doi.org/10.5430/ijh.v3n2p30

Thompson, R. J., Payne, S. C., Alexander, A. L., Gaskins, V. A., & Henning, J. B. (2022). A taxonomy of employee motives for telework. *Occupational Health Science, 6*(2). https://doi.org/10.1007/s41542-021-00094-5

Ton, D., Arendsen, K., De Bruyn, M., Severens, V., Van Hagen, M., Van Oort, N., & Duives, D. (2022). Teleworking during COVID-19 in the Netherlands: Understanding behaviour, attitudes, and future intentions of train travellers. *Transportation Research Part a: Policy and Practice, 159*, 55–73. https://doi.org/10.1016/j.tra.2022.03.019

Tremblay, D. G., & Genin, É. (2007). The demand for telework of IT self-employed workers. *The Journal of E-Working, 1*(2), 98–115.

Urien, B. (2023). Teleworkability, preferences for telework, and well-being: A systematic review. *Sustainability, 15*(13), 10631. https://doi.org/10.3390/su151310631

Uru, F. O., Gozukara, E., & Tezcan, L. (2022). The moderating roles of remote, hybrid, and onsite working on the relationship between work engagement and organizational identification during the COVID-19 pandemic. *Sustainability*. https://doi.org/10.3390/su142416828

Valero, D., Hirschi, A., & Strauss, K. (2015). Hope in adolescent careers: Mediating effects of work motivation on career outcomes in Swiss apprentices. *Journal of Career Development, 42*(5), 381–395. https://doi.org/10.1177/0894845314566866

Van den Broeck, A., Howard, J. L., Van Vaerenbergh, Y., Leroy, H., & Gagné, M. (2021). Beyond intrinsic and extrinsic motivation: A meta-analysis on self-determination theory's multidimensional conceptualization of work motivation. *Organizational Psychology Review, 11*(3), 240–273. https://doi.org/10.1177/20413866211006173

van Hooff, M. L., & van Hooft, E. A. (2017). Boredom at work: Towards a dynamic spillover model of need satisfaction, work motivation, and work-related boredom. *European Journal of Work and Organizational Psychology, 26*(1), 133–148. https://doi.org/10.1080/1359432X.2016.1241769

Vleeshouwers, J., Fløvik, L., Christensen, J. O., Johannessen, H. A., Bakke Finne, L., Mohr, B., & Lunde, L. K. (2022). The relationship between telework from home and the psychosocial work environment: A systematic review. *International Archives of Occupational and Environmental Health, 95*(10), 2025–2051. https://doi.org/10.1007/s00420-022-01901-4

Wang, H., Xiao, Y., Wang, H., Zhang, H., & Chen, X. (2023). "Who knows me understands my needs": The effect of home-based telework on work engagement. *Psychology Research and Behavior Management*. https://doi.org/10.2147/PRBM.S402159

Wang, X., Kim, S. H., & Mokhtarian, P. L. (2024). Identifying teleworking-related motives and comparing telework frequency expectations in the post-pandemic world: A latent class choice modeling approach. *Transportation Research Part a: Policy and Practice, 186*, 104070. https://doi.org/10.1016/j.tra.2024.104070

Wang, B., Liu, Y., Qian, J., & Parker, S. K. (2021). Achieving effective remote working during the COVID-19 pandemic: A work design perspective. *Applied Psychology, 70*(1). https://doi.org/10.1111/apps.12290

Widar, L., Wiitavaara, B., Boman, E., & Heiden, M. (2021). Psychophysiological reactivity, postures and movements among academic staff: A comparison between teleworking days and office days. *International Journal of Environmental Research and Public Health, 18*(18), 9537. https://doi.org/10.3390/ijerph18189537

Wollard, K. K., & Shuck, B. (2011). Antecedents to employee engagement: A structured review of the literature. *Advances in Developing Human Resources, 13*(4), 429–446. https://doi.org/10.1177/1523422311431220

Yang, D., Kelly, E. L., Kubzansky, L. D., & Berkman, L. (2023). Working from home and worker well-being: New evidence from Germany. *Industrial & Labor Relations Review, 76*(3), 504–531. https://doi.org/10.1177/00197939221148716

Zhang, C. (2019). Streams and future directions of research on work motivation based on the self-determination theory. *Advances in Psychological Science, 27*(8), 1489. https://doi.org/10.3724/SP.J.1042.2019.01489

Zhang, S., Moeckel, R., Moreno, A. T., Shuai, B., & Gao, J. (2020). A work-life conflict perspective on telework. *Transportation Research Part a: Policy and Practice, 141*, 51–68. https://doi.org/10.1016/j.tra.2020.09.007

Chapter 12
Expatriate Workers: Impact of Change on Adaptation to the Destination Country

Ana Francisca Oliveira Ribeiro and Isabel S. Silva

Abstract The globalization of markets promotes the internationalization of companies, offering a competitive advantage by facilitating business growth and consolidation. Consequently, it has become essential to identify employees capable of meeting emerging organizational needs. This study aimed to understand the challenges faced by expatriate workers in a company and to propose strategies to support their adaptation to the host country. An online questionnaire was completed by 157 expatriate workers with access to email. The results indicate that autonomous motivation is a critical factor for improved sociocultural adaptation. No significant differences were observed in the impact of the three types of support analyzed on sociocultural adaptation. In addition, the variables of the participants' gender and the continent where they work were also studied. Social relationships, social communication, and work performance were found to be positively and significantly associated with organizational support. Support from colleagues, managers, and the organization appears to foster expatriates' adaptation to the host country. These findings contribute to expanding knowledge and guiding the development of new organizational strategies tailored to the challenges and needs of expatriate workers and their families.

Keywords Sociocultural adaptation · Expatriation · Organizational strategies

12.1 Introduction

Technology can be characterized as a powerful force. In recent years, technological advances have transformed transportation, communication, travel, and modernized isolated regions and impoverished populations (Crooke et al., 2021). Along with

A. F. O. Ribeiro · I. S. Silva (✉)
School of Psychology, University of Minho, Braga, Portugal
e-mail: isilva@psi.uminho.pt

I. S. Silva
CICS.NOVA.UMinho, School of Psychology, Braga, Portugal

C. Machado (ed.), *Human Resource Management in a Disrupted World*, Lecture Notes on Multidisciplinary Industrial Engineering,
https://doi.org/10.1007/978-3-031-89948-5_13

these advancements, Bahadur and colleagues (1982) also highlight the significant impact on market realities, marking the onset of globalization.

The globalization of markets is a phenomenon that has garnered significant attention and sparked extensive debate within the economic sphere (Steger et al., 2023). A globalized market is characterized by its complexity and high connectivity. It can be defined as an increase in interdependencies among market activities and the outcomes of other dispersed national markets (Braun et al., 2021). The growing globalization of markets, which defines contemporary reality, fosters the internationalization of companies, providing them with a competitive advantage. This workforce relocation to other markets supports the growth and consolidation of diverse businesses (Braun et al., 2021). Consequently, it is vital to identify employees capable of addressing organizational needs (Murray & Holmes, 2021).

In Portugal, there is currently evidence of an emigration wave comparable in scale to the significant outflow observed during the 1960s and 1970s (Pires et al., 2016). However, the two waves differ notably, particularly in terms of the qualifications of the emigrating professionals (Pires, 2021). Unlike the first wave, which involved a less qualified workforce, the second wave consists predominantly of highly skilled professionals. This recent wave reflects individuals choosing to pursue professional careers abroad due to insufficient local opportunities that fail to meet their expectations (Koser & Salt, 1997).

According to Mauro (2021), the increasing availability of international job opportunities reflects companies' allocation of talent to diverse cultural contexts. To remain competitive, companies must become smaller, smarter, and more responsive to rapidly changing market conditions, anticipating environmental needs (Mauro, 2021). These changes are not limited to the workplace; employees must also develop greater flexibility and adaptability (Hall & Moss, 1998). Indeed, innovations and new technological applications are significantly influencing and transforming the structure of human resources (HR) departments. Today, HR management and practices play a pivotal role, as globalization and its associated challenges have reshaped HR functions, making them a critical component of organizational processes (Mantzaris & Myloni, 2023). However, it remains unclear how organizations and their employees can adapt effectively and productively to these dynamic changes (Anwar & Abdullah, 2021). Consequently, effective HR management emerges as a new and essential challenge for organizations.

Recognizing that relocating workers abroad impacts not only their professional lives but also their personal lives is crucial. Cultural adaptation to the host country becomes essential to avoid issues such as culture shock or maladaptation (Inverno, 2011). It is important to acknowledge that leaving one's home country represents a profound life transition, accompanied by numerous challenges for individuals and their families. As Bonache (2005) noted, these moves must offer some tangible benefits in return. In most cases, such relocations are motivated by aspirations for both professional and personal advancement (Boneva & Frieze, 2001). Among the various reasons for emigrating, the hope of improving quality of life and happiness is often the most influential factor (Fan et al., 2022). Ganguli and colleagues (2022) further conclude that workers increasingly seek a balance between their careers, family lives, and leisure activities. However, not everyone successfully achieves this balance. Evidence suggests that many workers face significant challenges when working and living abroad as expatriates (De Paul & Bikos, 2015).

12.1.1 Expatriation

From the literature reviewed (Chen & Shaffer, 2018), empirical studies suggest that the historical conceptualization of the term "expatriate" is closely linked to the needs of the business market. Expatriate job opportunities are shaped by the organizational work context abroad (Mäkelä et al., 2019) and by expatriates' ability to help organizations achieve their strategic objectives (Selmer & McNulty, 2017). Furthermore, evidence shows a growing number of expatriates, with projections indicating that this trend will continue to increase in the future (Mello et al., 2023).

An "expatriate" is defined as an employee of an organization sent to a unit in a foreign country to perform a specific job or fulfill an organizational objective for a designated period of time (Aycan, 1997). This definition implies two key aspects: (1) that expatriates are officially assigned by their organizations and (2) that the employment relationship between the worker and the organization is central and indispensable. This distinction separates expatriates from other groups, such as tourists or volunteers (Selmer & McNulty, 2017).

The process of expatriation has existed for centuries. According to Osland (2000), colonization can be regarded as the earliest form of expatriation. States or kingdoms would send colonizers—individuals leaving their homeland—to work in colonies for specified periods (Osland, 2000). In modern times, expatriation continues to play a significant role, offering numerous advantages (Colakoglu & Caligiuri, 2008). These advantages, as highlighted by the authors, include: (a) Improved coordination and communication between a company's headquarters and its subsidiaries (e.g., an expatriate previously employed at headquarters can more effectively transmit corporate objectives, goals, and vision to the subsidiary); (b) Enhanced competitiveness in international markets; (c) Opportunities to disseminate knowledge across borders. Expatriation, therefore, can be seen as a tool for acquiring, sharing, and spreading knowledge, ultimately contributing to competitive advantages for organizations (Colakoglu & Caligiuri, 2008).

12.1.1.1 Stages of the Expatriation Process

From the literature, it is possible to conclude that the expatriation process can be characterized by three distinct phases (McNulty & Brewster, 2020). The first phase is pre-planning, which involves assigning the expatriate to a position in the destination country. This process becomes increasingly complex as the cultural distance between the home country and the host country widens. The second phase is expatriation, during which the expatriate is stationed in the foreign country. This phase includes maintaining a connection with the company's headquarters and ensuring regular communication to support the expatriate's well-being and professional integration. The final phase is repatriation, where the expatriate and the organization prepare for the return to the home country (Inverno, 2011). Proper management of these phases not only facilitates the expatriate worker's adaptation to the new cultural and social environment but also ensures a return on the company's investment in expatriates (Colakoglu & Caligiuri, 2008).

12.1.1.2 Perspectives on the Concept of Expatriation

The concept of expatriation encompasses a wide range of work experiences that differ significantly from those an employee may be accustomed to. However, it is important to pay closer attention to this concept, as it involves various forms and classifications (Reiche, 2011).

According to Hussain and Zhang (2023), expatriates can be classified based on the initiative behind their relocation. Self-initiated expatriates are individuals who voluntarily decide to move to a country other than their own. Generally, these expatriates seek to improve their quality of life or wish to experience working in a foreign country (Chen, 2012). Research by Selmer et al. (2022) suggests that such individuals often aim for personal development and cultural enrichment.

In contrast, company-initiated expatriates are employees who leave their home country with the full support of their organization, also on a voluntary basis (Chen, 2019). This group typically agrees to relocate for financial incentives or career advancement opportunities. The concept of expatriation can also be categorized by the nature and duration of the assignment. At the organizational level, expatriates can be divided into two main categories (Kour & Jyoti, 2022):

Home-country expatriates: Employees originating from the country where the company's headquarters are located and working in a subsidiary in another country. Third-country expatriates: Employees who do not originate from the country where the company's headquarters are located but work in a third country covered by the organization. This group often includes employees transferred between subsidiaries located in countries other than the headquarters (Chen, 2019). Regarding the duration of expatriation, it can be classified into five types (Fan et al., 2022): Long-term expatriation: Assignments lasting more than one year. Short-term expatriation: Assignments that do not exceed one year. Commuter expatriation: Regular cross-border transfers, typically short-term. Extended business trips: Less frequent, short-term transfers for specific purposes. Permanent moves: Relocations in which expatriates remain in the host country without a set return date.

12.1.1.3 Success Factors in the Expatriation Process

In today's global landscape, the expatriation process increasingly demands that workers be prepared to navigate differences in both national and organizational cultures. In this context, it is essential for workers to develop personal skills that facilitate cultural adaptation. According to Li (2012), the following have been identified as critical factors for successful expatriation: The ability to commit fully to tasks and work responsibilities. Respect for the local culture. Learning from the experiences of other expatriates. The capacity to progress as a minority, as expatriates often represent the values and culture of their home company.

Achieving a balance between personal and professional life. Recognizing and respecting one's own limitations. Among various studies, Zhou's (2020) empirical research on factors promoting expatriation success reveals that age has a significant negative impact on the process. Specifically, the older the employee at the end of the expatriation assignment, the lower the likelihood of expatriation success.

Olsen and Martins (2009) identify additional factors that influence the expatriation process. Gender is highlighted as a significant component, as the host country may offer varying levels of support and employ personalized approaches depending on the expatriate's gender. Similarly, factors such as race and ethnicity can affect expatriation success, depending on how the organization values and appreciates the expatriate's ethnic and racial background. Finally, the authors note that cultural distance—the disparity between the expatriate's culture and that of the host country—plays a crucial role. Greater cultural distance increases the likelihood of exclusionary attitudes and a sense of alienation from the host country (Olsen & Martins, 2009).

12.1.2 Adapting to the Country of Origin

According to Black (1988), expatriates tend to integrate into the culture of their destination country through a process marked by uncertainty and transformation. By mimicking and practicing culturally acceptable behaviors, they can reduce fear and confusion during the adaptation process. Expatriates who successfully adapt to the culture of the destination country are typically characterized as being open and receptive to the customs of that society (Black, 1988). Conversely, another group of expatriates either fails or refuses to adapt to the habits, customs, and norms of the host country, often viewing foreign cultures as inferior to their own (Church, 1982).

Cagliuri (2000) identifies three aspects that are critical to the success of an expatriate's adaptation: General adjustment: Refers to the way individuals adapt to living abroad. Interaction adjustment: Involves adapting to new social contacts and relationships. Work adjustment: Relates to the expatriate's integration into the workplace, a key factor in successful adaptation. Evidence suggests that expatriates benefit significantly from assignments abroad when they receive language and cultural training prior to departure. Such training helps them better understand the customs, norms, and behaviors of the destination country's culture and its inhabitants (Selmer & Lauring, 2015). Furthermore, fostering interpersonal interactions between expatriates and host-country employees strengthens ties and eases the social transition.

Therefore, it is vital for subsidiaries in the destination country to promote such interactions. For instance, organizations could arrange activities involving expatriates and the close relatives of local employees, creating a friendly and inclusive environment.

Critical success factors also play a vital role in expatriation. These are the competencies and activities that, when well-executed, drive organizational success. Conversely, critical flaws can impede success or lead to failure. Key success factors include:

(i) a clear definition of expectations throughout the process,
(ii) consideration of the destination environment and context,
(iii) frequent performance evaluations,
(iv) the establishment of equity parameters (e.g., between source and destination plans),
(v) a well-defined career progression plan for expatriates, and
(vi) a robust system for evaluating the expatriation process, utilizing key performance indicators (KPIs) (Martin & Bartol, 2003).

Given these considerations, the aim is to address a request from the organization where the internship took place. The goals are twofold: To identify the main challenges faced by expatriate employees in adapting to their destination countries. To understand the strategies that facilitate adaptation to the expatriate context. The organization employs approximately 700 expatriate workers, of whom 255 have access to email—a necessary condition for data collection, as detailed below. These employees are spread across 16 countries and occupy a wide variety of roles. The organization's objective is to create optimal conditions for expatriate workers to adapt while minimizing the impact of relocation. This ensures expatriates' well-being abroad and reduces the risk of costly returns due to dissatisfaction with the expatriate location.

12.2 Methodology

12.2.1 Sample

The sample consisted of 157 Portuguese participants, the majority of whom were male (n = 133, 84.7%), married (n = 98, 68%), having been in the destination country for more than 5 years (n = 98, 62.4%). With regard to areas of activity, it was found that the majority of employees belonged to the Management functional group (n = 53, 33.8%), and the Technical group (n = 43, 27.4%). In addition, more than half of the participants work on the African continent (n = 93, 59.2%). Table 12.1 contains more detailed information about the participants.

Table 12.1 Sociodemographic characterization of the sample

Variable	*n*	%
Gender		
Female	24	15.3
Male	133	84.7
Marital status		
Divorced	14	8.9
Single	45	28.7
Married	98	62.4
Widowed	0	0.0
Duration of expatriation		
Less than 1 year	11	7.0
Between 3 and 5 years	14	8.9
Between 1 and 2 years	17	10.8
2–3 years	17	10.8
More than 5 years	98	62.4
Work area		
Operational	3	1.9
Specialization	5	3.2
Administrative	6	3.8
Top management	15	9.6
Coordination	32	20.4
Technical	43	27.4
Management	53	33.8
Continent of work		
Asia	4	2.5
America	8	5.1
Europe	52	33.1
Africa	93	59.2

12.2.2 Instruments

The research protocol includes several instruments which are described below. Initially, the Sociodemographic Questionnaire was presented with the aim of characterizing the sample, obtaining information on sociodemographic characteristics such as gender, nationality, marital status and destination country.

Next, we used the Portuguese version of The Multidimensional Work Motivation Scale, originally developed by Gagné et al. (2015). It was adapted for the Portuguese population by Santos and colleagues (2021), and modified for the expatriate context by Santos (2015). This scale is made up of twenty items, which integrate two dimensions, (1) autonomous motivation (e.g. "*Why I chose to work in a foreign country*") and (2) controlled motivation (e.g. "*Because working abroad gives me a feeling of success*"). These items are assessed on a 7-point Linkert scale, where 1 corresponds to "*None*" and 7 corresponds to "*Completely*". The internal consistency values are $\alpha = 0.82$ for autonomous motivation and $\alpha = 0.70$ for controlled motivation.

Next, the "Job Content Questionnaire" by Karasek (1985) and adapted for the Portuguese population by Malveiro (2016) was presented. The scale consists of 18 items, which assess the characteristics of the job, namely "support from supervisors" and "support from colleagues", "autonomy" and "psychological demands". In this protocol, only the items relating to support from supervisors (e.g. "*My supervisor cares about the well-being of his subordinates*") and colleagues (e.g. "*My co-workers take an interest in me*") were considered, and only 8 items were used, 4 per sub-scale. The items are rated on a 5-point Likert scale, with 1 corresponding to "*Strongly Disagree*" and 5 to "*Strongly Agree*". It should be noted that the items in question relate to the instrumental and socio-emotional support provided by the supervisor to the workers. The internal consistency is $\alpha = 0.81$ (Malveiro, 2016).

From the point of view of support, we also used the Organizational Support Perception Scale, originally developed by Eisenberger, Huntington, Hutchison and Sowa (1986) and adapted for the Portuguese population by Santos and Gonçalves (2013). It consists of 8 items (e.g.: "*The organization/institution values my contribution to institutional well-being.*") which are evaluated by on a 7-point Linkert scale, where 1 corresponds to "*Strongly Disagree*" and 7 corresponds to "*Strongly Agree*".

The final part of the protocol also includes the Revised Sociocultural Adaptation Scale (RSCAS) by Ward (2013). This is made up of twenty items (e.g. "Seeing both sides of an intercultural issue") which cover social situations experienced in the host society. The items are assessed on a 5-point Linkert scale, ranging from 1 ("no difficulty") to 5 ("extremely difficult"), in which higher scores indicate greater difficulties in cultural adaptation; with a Cronbach's alpha of 0.885 (Ward, 2013). As there has been no adaptation of the measure for the Portuguese context to date, an analysis of its psychometric properties will be carried out as part of this project.

12.2.3 Procedures

First, contact was made with the authors of the previously mentioned scales to request permission for their use. Regarding the Revised Sociocultural Adaptation Scale, its translation and adaptation process adhered to the normative procedures described in the literature (Orlando, 2000). The initial Portuguese translation was carried out by the Portuguese research team, ensuring that the original meaning of the items, as well as their accessibility and comprehensibility, were preserved. Subsequently, the Portuguese version was back-translated by a native English speaker. The two versions were then compared by the research team, and minor adjustments were made to achieve a consensual translation.

The project was then submitted to the Ethics Subcommittee for Social and Human Sciences at the University of Minho and received a favorable response (CEICSH

046/2022). Following this, contact was established with an organization employing a significant number of expatriate workers to present the research project and request authorization for its dissemination and data collection. Permission was granted by an organization in the Engineering and Construction sector, which has approximately 250 expatriate employees with access to email.

Data collection began in April 2024 and spanned three weeks. It was conducted online via the Google Forms platform. To promote the questionnaire, a request for approval was made to department directors, allowing the questionnaire link to be shared through email. Once authorized, the link to the research questionnaire was sent to all expatriate employees with email access, along with an introduction to the study and a formal participation request.

Participation in the study commenced with informed consent, which detailed the research topic, objectives, and ethical considerations, such as the voluntary nature of participation, anonymity, and confidentiality. Participants could only proceed with completing the protocol after providing their consent.

12.2.4 Data Analysis

The data was analyzed using the Statistical Package for Social Sciences software (SPSS; version 27.0 for Windows). Initially, descriptive analyses were carried out for all the variables involved and the assumptions for the parametric tests were checked.

Tests were carried out for differences in means according to certain variables (marital status, continent of destination country and functional group) and the dimensions of socio-cultural adaptation using parametric tests (t-Student and One-Way ANOVA) and non-parametric tests (Mann–Whitney and Kruskal Wallis). Correlations were also carried out, using Pearson's test when the assumptions were met and Spearman's when they were not. If the assumptions were not met and the results of the two types of test agreed, we opted to report the results of the parametric tests, following the recommendations of Fife-Schaw (2006).

12.3 Results

In this study, the impact of change on the adaptation of expatriate workers was analyzed using the variables motivation for expatriation, organizational support, support from colleagues and support from managers. The aim was also to see if there were any differences in terms of socio-cultural adaptation according to marital status, continent of destination and functional group.

Table 12.2 shows a descriptive analysis of all the variables present in the project. The range, minimum, maximum, mean and standard deviation for each of the variables are shown. The highest average was for Social Communication and the lowest

Table 12.2 Range, minimum, maximum, mean and standard deviation of all variables

	Ampl	Min	Max	Mean	Standard deviation
Social adaptation	4.00	1.00	5.00	3.64	0.64
Social communication	3.67	2.67	6.33	4.88	0.69
Social relations	4.00	1.00	5.00	3.56	0.73
Language proficiency	4.00	1.00	5.00	3.85	0.83
Performance at work	3.00	2.00	5.00	4.17	0.48
Peer support	3.00	1.00	4.00	2.13	0.58
Organizational support	3.50	3.00	6.50	4.76	0.68
Management support	3.75	1.00	4.75	2.06	0.65
Motivation for expatriation	3.89	1.95	5.84	3.80	0.52

for the variable Support from Managers. Similarly, there was a greater range in the Social Adaptation, Social Relations and Language Proficiency variables.

12.3.1 Sociocultural Adaptation of Expatriate Workers Scale

In addition, the psychometric properties of the Sociocultural Adaptation of Expatriate Workers scale were also studied. In order to study the validity of the workers' sociocultural adaptation scale, an exploratory factor analysis of the main components was carried out with varimax rotation, and the Kraiser normalization criterion was considered. The analysis of fidelity (internal consistency) was carried out by calculating the Cronbach's Alpha coefficient.

In relation to this, the KMO measure (Kaiser–Meyer–Olkin measure of sampling adequacy) verified that the sample is suitable for analysis, with a value of 0.819, which is above the acceptable limit of 0.50 (Field, 2005). On the other hand, Bartlett's test of sphericity $\chi^2(190) = 1398.56$, $p < 0.001$, indicated that the correlation between the items was large enough to allow the analysis to continue. As with the original scale, 5 components were extracted, called factors, which explained 63.77% of the total variance. Table 12.3 shows the saturation weights of the 21 items of the Sociocultural Adaptation scale for the 5 factors obtained.

As can be seen in Table 12.3, the first factor explains 16.13% of the total variance, saturating items 2, 7, 15, 16 and 17. Next, the second factor, which explains 14.05% of the total variance, was called Social Adaptation, because the items that make it up present a set of behaviors that imply adaptation to the society in which the individual is inserted (for example, "Adapt to the amount of noise in my neighborhood."). The third factor was called Social Relations because it involves actions that imply that the individual relates to others in the society in which they live (for example, "Building and preserving relationships"). In addition, the fourth factor includes items relating to

Table 12.3 Results of the exploratory factor analysis of the sociocultural adaptation scale

Items	h2*	Factor 1	Factor 2	Factor 3	Factor 4	Factor 5
1. Building and preserving relationships	0.668	0.455	0.040	**0.667**	0.063	0.104
2. Manage my work/ school	0.580	**0.611**	0.140	0.252	0.333	0.109
3. Interact at social events	0.692	0.267	0.097	**0.735**	0.013	0.266
4. Keeping my hobbies and interests	0.626	−0.007	0.235	**0.755**	−0.026	−0.012
5. Adapt to the amount of noise in my neighborhood	0.623	−0.183	**0.740**	0.152	0.097	−0.095
6. Interpret and respond accurately to the gestures and facial expressions of others	0.427	0.314	0.342	0.088	0.214	0.3**98**
7. Working effectively with other work colleagues	0.571	**0.698**	0.017	0.121	0.026	0.263
8. Provide requested communication services	0.668	0.054	**0.588**	0.287	−0.065	0.482
9. Adapt to population density	0.730	0.286	**0.772**	0.086	−0.53	0.206
10. Perceiving and speaking	0.842	0.097	0.033	0.028	**0.911**	−0.001
11. Varying the form of my speech in one way culturally appropriate	0.420	0.118	0.332	0.022	**0.484**	0.247
12. Receiving feedback from other co-workers to help me to improve my work performance	0.578	0.190	−0.168	0.300	0.026	0.6**50**
13. Interpreting and responding accurately to the emotions of other people	0.797	0.181	0.261	0.049	0.125	0.8**23**

(continued)

Table 12.3 (continued)

Items	h2*	Factor 1	Factor 2	Factor 3	Factor 4	Factor 5
14. Being present, or participate in community activities	0.731	0.019	−0.168	0.300	0.026	**0.650**
15. Finding my way	0.607	**0.705**	0.309	−0.078	0.056	0.067
16. Interacting with opposite sex	0.679	**0.773**	0.161	0.234	0.010	−0.013
17. Express my ideas in a culturally appropriate way to other work colleagues	0.695	**0.669**	0.338	0.116	0.122	0.324
18. Dealing with bureaucracy	0.420	0.166	**0.504**	0.333	0.161	0.037
19. Adapting to the rhythm of life	0.548	0.447	**0.492**	0.158	0.267	0.096
20. Reading and writing	0.853	0.077	−0.010	−0.045	**0.918**	0.034
Variance explained by factor		**16.13**	**14.05**	**12.38**	**11.10**	**10.19**

Note * h2—communalities
Bold numbers means that p <.05

the language of the destination country, and the name of the original scale, Language Proficiency, has been retained.

The main difference between this factor and the original scale is based on the number of items. Specifically, in this adaptation the factor consists of three items (10, 11 and 20) and in the original, only two (10 and 20). Finally, the last factor was called Social Communication, since it includes items related to the individual's verbal or non-verbal communication with other members of the adapting society (for example, "Being present or participating in community activities"). In naming the five dimensions extracted, the name given to the dimensions in the original version was considered.

12.3.2 Cronbach's Alpha Coefficient

Next, the internal consistency (Cronbach's Alpha coefficient) was analyzed for each of the subscales, and it was found that this coefficient ranged from 0.66 for social communication to 0.82 for work performance. The coefficients for all the sub-scales can be seen in the following Table 12.4. With the exception of the social communication subscale, the other subscales have a Cronbach's Alpha value of over 0.70,

Table 12.4 Descriptive statistics, matrix of correlations between the dimensions of the sociocultural adaptation scale and cronbach's alpha for each dimension (shown diagonally)

Dimension	Average	Standard deviation	1	2	3	4	5
1.Performance at work	4.17	0.404	(0.82)				
2. Social adaptation	3.64	0.772	0.574**	(0.78)			
3.Social relations	3.56	0.835	0.456**	0.470**	(0.71)		
4. Language proficiency	3.85	1.05	0.302**	0.253**	0.099	(0.75)	
5. Social communication	4.88	0.537	0.518**	0.607**	0.593**	0.246**	(0.66)

Note ** $p < 0.01$

and therefore, according to Field (2005), acceptable or good reliability values. With regard to the social communication subscale, as the value obtained is close to the acceptable value, it was decided to consider including it in the following analyses. In any case, it is important to bear in mind that the results obtained will be read with due reservations. Table 12.4 also shows the results obtained between the correlations of the different sub-scales of sociocultural adaptation. All the scales are significantly correlated with each other, with the exception of the correlation between social relationships and language proficiency.

12.3.3 *Relationship Between the Variables*

To study the relationship between the variable's sociocultural adaptation, motivation for expatriation, organizational support, support from managers and support from colleagues, a Pearson correlation test was carried out. The results obtained (Table 12.5) show statistically significant correlations, both positive and negative, between the variables included in the project.

In order to study the differences between sociocultural adaptation according to the gender of the participants, the Student's t-test was used. From the results obtained, it can be seen that there are no statistically significant differences in the sociocultural adaptation of the female and male participants ($t = 1.38$; $p > 0.05$).

The t-Student test was also used to study the differences between sociocultural adaptation according to the country of expatriation. In this analysis, only the European continent was compared with the African continent, since the number of expatriate participants in Asia ($n = 8$) and America ($n = 4$) was very small. From the results obtained, it can be seen that there are statistically significant differences in the sociocultural adaptation of the participants depending on the destination continent. In

Table 12.5 Pearson correlations between variables

Dimension	Organizational support	Management support	Peer support	Motivation for expatriation
Organizational Support	–	–		
Management support	−0.356**	–		
Peer support	−0.317**	0.370**	–	
Motivation for expatriation	0.157*	−0.183*	−0.172*	–
Social adaptation	0.101	−0.061	−0.011	0.228**
Social communication	0.205*	0.011	−0.153	0.202*
Social relations	0.168*	−0.166*	−0.071	0.144
Language proficiency	0.768	−0.094	−0.159*	0.151
Performance at work	0.179*	−0.078	−0.074	0.289**

Note * $p < 0.05$; ** $p < 0.01$

Table 12.6 Results for the independent samples T-test continent of destination country

	Europe ($n = 52$) Mean (SD)	Africa ($n = 93$) Mean (SD)	t (143)
Socio-cultural adaptation	3.92 (0.43)	3.77 (0.40)	2.03*

Note * $p < 0.05$

other words, it can be seen that participants from the European continent show greater socio-cultural adaptation compared to the other continents (Table 12.6).

The One-Way ANOVA test was used to study the differences between the participants' sociocultural adaptation according to their marital status and functional group. The results show that there are no statistically significant differences in the socio-cultural adaptation of the participants according to their marital status ($F = 1.91$; $p > 0.05$) and their functional group ($F = 2.17$; $p > 0.05$).

12.3.4 Characterization of the Motivation for Expatriation

Following the results of the correlations obtained above (Table 12.5), the need arose to go into greater depth knowledge about some of the variables under study, in order to achieve more beneficial results for its objective. Thus, the motivation variable

Table 12.7 Dimensions of motivation for expatriation

	M	DP	Median	Min	Max
Motivation					
Autonomous	4.12	0.59	4.00	2.56	6.33
Controlled	3.05	0.62	3.50	1.40	5.30

can also be divided into two dimensions, autonomous motivation and controlled motivation (Table 12.7).

A frequency analysis was carried out on the dimensions of this variable, obtaining the mean, standard deviation, median, maximum and minimum. Table 12.7 shows the characterization of the two dimensions of this scale. The results show a higher average value for autonomous motivation compared to controlled motivation.

12.3.5 *Predictors of Adaptation*

In order to explore the motivational predictors of sociocultural adaptation, a multiple linear regression was used. The results show a statistically significant model $F(2, 154) = 10.793$, $p < 0.001$. The contribution of each is shown in Table 12.8 below.

In addition, a multiple linear regression was also carried out to see which of the supports (organizational, management and colleagues) corresponded to a better predictor of the participants' sociocultural adaptation. However, the results obtained indicate that the model is not statistically significant, $F(3, 153) = 1.949$, $p > 0.05$.

12.4 Discussion

The ongoing trend of globalization continues to shape how organizations operate, with significant implications for their business practices in the foreseeable future. As companies expand their activities globally, the demand for employees willing and

Table 12.8 Motivational predictors of sociocultural adaptation

	Non-standardized coefficients		Standardized coefficients		
	B	Error	Beta	t	Sig
(Constant)	2.653	0.263		10.078	0.000
Autonomous motivation	0.295	0.072	0.365	4.116	0.000
Controlled motivation	−0.021	0.064	−0.030	−0.333	0.74

prepared to take on international assignments is increasing (Varma & Russel, 2016). Against this backdrop, this project sought to address two primary objectives. First, it aimed to identify the main challenges expatriate workers face when adapting to their host countries. Second, it explored strategies that organizations could develop and implement to facilitate and enhance the adaptation process for expatriate employees.

12.4.1 Discussion of the Results and Implications for the Intervention

After analyzing the results, it can be seen that the socio-cultural adaptation of expatriate workers will be influenced by their motivation. According to Ryan and Deci (2000), motivation corresponds to energy, direction, persistence and intention, and is related to productivity. In other words, human beings only produce when they are motivated (Ryan & Deci, 2000). It is therefore possible to conclude that the greater the participants' autonomous motivation for expatriation, the greater their sociocultural adaptation competence. This result corroborates the findings of Lang's study (2015), in which he argues that sociocultural context factors such as competence, autonomy and familiarity enhance and sustain motivation, more specifically intrinsic motivation (autonomous motivation).

In addition, according to Vianen et al. (2004), it was possible to analyze the participants' cultural adjustment, i.e. the degree of "psychological comfort" with which an expatriate deals with the different cultural and social aspects of the host country. According to the authors cited above, this degree of comfort is made up of (1) performance at work, (2) interaction with the locals, and (3) organizational supervision. From the results in Table 12.2, it can be seen that work performance has a higher average value than the other factors. In other words, the participants in this project show a high level of competence in terms of work performance, which leads to an increase in the degree of comfort experienced by the expatriate worker (Vianen et al., 2004).

On the other hand, the second aspect that influences expatriates' adjustment is interaction with locals (i.e. social relations). After analyzing the results, the average for this factor in Table 12.3 is the lowest compared to the other factors. In this way, it can be seen that relationships with other people and social interaction need to be worked on and improved with this group of people. In fact, communication is the most important means in the context of interculturality, as it facilitates dialog, exchange and mutual understanding between individuals from different cultures (Sarmento, 2015).

In order to improve this variable, the organization can promote mentoring programs, where experienced locals guide the expatriates and help with their integration. Mentors can provide cultural and professional insights, making it easier for expats to adapt to the new environment (Caligiuri, 2000). They can also promote open communication and hold regular feedback sessions. This can help to identify and resolve integration problems quickly. Encouraging both expatriates and locals to voice their concerns and suggestions contributes to a more positive environment (Chen & Shaffer, 2018). Finally, Andresen et al. (2012) also argue that another strategy that enhances interaction focuses on including expatriates in mixed project teams with local employees. This promotes collaboration and knowledge sharing. Working together towards common goals can strengthen working and personal relationships (Andresen et al., 2012).

With regard to support from the organization, the results obtained (Table 12.2) show that this variable has a high average compared to the others. In this way, it is possible to see that the participants report feeling support from the organization. In fact, Lee and Qomariyah (2015) demonstrated the importance of organizational support for expatriates' optimal adjustment to their mission. Employees believe that the company should play a key role in their adaptation, contributing to their well-being in the host country (Briscoe et al., 2008). This evidence leads to the conclusion that the participants have a good degree of cultural adjustment, i.e. a good level of interaction with the locals and a good perception of performance, without the influence of stress crises (Lee & Qomariyah, 2015).

The results showed that this variable was positively and significantly related to work performance, social communication, social interaction and motivation for expatriation. In other words, the greater the expatriates' organizational support, the greater their performance, communication, interaction and motivation. Furthermore, this variable is not significantly correlated with social adaptation. Therefore, in order to enhance organizational support for expatriates, Varma and Russel (2016) argue that organizations need to guarantee adequate training before expatriation and create conditions for expatriates to receive adequate and specific information about their role in the destination country. For example, offer training on cultural differences, local customs, social norms and business practices in the destination country. Also, offer help in learning the foreign language, in order to facilitate communication with the society of the destination country (Dowling et al., 2013).

With regard to the variable support from managers, the average obtained for this variable was lower. Furthermore, in terms of correlations (Table 12.4), it was only positively and significantly related to support from colleagues. In turn, it is negatively and significantly related to motivation for expatriation and social relationships. In this sense, the study by Varma and Russel (2016) reveals that, as indicated above, expatriate support should provide direct and indirect support to expatriates and their families in terms of language training and social relationships (Bhatti et al., 2013).

In addition, Haslberger and Brewster (2009) advocate the importance of continuous communication with expatriates during the international assignment. In other words, holding regular meetings to assess the worker's adaptation and identify areas where they may need more support. In addition, feedback also stands out as a powerful tool, which can help the expatriate adapt, through continuous and constructive feedback on the worker's performance and adaptation (Haslberger and Brewster, 2009).

In addition, other aspects such as support from managers in clarifying the expatriate worker's career development, help with self-development and the availability of materials needed for growth at work are also key to a good adaptation (Bhatti et al., 2013). On the other hand, logistical assistance (e.g. detailed guides on life in the new country, information on culture, transport, health and other essential services), family guidance, psychological counselling and social activities (such as exploring the new city through organized tours or tips on places to visit). This direct and indirect support will help expats to carry out their professional activities in a comfortable environment, interact easily with host country nationals and facilitate general adaptation (Bhatti et al., 2013).

Unlike most studies, no statistically significant differences were found in terms of the gender of the participants. In fact, one study from 2022 was in line with the evidence previously found in the literature. That is, male adaptation turns out to be more competent than female adaptation in aspects such as the need for support, the prevalence of gender roles in society, among others (Wurtz, 2022). Possibly, in this study, this result was not found because the sample of male participants was five times larger than the female sample.

Finally, the results also indicate that expatriate participants in Europe have a greater capacity to adapt, compared to expatriates in Africa. Since the participants were Portuguese (which was a criterion for inclusion in this study) the culture shock was not as great, or did not exist, when compared to other countries on different continents. In other words, in these cases there was not culture shock in its classic definition (i.e. the anxiety resulting from the loss of all the familiar signs and symbols of social interaction Oberg, 1960), but a mismatch, sustained over time, between the conduct, values and behaviors accepted in the destination society and the principles of the expatriate participant (González & Oliveira, 2011).

With regard to the organization's strategies for reducing culture shock, Ward and Kennedy (1999) argue that it is important for expatriates to maintain an open and flexible attitude in order to accept and adapt to new situations and customs. They should also avoid constantly comparing the new culture with their culture of origin (Ward & Kennedy, 1999). In addition, they should promote self-care practices, such as exercise, meditation and hobbies, with the aim of reducing stress and anxiety. Finally, it is important to seek psychological support or a counselor, so as to have someone to support them in this change and, if necessary, to deal with feelings of culture shock (Bennett, 1993).

This study highlights the importance of the variable's motivation, management support and organizational support as influencers of expatriates' sociocultural adaptation to their destination country. It was possible to see that these are intrinsically involved, and it was possible to meet the objectives. In other words, it was possible to understand some of the main difficulties faced by expatriate employees in adapting to their destination country, and it was also possible to understand some strategies that can help employees adapt to the expatriate context.

12.4.2 Limitations and Proposals for Future Work

Despite the various advances in expatriation research, the literature suggests more research, specifically in geographical contexts that are still unknown. This study not only responds to the organization's request, but also supports the existing literature in this area. Starting with cultural influence, where the evidence obtained reveals the need to pay greater attention to cultural differences in order to understand the specificities of expatriate management in each country. And concluding with more classic topics such as gender, experience and types of expatriation.

It is important to note that this study has some limitations. Starting with the fact that it is not possible to generalize these results since the sample is relatively small and the fact that this study is cross-sectional rather than longitudinal. As such, this evidence does not allow causal inferences to be made. In addition, the fact that it was only carried out in one company and in one context are also limitations.

The variables were characterized as external and non-controllable, which could lead to a "manipulation" of the participants' responses; in particular, social desirability and the potential denial or minimization of problems.

Despite the limitations mentioned above, we believe that the results obtained will allow us to discuss with the organization hosting the internship/project possibilities for intervention in promoting the adaptation of expatriate workers to their country of destination.

References

Andresen, M., Bergdolt, F., Margenfeld, J., & Dickmann, M. (2012). Addressing international mobility confusion: Developing definitions and differentiations for self-initiated and assigned expatriates as well as migrants. *The International Journal of Human Resource Management, 23*(11), 2295–2318. https://doi.org/10.1080/09585192.2011.610342

Anwar, G., & Abdullah, N. N. (2021). The impact of Human resource management practice on organizational performance. *International Journal of Engineering, Business and Management, 5*(1), 35–47. https://doi.org/10.22161/ijebm.5.1.4

Aycan, Z. (1997). Expatriate adjustment as a multifaceted phenomenon: Individual and organizational level predictors. *Journal of Human Resource Management, 8*(4), 434–456.

Bahadur, S., Bhatia, R., & Kacker, S. K. (1982). Nerve decompression for intratemporal facial nerve injuries. *Indian Journal of Otolaryngology, 34*(4), 64–68. https://doi.org/10.1007/BF02992358

Bennett, M. J. (1993). Towards ethnorelativism: A developmental model of intercultural sensitivity. In R. M. Paige (Ed.), *Education for the intercultural experience* (pp. 21–71). Intercultural Press.

Bhatti, M. A., Kaur, S., & Battour, M. M. (2013). Effects of individual characteristics on expatriates' adjustment and job performance. *European Journal of Training and Development, 37*(6), 544–563. https://doi.org/10.1108/EJTD-02-2013-0013

Black, J. S. (1988). Work role transitions: A study of American expatriate managers in Japan. *Journal of International Business Studies, 19*(2), 277–294. https://doi.org/10.1057/palgrave.jibs.8490383

Bonache, J. (2005). Job satisfaction among expatriates, repatriates and domestic employees: The perceived impact of international assignments on work-related variables. *Personnel Review, 34*(1), 110–124. https://doi.org/10.1108/00483480510571905

Boneva, B. S., & Frieze, I. H. (2001). Toward a concept of a migrant personality. *Journal of Social Issues, 57*(3), 477–491.

Braun, B., Krampf, A., & Murau, S. (2021). Financial globalization as positive integration: Monetary technocrats and the Eurodollar market in the 1970s. *Review of International Political Economy, 28*(4), 794–819. https://doi.org/10.1080/09692290.2020.1740291

Briscoe, J. P., Hall, D. T., & DeMuth, R. L. F. (2008). Protean and boundaryless careers: An empirical exploration. *Journal of Vocational Behavior, 73*(3), 341–352. https://doi.org/10.1016/j.jvb.2008.07.003

Caligiuri, P. M. (2000). The Big Five personality characteristics as predictors of expatriates' desire to terminate the assignment and supervisor-rated performance. *Personnel Psychology, 53*(1), 67–88. https://doi.org/10.1111/j.1744-6570.2000.tb00194.x

Chen, M. (2019). The impact of expatriates' cross-cultural adjustment on work stress and job involvement in the high-tech industry. *Frontiers in Psychology, 10*, 1–10. https://doi.org/10.3389/fpsyg.2019.02228

Chen, Y.-P., & Shaffer, M. (2018). The influence of expatriate spouses' coping strategies on expatriate and spouse adjustment. *Journal of Global Mobility, 6*(1), 20–39. https://doi.org/10.1108/JGM-04-2017-0012

Chen, Y. P. (2012). *A three-stage process model of self-initiated expatriate career transitions: A self-determination theory perspective*. https://doi.org/10.2307/j.ctv2x1npz7.6

Church, A. T. (1982). Sojourner adjustment. *Psychological Bulletin, 91*(3), 540–572. https://doi.org/10.1037/0033-2909.91.3.540

Colakoglu, S., & Caligiuri, P. (2008). Cultural distance, expatriate staffing and subsidiary performance: The case of US subsidiaries of multinational corporations. *The International Journal of Human Resource Management, 19*(2), 223–239.

Crooke, S. T., Liang, X. H., Baker, B. F., & Crooke, R. M. (2021). Antisense technology: A review. *Journal of Biological Chemistry, 296*, 100416. https://doi.org/10.1016/j.jbc.2021.100416

De Paul, N. F., & Bikos, L. H. (2015). Perceived organizational support: A meaningful contributor to expatriate development professionals' psychological well-being. *International Journal of Intercultural Relations, 49*, 25–32. https://doi.org/10.1016/j.ijintrel.2015.06.004

Dos Santos, J. V., & Gonçalves, G. (2013). Contribuição para a adaptação portuguesa da escala de Percepção de Suporte Organizacional de Eisenberger, Huntington, Hutchison e Sowa (1986). *Laboratório de Psicologia, 8*(2), 213–223. https://doi.org/10.14417/lp.642

Dowling, P. J., Festing, M., & Engle, A. D. (2013). *International human resource management* (6th ed.). Cengage Learning.

Fan, D., Wu, S., Su, Y., & Li, Y. (2022). Managing expatriates to achieve mutual benefits: An integrative model and analysis. *Journal of International Management, 28*(2). https://doi.org/10.1016/j.intman.2021.100882

Field, A. P. (2005). Is the meta-analysis of correlation coefficients accurate when population correlations vary. *Psychological Methods, 10*(4), 444.

Fife-Schaw, C. (2006). Questionnaire design. In G. M. Breakwell, S. Hammond, C. Fife Schaw, & J. A. Smith (Eds.), *Research methods in psychology* (3rd ed., pp. 161–176). Sage Publications.

Gagné, M., Forest, J., Vansteenkiste, M., Crevier-Braud, L., van den Broeck, A., Aspeli, A. K., Westbye, C. (2015). The multidimensional work motivation scale: Validation evidence in seven languages and nine countries. *European Journal of Work and Organizational Psychology, 24*(2), 178–196. https://doi.org/10.1080/1359432X.2013.877892

Ganguli, R., Padhy, S. C., & Saxena, T. (2022). The characteristics and preferences of Gen Z: A review. *IUP Journal of Organizational Behavior, 21*(2), 79–99.

Gittins, T., Lang, R., & Sass, M. (2015). The effect of return migration driven social capital on SME internationalisation: A comparative case study of IT sector entrepreneurs in Central and Eastern Europe. *Review of Managerial Science, 9*, 385–409.

González, J. M. R., & Oliveira, J. A. de. (2011). Os efeitos da expatriação sobre a identidade: estudo de caso. *Cadernos EBAPE.BR, 9*(4), 1122–1135. https://doi.org/10.1590/s1679-39512011000400011

Hall, D. T., & Moss, J. E. (1998). The new protean career contract: Helping organizations and employees adapt. *Organizational Dynamics, 26*(3), 22–37. https://doi.org/10.1016/s0090-2616(98)90012-2
Haslberger, A., & Brewster, C. (2009). Capital gains: Expatriate adjustment and the psychological contract in international careers. *Human Resource Management, 48*(3), 379–397. https://doi.org/10.1002/hrm.20286
Hussain, T., & Zhang, Y. (2023). The influences of cross-cultural adjustment and motivation on self-initiated expatriates' innovative work behavior. *Personnel Review, 52*(4), 1255–1272.
Inverno, A. M. F. S. (2011). *O impacto cultural na expatriação e a influência das práticas de selecção e de formação na adaptação ao novo contexto cultural.*
Karasek, R. A. (1985). The job content questionnaire (JCQ): An instrument for internationally comparative assessments of psychosocial job characteristics. *Journal of Occupational Health Psychology, 10*(4), 322–335.
Koser, K., & Salt, J. (1997). The geography of highly skilled international migration. *International Journal of Population Geography, 3*(4), 285–303.
Kour, S., & Jyoti, J. (2022). Cross-cultural training and adjustment through the lens of cultural intelligence and type of expatriates. *Employee Relations, 44*(1), 1–36. https://doi.org/10.1108/ER-07-2020-0355
Lee, L. Y., & Qomariyah, A. (2015). The effects of work-role demands on cross-cultural adjustment and expatriate effectiveness: A meta-analysis. *Anthropologist, 22*(3), 636–649.
Li, N. N. (2012). *Seven secrets to success as an expat executive.* Forbes.
Mäkelä, L., Kangas, H., & Suutari, V. (2019). Satisfaction with an expatriate job: The role of physical and functional distance between expatriate and supervisor. *Journal of Global Mobility, 7*(3), 255–268. https://doi.org/10.1108/JGM-04-2019-0025
Malveiro, S. V. (2016). *Relação entre as características do trabalho e o engagement à luz da teoria da Autodeterminação.*
Mantzaris, K., & Myloni, B. (2023). Human vs technology: A cross-cultural comparison of HR professionals' perceptions. *International Journal of Manpower, 44*(1), 58–76. https://doi.org/10.1108/IJM-05-2020-0197
Martin, D.C., & Bartol, K. M. (2003).Factors influencing expatriate performance appraisal system success: An organizational perspective. *Journal of International Management.*
McNulty, Y., & Brewster, C. (2020). The changing context of expatriation and its impact on crosscultural management. *The SAGE handbook of contemporary cross-cultural management* (pp. 424–436).
Mello, R., Suutari, V., & Dickmann, M. (2023). Taking stock of expatriates' career success after international assignments: A review and future research agenda. *Human Resource Management Review, 33*(1), 100913. https://doi.org/10.1016/j.hrmr.2022.100913
Murray, W. C., & Holmes, M. R. (2021). Impacts of employee empowerment and organizational commitment on workforce sustainability. *Sustainability (Switzerland), 13*(6). https://doi.org/10.3390/su13063163
Oberg, K. (1960). Cultural shock: Adjustment to new cultural environments. *Practical Anthropology, 4*, 177–182.
Olsen, J. E., & Martins, L. L. (2009). The effects of expatriate demographic characteristics on adjustment: A social identity approach. *The University of Michigan and in Alliance with the Society of Human Resources Management, 48*(2), 311–328.
Osland, J. S. (2000). The journey inward: Expatriate hero tales and paradoxes. *Human Resource Management Review, 39*(2–3), 227–238.
Pires, R. P., Pereira, C., Azevedo, J., Inês Espírito-Santo, I. V. (2016). Emigração Portuguesa. Relatório Estatístico. *Observatório Da Emigração*, 280. https://doi.org/10.15847/CIESOEMRE032016
Pires, R. P. (2021). A emigração portuguesa no século XXI. In J. Peixoto, J. Malheiros & P. Marques (Eds.), *Migrações e Mobilidade: Novos Desafios e Perspetivas* (pp. 45–67). Imprensa de Ciências Sociais. https://journals.openedition.org/cs/

Reiche, S. (2011). *Expatriation close-up–calling things by their right names*. Blog.Iese. https://blog.iese.edu/expatriatus/2011/08/08/expatriation-close-up-calling-things-by-their-right-names/
Rosaria Mauro, M. (2021). Commercio internazionale e tutela dei diritti dei lavoratori: alla ricerca di una globalizzazione sostenibile. *E-Revista Internacional De La Proteccion Social, 2*(6), 372–401. https://doi.org/10.12795/e-rips.2021.i02.17
Ryan, R. M., & Deci, E. L. (2000). Intrinsic and extrinsic motivations: Classic definitions and new directions. *Contemporary Educational Psychology, 25*(1), 54–67.
Santos, G. A. (2015). *A motivação dos expatriados e o seu bem-estar no trabalho: estudo exploratório numa empresa de construção EM ÁFRICA*. Universidade de Lisboa.
Dos Santos, N. R., Mónico, L., Pais, L., Gagné, M., Forest, J., Cabral, P. M. F., & Ferraro, T. (2022). The multidimensional work motivation scale: Psychometric studies in Portugal and Brazil. *Management Research: The Journal of the Iberoamerican Academy of Management, 20*(2), 89–110. https://doi.org/10.1108/MRJIAM-07-2021-1206
Sarmento, C. (2015). *Semióticas da comunicação intercultural: Da teoria às práticas*. Centro de Estudos Interculturais, ISCAP-P.PORTO.
Selmer, J., & McNulty, Y. (ed.). (2017). *Research handbook of expatriates* (p. 576). Edward Elgar Publishing Limited.
Selmer, J., & Lauring, J. (2015). Host country language ability and expatriate adjustment: The moderating effect of language difficulty. *The International Journal of Human Resource Management, 26*(3), 401–420. https://doi.org/10.1080/09585192.2011.561238
Selmer, J., Suutari, V., & Brewster, C. (2022). Self-initiated expatriates. In *Expatriates and managing global mobility* (pp. 114–136).
Steger, M. B., Benedikter, R., Pechlaner, H., & Kofler, I. (2023). *Globalization*. University of California Press.
Van Vianen, A. E. M., De Pater, I. E., Kristof-Brown, A. L., & Johnson, E. C. (2004). Fitting in: Surface- and deep-level cultural differences and expatriates' adjustment. *Academy of Management Journal, 47*(5), 697–709. https://doi.org/10.5465/20159612
Varma, A., & Russell, L. (2016). Women and expatriate assignments: Exploring the role of perceived organizational support. *Employee Relations, 38*(2), 200–223. https://doi.org/10.1108/ER-07-2015-0154
Ward, C., & Kennedy, A. (1999). The measurement of sociocultural adaptation. *International Journal of Intercultural Relations, 23*(4), 659–677. https://doi.org/10.1016/S0147-1767(99)00014-0
Ward, W. (2013). *SCAS-R revised sociocultural adaptation scale factor loadings for exploratory factor analysis with oblique Promax rotation of the SCAS-R*.
Wurtz, O. (2022). A transactional stress and coping perspective on expatriation: New insights on the roles of age, gender and expatriate type. *Journal of Global Mobility, 10*(3), 351–372. https://doi.org/10.1108/JGM-10-2021-0086
Zhou, Y. (2020). An empirical study on expatriation success of MNCs: Evidence from Japan. *International Journal of Manpower, 42*(3), 394–408. https://doi.org/10.1108/IJM-02-2019-0068

Part VII
Thematic Area: Sustainability and Corporate Social Responsibility

Chapter 13
Work Time Reduction: A Critical Analysis of the Main Arguments for Adopting a 4-Day Workweek

Ana Oliveira, Sofia Alexandra Cruz, Ana Isabel Couto, Jorge Cerdeira, Cristina Parente, and Carlos Manuel Gonçalves

Abstract Over the past five years, the private sector, policymakers, and civil society have developed a growing interest in reduced working hours. The rapid growth of 4-day workweek pilots in multiple firms across different countries has led to considerable scholarly attention regarding this topic. However, most of this discussion is taking place outside of the academic domain. The lack of representativeness of academic literature on the ongoing debate implies that a comprehensive scientific review of the arguments about the 4-day workweek is still missing. This chapter fills this gap by critically assessing the most frequently cited arguments for adopting shorter weeks—increasing productivity, improving well-being, correcting gender imbalances, and reducing environmental impact—showing that some of these arguments lack sound scientific support. It frames the 4-day workweek discussion in the general debate about reducing working time in different periods in time and discusses the results of the main 4-day workweek pilots carried out over the most recent years.

Keywords Working time · Work time reduction · 4-day workweek · 4-day workweek arguments

---

A. Oliveira
Utrecht School of Economics, Utrecht University, Utrecht, Netherlands

S. A. Cruz · A. I. Couto (✉)
Faculty of Economics of University of Porto, Institute of Sociology of the University of Porto, Porto, Portugal
e-mail: aicouto@fep.up.pt

A. I. Couto
CEOS.PP, ISCAP, Polytechnic of Porto, Porto, Portugal

J. Cerdeira · C. Parente · C. M. Gonçalves
Faculty of Arts and Humanities of University of Porto, Institute of Sociology of the University of Porto, Porto, Portugal

J. Cerdeira
Center for Economics and Finance at the University of Porto, Porto, Portugal

C. Machado (ed.), *Human Resource Management in a Disrupted World*, Lecture Notes on Multidisciplinary Industrial Engineering,
https://doi.org/10.1007/978-3-031-89948-5_14

13.1 Brief Introduction

Shorter working weeks and, in particular, implementing a 4-day workweek (4DWW) of 30–32 working hours as a norm have received much attention in the wake of COVID-19, the emergence of the so-called "Great Resignation" phenomena, and growing difficulties in hiring and retaining workers.[1] Academics, political parties, policymakers, and think tanks have been increasingly interested in such debate, reflected in extensive media coverage, new research publications, and a growing number of management-led trials.

In this chapter,[2] we aim to understand how the claims for the (4DWW) have emerged in the context of the evolution of working time since the early twentieth century and how they are framed within the wider debate on working time reduction. We critically review the main arguments for the 4DWW, confronting them with empirical evidence and compiling the results of the most well-known 4DWW pilots.

This chapter has three aims. First, it frames the 4DWW discussion within the general debate on reducing working time in different periods. Second, it explores the most frequently cited arguments in favour of the 4DWW—increasing productivity, improving well-being, correcting gender imbalances, and reducing environmental impact (Campbell, 2023; Eurofound, 2024; Harper & Martin, 2018; Jahal et al., 2023; Veal, 2023)—and confronts them with the existing robust empirical evidence on the effects of reducing working time. Finally, it summarises the design and reported outcomes of the foremost 4DWW pilots conducted over the most recent years and the limitations of these experiments.

13.2 Reducing Working Time in Different Periods in Time

Despite the recent buzz around the 4DWW, this is not the first time that it has been debated. The first reported shortened week dates back to 1930, when 8-h shifts at Kellogg's manufacturing plant in Michigan were reduced to 6 working hours a day, with no wage cuts (Hunnicutt, 1992). Kellogg's primary motivation for introducing shorter shifts was to create more jobs in Battle Creek during the Great Depression. Moreover, the company tried to retain its workforce in a highly routinised job by increasing workers' motivation and satisfaction.[3]

Kellogg's management-led experience of shorter regular working hours became a relevant case study in a context in which the 40-h working week had recently been implemented in only some large plants and a few US states. The path to a statutory maximum of 40 "normal" weekly hours was still under construction. The Fair Labor

[1] https://time.com/charter/6241211/four-day-workweek/.

[2] The research that supports this chapter was conducted in July 2023.

[3] Changes in management in the late 30 s initiated four decades of constant attempts to increase the number of daily working hours back to eight. In 1984, under threat of reallocation and job displacement, workers who were still working under six-hour shifts resigned (Hunnicutt, 1992).

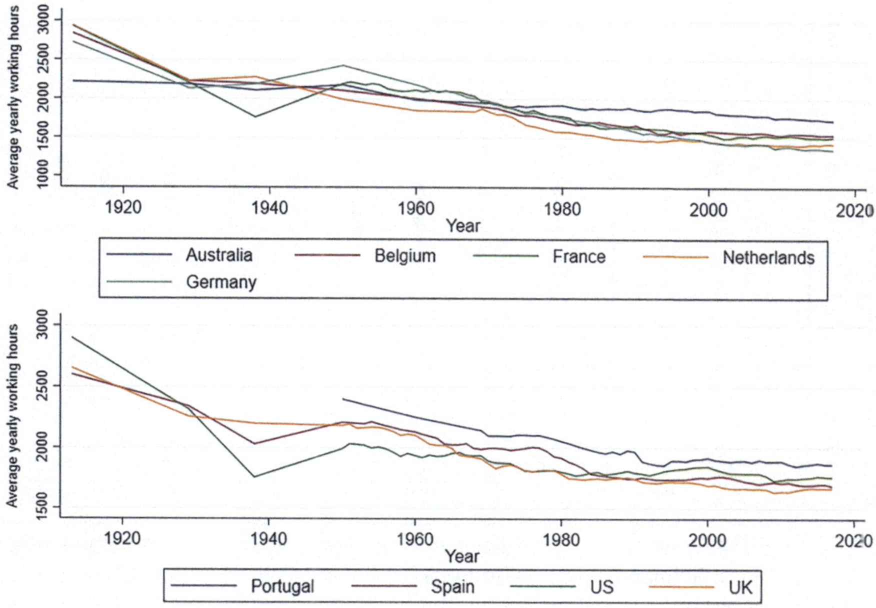

Fig. 13.1 Average yearly working hours. *Note* Based on Giattino et al. (2020) and our world data

Standards Act of 1940 made the 40-h week the law in the USA. By then, the average working week was still 46 h long in the UK (Bank of England, 2023) and over 47 h in manufacturing plants in the Netherlands and Sweden (Gilmore, 2021). Workers in these countries had to wait until the mid-fifties to reach 40 weekly working hours. Figure 13.1 shows the evolution of yearly working hours for a selection of European countries and the US.

Working hours fell significantly before the Second World War: in countries such as the Netherlands and France, average yearly hours fell by 750 between 1913 and 1929. Equally significant changes occurred between 1929 and 1938. The ILO Hours of Work (Industry) Convention, 1919 (No.1), ratified in the late 1920s and early 1930s by countries such as Belgium (1926), France (1927), Portugal (1928), Spain (1929) or Canada (1935), represented an important milestone towards the eight-hour working day in the industrial sector during this period.

The war and post-war efforts restored longer working periods. Of all the countries considered, only the Netherlands and Belgium registered a decrease in the number of hours worked between 1938 and 1950. In all other countries, those increased, sometimes reversing all the progress made in the previous decade.

The pre-war downward trend in working time was resumed in the fifties. Between 1950 and 1970, several European countries registered double-digit reductions in yearly working hours: 10% in Belgium and Spain, 11% in France, 14% in the UK, and 19% in Germany (Fig. 13.2). The increase in the number of vacation days, sick leave, and parental leave were the main drivers of this decrease over this period,

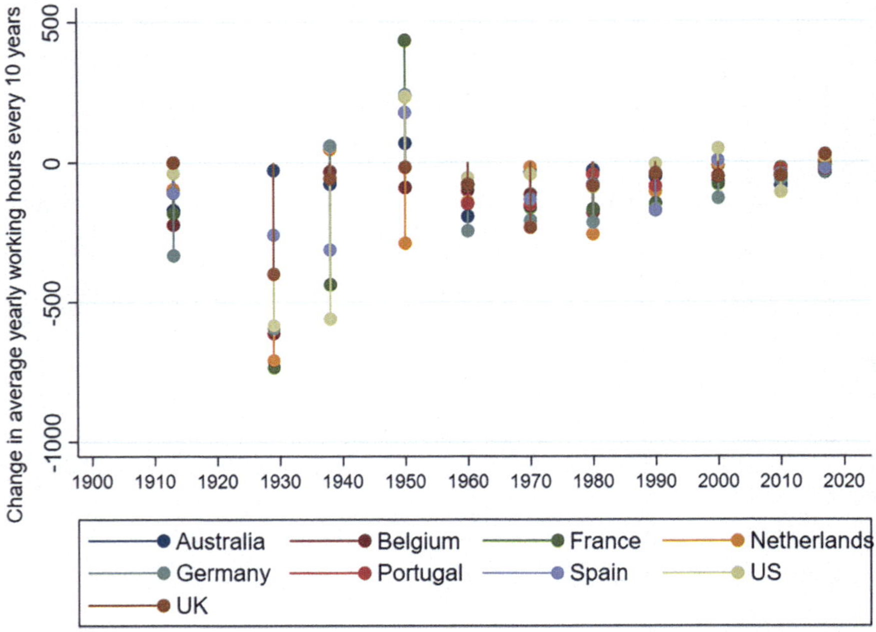

Fig. 13.2 10-year change in average yearly working hours. *Note* Based on Giattino et al. (2020) and our world data

although shorter working days have also been reported, especially in the US (Costa, 2000).

In the 1970s, some US companies tried compressed 4-day working weeks (40 weekly hours in 4 days). A meta-analysis of the literature examining the impacts of compressed working weeks revealed that workers were generally satisfied with the new working schedule, despite reporting increased fatigue (Moores, 1990). Similar pilots were replicated in the UK and Australia. However, compressed working weeks did not become the norm in all companies and there is no evidence that pilot companies adopted this schedule after the experiment ended (Veal, 2023).

As in the 1920s and 1930s, the potential impact on employment (work-sharing) was the most referred argument in favour of reduced working hours during the 1970s, when the US unemployment rate reached 8.8% (US Bureau of Labor Statistics, 2023). Work sharing was again a popular argument to introduce the 35-h standard working week in France in the mid-nineties, motivating a series of economic papers to determine the relationship between employment and working time reductions.[4]

From the 1990s onwards, the decrease in yearly working hours stalled. In the UK, where yearly working hours fell by 17% between 1960 and 1990, only a 3% reduction was registered in the following thirty years; in the US, the 2.5% reduction between

[4] The results are ambiguous, but reducing working hours may increase employment under certain conditions. A compilation of empirical evidence is described in Estevão et al. (2008).

1990 and 2017 contrasted with the 7%-change in the previous period; in Germany, where the most recent change is considerable (−14.5%), it was still far below the change in working time achieved between the 1960s and the 1990s (−27%) (Giattino et al., 2020).

The literature documents two episodes of national reductions in the statutory working week. In 1996, the Portuguese Government reduced the standard working week from 44 to 40 h, affecting around 50% of all private sector employees (Raposo & Ours, 2010). The policy aimed converge with the European working time standards and companies had two years to implement the new working schedule. In 1998, the French Government passed a law reducing the 40-h working week to 35 h, which came into force for all companies in 2002. High unemployment and the expectation that the policy would encourage work-sharing were at the core of the political debate in France at the time (Estevão et al., 2008). However, the impact on employment was not statistically significant (Estevão et al., 2008).

Two countries have introduced compressed 4-day workweeks into their labour law. With the Working Time Act of 1997 and amended in 2007, Austria opened up the possibility of extending the daily working time to 10 h a day to accommodate a 4-day workweek. Belgium, in 2001, ruled for flexible schedules in the private sector of a 9-h working day in 4 days (Cabrita & Boehmer, 2016). The most recent campaigns for the 4-day workweek have instead claimed for an effective reduction in the standard working week, in most cases to 30 or 32 h.

13.3 Arguments for a 4-Day Workweek: Discussion and Empirical Evidence

The arguments in favour of the 4-day workweek (4DWW) cover many fields and are ideologically diversified (Boulin, 2024; Burchell et al., 2024; Sanchis, 2023). The more ancient literature placed more emphasis on the revolutionary ideal of human emancipation, considering the 4DWW as a way of freeing people from the burden of work and giving them more time for leisure and creative activities (Hunnicutt, 1992). Proponents of the 4DWW now emphasise other potential outcomes rather than an increase in leisure time (Veal, 2023).

The arguments in favour of implementing the 4-day workweek can be clustered into the following four themes: (i) increasing productivity, (ii) promoting general well-being, (iii) improve gender imbalances and (iv) reducing environmental impact. These arguments will be discussed in detail in the following sub-sections.

13.3.1 *Increase of the Productivity*

Highly prominent advocates of the 4DWW implementation have defended its role in increasing workers' productivity. In the book "Overworked American: The Unexpected Decline of Leisure", Juliet Schor mentions that "reduced hours are highly correlated with lower emissions, and they also yield improvements in worker well-being, gender equity, and productivity" (Schor et al., 2022); in 2019, the think tank Autonomy states: "there are a number of reasons why a reduction of working hours is desirable and possible with productivity in mind." (Stronge & Harper, 2019); in "Friday is the New Saturday" (2021), Pedro Gomes argues that "the four-day week would reduce this pressure [on mental health] and raise workers' productivity." Similar claims can be found in countless news and reports, often referring to the outstanding results reported after the end of pioneering 4DWW pilots.

In most of these documents, we find references to the impact of long working hours on labour productivity. Long working hours is still common in many countries (ILO, 2023), especially among high-skilled workers (Cortes & Pan, 2019; Kodz et al., 2023). Long working hours are culturally associated with a prototype of a worker who puts their job above everything else, is highly committed to their job, and is highly productive (Kim et al., 2018; Rosengren, 2019; Rutherford, 2001). In certain occupations and workplaces, the ideal worker is always available, can travel and change work locations, and prioritises work over family time and leisure (Weeden et al., 2016; Williams, 2010).

Some authors claim that the association between long working hours and high individual productivity is inaccurate and may involve a performative act to meet hierarchical expectations and seek a career promotion (Heejung, 2022; Weeden et al., 2016). Despite the expectations of employers and employees, empirical evidence shows that increasing individual working hours does not proportionally increase the outcome and decreases hourly productivity (Golden, 2011; Shepard & Clifton, 2000). The negative relationship between long working hours and labour productivity stems from the impact of the former on health outcomes, as well as fatigue and stress (Collewet & Sayermann, 2017; Wong et al., 2019). Rivera et al. (2020) show that people working more than 40 h per week are more likely to suffer from coronary heart disease, and pregnant women are more likely to have premature births and low birth weight. Long working hours are also associated with a higher incidence of musculoskeletal disorders (Johnson & Lipscomb, 2006), cerebrocardiovascular disease (Wong et al., 2019), stress and fatigue (Johnson & Lipscomb, 2006; Pencavel, 2015), depression, anxiety and sleep condition (Afonso et al., 2017; Bannai & Tamakoshi, 2014; Milner et al., 2015; Wong et al., 2019). Long working hours are also related to higher odds of suffering a work-related injury (Dembe et al., 2005; Dong, 2005) and being overweight (Li et al., 2021; Virtanen et al., 2022).

Reducing overtime may have had a positive impact on labour productivity. Firstly, it helps prevent the occurrence of the health hazards described above (Golden et al., 2011; Sparks et al., 2001), thus reducing absenteeism. Secondly, teams whose members experience less fatigue are, on average, more productive (Shangguan et al., 2021) and less likely to make mistakes at work (Landrigan et al., 2004).

Apart from Schor et al. (2022), Gomes (2021) and Autonomy (2019), many other authors (Barnes, 2002; Chung, 2022; Pang, 2020; Spencer, 2022) reference existent empirical evidence on the effects of reducing long working hours (50 or more hours) to argue for a reduction in the standard working week (35–40 h). Such reference bias is likely related to the scarcity of robust and consensual evidence on the causal relationship between regular working hours and labour productivity. Indeed, many studies that aim to prove causality use industry-level data only, which suffers from endogeneity when measuring both working time and output (Collewet & Sayermann, 2017).

One of the few exceptions is Collewet and Sayermann (2017). The authors look at the productivity variation along an 8-h working day and find that for a 1% increase in working time, individual output increases by only 0.9%. Fatigue explains this loss of productivity. Accordingly, one may expect that reducing the length of the working day increases hourly productivity, but it is unknown whether the same applies to reducing the duration of the working week.

Only four other papers examine the relationship between part-time work and company productivity. The results are contradictory. Specchia and Vandenberghe (2013) and Decicienti et al. (2015) found a negative impact of employing a larger share of part-time workers on the productivity of Belgian and Italian companies, respectively. Garnero et al. (2014) and Künn-Nelen et al. (2013), however, found that Belgian and Dutch part-time workers are relatively more productive than full-timer workers. Across the four studies, results vary by gender, part-time duration, and sector.

In a systematic review of studies examining the impact of reduced working hours on health outcomes (while keeping the same wage and workload), Voglino et al. (2022) show that there is a positive relationship between a shorter working week and working life quality, improved sleep habits and reduced stress. Although these positive outcomes seem to depend on the organisational aspects of the workplace and how leisure time is spent (Voglino et al., 2022), shorter working hours contribute to preventing some of the underlying causes of absenteeism (Van De Poll et al., 2020), thus increasing the number of productive days.

David Spencer (2022) argues against placing productivity at the centre of the 4DWW debate. The author states that whenever working time has been reduced on the initiative of the employers to increase productivity, 4-day workweek experiences have either failed (such as at Microsoft Japan, whose pilot was never extended), or back-fired on workers' well-being by keeping the same workload but reducing the time to complete it and increasing work intensity. Work intensification and higher stress levels were reported in 4DWW trials when resources and workload remained unchanged despite the reduced working time (Delaney & Casey, 2022). Similar outcomes were reported in some British companies participating in the 2022 pilot.[5]

[5] https://www.bbc.com/worklife/article/20230319-four-day-workweek-trial-the-firms-where-it-didnt-work.

The "pure" effect of reducing working hours on company productivity—by reducing unproductive time, reducing fatigue, or introducing more efficient management methods (e.g., shorter meetings)—may be difficult to quantify. The lack of robust evidence indicates this. Comparatively, the effect of shorter statutory working hours on absenteeism and stress-related sick leaves is more debated and the existing literature, while also scarce, is more consensual. Overall, we find that the case for the 4DWW as a means of increasing productivity is not well supported by empirical evidence and is based mainly on the literature focused on long working hours.

13.3.2 Promotion of General Well-Being

In the article "A Social Policy Case for the 4-day workweek" (2021), Chung states that "providing workers more capacity to spend time with others can also help improve the well-being of others—such as their children and other family members". Similar claims can be found on the 4-day workweek Global webpage, while most 4DWW pilots have examined the experiment's impact on well-being indicators. It is, thus, safe to say that concerns about physical and psychological well-being have been at the core of the latest 4DWW debate.

Similarly to the productivity argument, many 4DWW advocates refer to the literature focused on long working hours and their consequences for workers' well-being (Gomes, 2021; Heejung, 2022; Lewis & Stronge, 2021). Based on this literature, they suggest that reducing statutory working hours will prevent the negative well-being outcomes observed when employees work long hours, often assuming a linear relationship between the two.

Zheng et al. (2023) show precisely that the relationship between job satisfaction and working hours is non-linear and point out the existence of heterogeneous effects across genders and types of employment.[6] Dong (2005) and Kamerade et al. (2019) also demonstrate that life satisfaction increases as working hours decrease non-linearly. Although these results may sound trivial, they highlight the importance of treating long and regular working hours differently.

The relationship between long weekly working hours, job satisfaction, and health outcomes is well documented. Some of the extensive literature on long working hours and health outcomes—e.g., cardiovascular disease, sleep quality, obesity, stress, and mental health conditions such as depressive and anxiety symptoms—was already described above. Several other papers also provide evidence of the negative effects of long working hours on overall well-being and jobs (e.g., Anfrave & Charlwood, 2015; Zheng et al., 2023).

Although it is reasonable to expect that fewer working hours increase leisure time and free up time for physical activity, which has been shown to positively

[6] In particular, they show that job satisfaction decreases as the number of hours worked increases for married women (but not for unmarried women) and hat job satisfaction decreases for self-employed women only after 12 daily working hours.

affect people's health, the relationship between shorter statutory working weeks and workers' physical and mental well-being is considerably less documented.[7] A few exceptions report the Swedish experiences of working time reductions in various companies from the private (Schwarz & Hasson, 2011; Schwarz et al., 2008) and the public sectors (Schiller et al., 2017) and well-being outcomes. Health benefits were found whenever physical activity replaced working time (Schwarz et al., 2008), and less stress and longer sleep duration were reported by public servants who benefited from shorter working hours (Schiller et al., 2017).

Lepinteur (2019) evaluates the nationwide reduction of the statutory weekly working hours in France and Portugal in the late nineties. The author shows that implementing a shorter working week had positive, robust, and non-transitory effects on overall job satisfaction due to higher satisfaction with working hours and perceived working conditions. The paper also explores heterogeneous outcomes across sectors and gender: men employed in manufacturing and agriculture were relatively more satisfied with their jobs after France introduced the 35-h working week compared to the remaining workforce; in Portugal, the reduction in weekly working hours seemed to increase job satisfaction in all sectors, particularly among women. Berniell and Bientenbeck (2020) look at the impact of the French reform on health-related outcomes and demonstrate that four years after the reform, workers whose working week was reduced to 35 h were less likely to smoke, had a lower BMI and reported better health than workers who were not subject to the reform.

Despite the focus of the 4DWW literature on the hazards of long working hours for workers' general well-being and satisfaction, the non-linear pattern between working time and wellbeing-related variables confirms the need to use the typical working week as the baseline scenario to test whether a 4DWW improves these variables. Studies such as Lepinteur (2019) and Berniell and Bientenbeck (2020) prove the benefits of reducing working time below standard working hours to these outcomes.

13.3.3 Improvement of Gender Imbalances

In 2019, the World Economic Forum claimed that the 4DWW would be a game changer in achieving gender equality.[8] Women's rights organisations, such as UNICEF, the ILO or the UK Women's Budget Group, have appealed to the 4DWW adoption as a strategy to improve women's access to leisure time and more equal wages. Likewise, prominent 4DWW advocates mention the contribution of shorter working weeks to narrow gender gaps at home and at the workplace. It is safe to

[7] Voglino et al. (2022) conducted a systematic review of research publications on quantitative health outcomes of reducing working hours without loss of income, and the pool of publications was reduced to seven, all documenting experiences in Sweden.

[8] https://www.weforum.org/agenda/2019/01/why-the-four-day-work-week-is-good-news-for-gender-equality.

say that these organisations and personalities strongly believe in changing gendered social norms by changing working schedules.

Gender pay gaps are still significant in most countries, although they have been narrowing in recent decades. In OECD countries, women still earn 11.9% less than men, in EU countries 10%, and in the UK and US 15 and 17%, respectively. Social norms, and large inequalities in the distribution of care and household responsibilities between men and women, not only contribute to the persistence of gender gaps in the labour market but also prevent women's equal participation in several societal dimensions (e.g., men are twice as much involved in politics as women (EIGE, 2021).

Across the EU Member States, women spend an average, of 16.5 h a week on cooking and/or housework outside paid work, 67% longer than men (EIGE, 2021). In 2016, European mothers spent 38.8 h per week on caring and/or educational activities outside paid work, while fathers spent only 20.5 h (EIGE, 2021). In most OECD countries, women combine time spent on unpaid work with 40 weekly hours of paid work. Moreover, in countries where part-time work is prevalent (such as Germany or the Netherlands), women are more likely to be employed part-time.[9] Gender differences in the use of time and the duration of paid work harm women's career prospects by reducing the probability of being promoted, limiting mobility between companies, and benefiting from lower wage gains when switching between companies (OECD, 2021).

Gendered social norms reduce women's full-time employment and affect the relative hourly wages of male and female workers. Some companies disproportionately reward individuals who work long hours and outside regular working hours. First, the gender gaps in hourly wages increase because off-hours are paid at higher rates and men are over-represented among those rewarded (Goldin, 2014). Second, men are more likely to be promoted in these companies (Blau & Kahn, 2013). Third, companies with non-linear hourly wages that promote unpredictable (and deregulated) working hours often employ high-wage professionals such as law firms, financial services, and consulting (Goldin, 2014). As women with children tend to seek flexible and shorter working hours, they are less likely to be employed in these companies, widening the gender pay gap (Card et al., 2016). Overall, time is "a resource to which men have more access than women" (Rutherford, 2001, p. 274), and as organisations demand more of this resource, women's job market prospects are damaged.

Reducing weekly working hours may contribute to improving women's labour market outcomes. First, reducing the gap between part-time and full-time work may reduce the wage penalty associated with part-time work (Budig & England, 2001; Dias et al., 2018). Second, shorter full-time schedules may induce part-time women to move to full-time employment (Spiegelaere & Piasna, 2017) and may reduce the incentives to disproportionately reward men for time availability. Therefore,

---

[9] In Austria, almost 35% of employed women work less than 30 h a week, while less than 8% of men do; in Belgium, 26% of women and 8.1% of men; in Germany, 34% of women and 8.7% of men; in Netherlands, 51.7% of women work part-time (<30 h), and the rate drops to 8.7% among men (OECD, Incidence of FTPT employment, 2022).

the gender pay gap related to the number of hours worked may be narrowed by implementing a reduced working week. In Portugal, where the working week was reduced from a maximum of 44–40 h between 1996 and 1998, reducing the statutory working hours narrowed the gender employment gap between 0.7 and 1.9% points.[10] Although the gender employment gap in Portugal was already small, due to the low occurrence of part-time employment and the traditionally high female labour market participation rate, reducing the regular working week narrowed the gap even further. Consequently, a more significant impact may be expected in countries where female part-time employment is more prevalent.

Advocates of the 4DWW often refer to its potential to reduce the inequalities in the division of household tasks (4 Day Week Global, 2023; Chung, 2022; Coote et al., 2021). However, survey data descriptions collected after some 4-day-week pilots and more robust qualitative studies show that reducing working time does not necessarily mean a more equal redistribution of care and household tasks between spouses when men spend less time at work. Piasna and Spiegelaere (2021) mention the observation of increased male participation in unpaid work after the French reform from 40 to 35 weekly hours, but to a small extent and without structural changes in social norms. The report on the Icelandic 35-h-week pilot also shows modest results: in some cases, men started participating more in household tasks, but women did not perceive significant normative changes (Haraldsson & Kellam, 2021). No change in the distribution of household labour was found in the recent UK 4-week pilot either: 68% of the survey trial respondents reported no change in this regard (Lewis et al., 2023).

Consistent with these views, Flèche et al. (2020) demonstrate that (1) men who work significantly less than their female partners do not spend much more time on household work than men who work significantly more time than their female partners, and (2) women also spend similar amounts of time on unpaid work whether they work more or less than their partners.

Overall, gender differences in the distribution of time between paid work, unpaid work, and leisure time within the household seem to depend much more on social norms and individual views of these social norms (Flèche et al., 2020). Therefore, despite the high hopes that shortening the working week will change social norms and ease women's burden of unpaid work, there is little empirical evidence to support this. While social norms are remarkably persistent (Giuliano, 2017), there are public policies able to change them, with significant intergenerational effects.[11]

[10] Controlling for age, tenure, education, and 1-digit occupations. The first estimate (0.7 p.p.) is obtained by adding firm fixed effects and the second by adding industry fixed effects. Data from *Quadros de Pessoal*, authors' analysis.

[11] Farré et al. (2023) analyse the introduction of paternity leave in Spain and found that children born after the reform were more likely to have gender egalitarian attitudes. Bünning (2015) found that German fathers who took at least two months' leave were more involved in subsequent childcare.

13.3.4 Reduction of Environmental Impact

Perspectives on the 4-day week and its environmental impact are also divergent. Orthodox macroeconomics wrongly assumes that economies will grow indefinitely, because ecosystems, although self-regenerating, also deteriorate when overexploited (Pullinger, 2014; Stagl, 2014). Thus, it is understood that the environmental impact can be reduced if fewer working hours result in a decrease in demand for transport and consumption of water and light in the workplace (Ashford & Kallis, 2013). Furthermore, if productivity gains are invested in leisure instead of capital, there can be a fall in output growth, with a consequent reduction in consumption, while at the same time increasing well-being, in a perspective known as sustainable degrowth (Fitzgerald et al., 2018; Knight et al., 2012; Pullinger, 2014). In this way, it is increasingly assumed that technological change alone will not achieve sustainable consumption, so the path to sustainability must be based on stabilising consumption by reducing working hours (Marston, 2016; Schor et al., 2022). This idea is supported by two effects: the scale effect, which refers to the impact of working hours on the size or scale of the economy through GDP growth, and the composition effect, which measures the influence that the availability of time has on the carbon intensity of household consumption (Fitzgerald et al., 2018). On the one hand, the scale effect means that, without mechanisms to reduce working hours, longer working hours generate a higher level of economic production (GDP) and, consequently, income and consumption (Fitzgerald et al., 2018; Knight et al., 2012). This creates a cycle called 'work and spend' which shapes workers' behaviour to lead them to consume more, because productivity increases translate into production and consumption, with the aggravating factor that consumption is environmentally degrading (Knight et al., 2012). On the other hand, the composition effect implies a decision-making component at the family level that incorporates income and time availability (Fitzgerald et al., 2018). In this sense, it states that families with shorter working hours and more leisure time will be more inclined to adopt less ecologically intensive lifestyles, because productivity increases are channelled into more free time rather than more income and consumption, unlike families with a shortage of time (Knight et al., 2012; Pullinger, 2014; Schor et al., 2022).

The objections raised regarding the relationship between the 4-day workweek and its environmental effects refer to the uncertainty of the environmental impacts. If, on the one hand, capital and energy were to replace workers, the environmental damage would increase. On the other hand, it is not known whether the extra leisure time will be redirected to less resource-intensive activities and, therefore, it is not certain that there will be a reduction in consumption (Kallis et al., 2013; Pullinger, 2014; Schor et al., 2022; Worstall, 2019). How leisure is used depends on people's consumption habits, their interests and priorities, the friends they have, the opportunities that are available and the money they have at their disposal (Coote et al., 2021; Schor et al., 2022). Given that there is still an association between success and greater economic ownership and, at the same time, between greater economic ownership and more leisure, the tendency is to buy more things, which increases the consumption of resource-intensive products (Coote et al., 2021; Worstall, 2019). Therefore, environmental benefits should not be taken for granted unless measures are taken to control resource-intensive consumption (Kallis et al., 2013; Pullinger, 2014; Stagl, 2014).

13.4 Empirical Evidence on 4-Day Workweek Pilots

Although the first experiments to implement a shorter working week belong back to the 1930s—with the six-hour shifts in Kellogg's manufacturing plant—the number of private companies piloting a 4DWW has boomed in the most recent years. Examples of completed pilots can be found for instance in Australia, Belgium, Brazil, Canada, Japan, New Zealand, Portugal, South Africa, Spain, UK, and the US. Some of these trials began as a single-company initiative—e.g., Perpetual Guardian in New Zealand or Microsoft Japan, while others were simultaneous experiments in multiple companies that voluntarily joined collective trials, such as in the UK and Portugal. In all cases, private companies self-selected for the experiment, adjusting the basic setting—reducing the number of hours and the number of weekly working days with no reduction in wages—to a working schedule fitting its needs the best (e.g., all workers in the company have the same day-off or half of workers has Mondays off, and the remaining half has Fridays).

Some of these pilots have been subject to significant media exposure, expanded the number of advocacy groups and 4DWW supporters, and have influenced other experiments. It is the case of the pilot run at Perpetual Guardian, a 250-worker company in New Zealand, which tested a shorter working week in 2018 and adopted it after the end of the trial, and whose CEO leads the international advocacy group "4 Day Week Global". It is also the case of the pilot in Iceland, held in the public sector, which did not implement 4DWW but only a reduction of working time from 40 to 35–36 weekly hours (distributed across five days); remarkably, the Icelandic experience is widely used as an example of the 4-day workweek success both by the media, advocacy organisations (e.g., 4-day workweek Global, Autonomy) and some policymakers. In 2019, the women's organisation Femma Wereldvrouwen conducted a 12-month trial, reducing the working time to 30 h a week (Mullens et al., 2021). More recently, 61 UK-based companies participated in the UK pilot over six months. The results of this pilot, which attracted a lot of media coverage, are described by the UK-based think tank Autonomy.[12] Moreover, in 2022, the organisation 4-day workweek Global conducted pilots in 33 companies operating entirely remotely or based in the US, Ireland, UK, Australia, and New Zealand.

Most of the reported 4DWW trials reveal outstanding results on productivity, workers' well-being, and job satisfaction. Perpetual Guardian reported a 20% productivity increase and a significant improvement in workers' perception of work-life

[12] The results are in: the UK's four-day week pilot". Autonomy, 2023. Available at https://autonomy.work/portfolio/uk4DWWpilotresults/.

balance (Cuello, 2023). Microsoft Japan, frequently referred to as an example of a successful 4DWW pilot, reported a 40% productivity increase (Chappell, 2019). Revenues also rose significantly in several companies participating in the 4-day work-week Global multi-country trial: compared to the period before the trial, 20 participating companies had, on average, registered a 37% increase in revenues (Schor et al., 2022). The UK report uses the same metric and states that 34 participating organisations (out of 61) had met, on average, a 35%-expansion in revenues (Lewis et al., 2023).

Most workers covered by the pilots reported lower stress levels, lower incidence of burnout symptoms, better sleep, and less fatigue (Booth, 2019; Lewis et al., 2023; Schor et al., 2022). Among other impacts, workers stated that it was easier to balance work and family life (Schor et al., 2022), and they were spending more time with their children (Delaney & Casey, 2022; Lewis et al., 2023). Consistent with other findings described in the previous section, surveyed men did not spend more time performing household chores when they started having a third day off (Lewis et al., 2023).

These and other results are described in further detail in Table 13.1, which holds a comprehensive overview of the main 4DWW pilots reported in publicly available sources in English, until July 2023. This excludes the experiment developed in Microsoft Japan, for which no results are consistently reported.

The outstanding outcomes described in the abovementioned reports contrast with concerns raised about the countervailing impacts of these experiments. Delaney and Casey (2022) conducted a qualitative case study on the Perpetual Guardian pilot and assessed the implications of this experiment as a management-led initiative, mainly motivated to increase labour productivity and workforce engagement (Delaney & Casey, 2022). The authors found that, despite the initial workers' involvement to reflect upon organisational changes (e.g., how to measure productivity, how to innovate working processes), the weight of maintaining the same workload to hold a shorter week—which required increasing individual productivity and work pace—counterbalanced the overall positive attitudes towards the trial. In particular, some staff felt compelled to perform some tasks over the day off (or being available to do it if necessary), increasing stress levels. Moreover, Delaney and Casey (2022) observed stricter management monitoring of employees and performance measurement and the absence of a "social rights and collective interests' perspective" (p. 185).

Spencer (2022) expresses similar heed. In particular, the author is concerned that if work time reduction is motivated by productivity gains, the 4DWW can backfire on workers and hinder wage progression, job quality among the low-skilled, and companies' rent-sharing. These adverse effects may be amplified if a 4DWW is achieved through compressed weeks or wage cuts.[13]

The results of the 4DWW pilots described above have also been questioned based on their robustness and external validity. Cuello (2023) screens three of the most cited pilots by the OECD quality assessment metric, which sets the standards for policy

[13] As Telefonica, a Spanish communication company, attempted to test an opt-in 4DWW with a 12% wage cut and found little support among employees.

Table 13.1 4DWW reported pilots

Pilot	Country	No. of companies	Year and pilot duration	Performance metric	Health and wellbeing	Job and life satisfaction	Work-life balance and use of time	Other reported impacts	References
Perpetual guardian	New Zealand	1	2018, 2 months	Productivity: +20%	Reduction of work stress: from 45 to 38% Some individuals experienced greater workloads		Increase in work-life balance scores: from 54 to 78%		Booth (2019), Delaney (2018)
4-day workweek Global	Ireland, US, UK, Australia, New Zealand, remote	33	2022, 6 months	Revenue change from comparison to the same period in 2021: +37.6% (weighted average among companies) Revenue, change from trial start and endpoint comparison to trial period: +37.6% (weighted average)	Reduction of work stress: −0.21 on a 1–5 scale Reduction in burnout symptoms: −0.44 on a 1–5 scale; Increase in time spent doing physical exercise: more 23 min/week Reduction in fatigue and sleep problems: −0.38 and −0.33 on a 1–4 scale	Increase in job satisfaction by 0.28 on a 1–10 scale Increase in life satisfaction: +0.89 on a 0–10 scale	Improvement in combining paid work with care duties: +0.72 on a 1–5 scale Improvement in combining paid work with social life: +0.78 on a 1–5 scale No change in time spent on caring responsibilities and housework No change in the gendered division of non-paid labour	Reduction in time spent communing: 1 h per week Increase in international traveling, energy use, and leisure travel	Schor et al. (2022)

(continued)

Table 13.1 (continued)

Pilot	Country	No. of companies	Year and pilot duration	Performance metric	Health and wellbeing	Job and life satisfaction	Work-life balance and use of time	Other reported impacts	References
UK Pilot	UK	61	2022, 6 months	Revenue change compared to the same period in 2021: +34.5% (weighted average across companies) Revenue, change from trial start and endpoint comparison to trial period: + 1.4% (weighted average)	Reduction of work stress: −0.33 on a 1–5 scale Reduction in burnout symptoms: −0.46 on a 1–5 scale Reduction in fatigue and sleep problems: 40% of workers report fewer sleep problems Negative emotions are experienced less frequently	Increase in job satisfaction by 0.57 on a 0–10 scale	Improvement in combining paid work with care duties: +0.82 on a 1–5 scale Improvement in combining paid work with social life: +0.88 on a 1–5 scale No change in the gendered division of non-paid labour	Increase in international traveling, energy use, and leisure travel Reduction in childcare costs	Lewis et al. (2023)
Femma Wereldvrouwen	Belgium	60	2019, 1 year		Reduction in stress and pressure in employees' private lives Higher levels of life satisfaction		Women spent more time on household-related activities but experienced less stress while performing those Some women became less satisfied with the division of unpaid labour	Some employees experienced less open interaction moments with colleagues Workers became more aware of their working time usage	Mullens and Ignace (2023) Mullens et al. (2021)

evaluation (OECD, 2023). Cuello categorizes the limitations with the methodologies adopted in each experiment into four groups: (1) concerns with addressing causality (omitted variable bias and selection bias), (2) issues with collecting accurate data (e.g., low-quality administrative data), (3) limited transparency (e.g., incomplete statistical information, misuse of accessible summaries), and (4) incomplete information on sample and intervention characteristics (e.g., lack of sample representativeness, small sample size). The author finds that all pilots analysed suffer severe methodological flaws across all four categories. In particular, the pilots failed to collect quality data, report all statistical information collected, disclose pre-registered plans, run robustness checks, gather a representative sample, address heterogeneous effects, analyse data from non-compliers, and provide information on costs and scalability of the intervention (Cuello, 2023). Some of these limitations imply, for instance, that macroeconomic contexts—which may be correlated with some of the experiment outcomes (e.g., productivity)—are not considered at all, and there is an overselling of positive results, while (partially) ignoring more modest (or inexistent) results.

Overall, one may conclude that future 4DWW pilots should rely on robust experimental design, collect high-quality administrative data, ensure the robustness of the results, and provide adequate reporting.

13.5 Discussion and Conclusion

Working hours significantly decreased in the decades following the Second World War. Extending paid holidays, sick leave, and parental leave, as well as an effective reduction in weekly working hours, contributed to lower yearly working hours. The downward trend stalled in the 1990s, and only a few countries reduced the standard weekly working hours. Sectoral collective bargaining agreements and company-specific practices have sometimes led to lower full-time schedules. At the same time, in some high-wage countries, workers—especially women—have opted for part-time work, thereby reducing their payslips proportionally.

The debate on reducing working weeks intensified after the COVID-19 pandemic. Changes in how (and from where) people work, the introduction of technological applications that make teamwork more effective (even at a distance) and can monitor employees more closely, and increasing difficulties in retaining workers led some companies to test 4-day working weeks. In most cases, there is an effective reduction in working hours (rather than compressed weeks) and no wage cuts. The acknowledgment that men and women used their time in lockdown differently highlighted the persistence of gender norms that have led to gender inequalities in the distribution of unpaid work and labour market outcomes. Labour organisations, women's rights advocacy groups, left-wing think tanks and politicians have, thus, campaigned for the 4DWW.

The arguments most often used in favour of the 4DWW have been grouped under four themes: increasing productivity, improving well-being, correcting gender imbalances, and reducing environmental impact. We noted that although the demand for a

4DWW is for shorter statutory working hours, its advocates often refer to the consequences of working long hours to make the case for the 4DWW. Existing literature proves that the hazards of working overtime may not necessarily apply to standard working hours. Therefore, we argue that the impact of shorter working hours should be assessed regarding the existing statutory schedules. When researching empirical evidence on the positive effects of shorter standard working weeks on productivity, workers' well-being and gender inequality, we conclude that it is somewhat limited and, sometimes, poorly robust.

The claims on the impact of the 4DWW on productivity gains are susceptible to the choice of working hours—standard or including overtime—as the baseline. The importance of this argument in the media debate requires a better understanding of the effective change in companies' outcomes when employees work fewer hours. Moreover, some studies indicate that making the 4DWW implementation dependent on productivity gains may constrain the well-being benefits of shorter working hours. Further studies assessing these countervailing effects of performing the same tasks in shorter periods are lacking.

We reviewed the reports describing the 4DWW pilots. Only a few trials were assessed and their experimental design suffered from significant methodological limitations. Often, the results of these trials are reported in the media without sufficient evaluation of their actual impact and external validity.

Overall, we conclude that the discussion on working time reductions is opportune, considering the stagnation of hours worked, the recent productivity gains driven by technological change and the positive effects on workers' well-being. However, 4DWW trials should be based on sound and robust policy evaluation methods and identify the advantages, disadvantages and challenges of implementing a 4DWW in the workplace.

Funding This work was funded by the Foundation for Science and Technology under Project UID/00727/2020. Oliveira thanks to Instituut Gak for support. Couto also thanks to the Portuguese national funds through FCT – Fundação para a Ciência e Tecnologia, under the project UIDB/05422/2020.

References

4 Day Week Global. (2023). From https://www.4dayweek.com

Afonso, P., Fonseca, M., & Pires, J. F. (2017). Impact of working hours on sleep and mental health. *Occupational Medicine, 67*, 377–382.

Anfrave, D., & Charlwood, A. (2015). What is the relationship between long working hours, over-employment, under-employment and the subjective well-being of workers? Longitudinal evidence from the UK. *Human Relations, 68*(9), 1491–1515.

Autonomy. (2019). The shorter working week: A radical and pragmatic proposal. https://autonomy.work/wp-content/uploads/2019/03/Shorter-working-week-docV6.pdf

Ashford, N., & Kallis, G. (2013). A four-day workweek: A policy for improving employment and environmental conditions in Europe. *The European Financial Review*, 53–58. www.europeanfinancialreview.com

Bank of England. (2023). Average weekly hours worked in the United Kingdom [AWHWUKA]. Retrieved from FRED, Federal Reserve Bank of St. Louis. https://fred.stlouisfed.org/series/AWHWUKA

Bannai, A., & Tamakoshi, M. (2014). The association between long working hours and health: A systematic review of epidemiological evidence. *Scandinavian Journal of Work Environment and Health, 40*(1), 5–18.

Barnes, A. (2002). *The 4 day week*. Piatkus.

Berniell, I., & Bientenbeck, J. (2020). The effect of working hours on health. *Economics and Human Biology, 29*, 100901.

Blau, F. D., & Kahn, L. M. (2013). Female labor supply: Why is the US falling behind? *American Economic Review, 103*(3), 251–56.

Booth, R. (2019). *Four-day week: Trial finds lower stress and increased productivity*. From The Guardian. https://www.theguardian.com/money/2019/feb/19/four-day-week-trial-study-finds-lower-stress-but-no-cut-in-output

Boulin, J.-Y. (2024). Semaine de Quatre Jours: Une Préfiguration de l'avenir de Temps de Travail? *Futuribles, 4*(461), 57–76.

Budig, M. J., & England, P. (2001). The wage penalty for motherhood. *American Sociological Review, 66*, 204–225.

Bünning, M. (2015). What happens after the 'Daddy Months'? Fathers' involvement in paid work, childcare, and housework after taking parental leave in Germany. *European Sociological Review, 31*(6), 738–748.

Burchell, B., Deakin, S., Rubery, J., & Spencer, D.A. (2024). The future of work and working time: Introduction to special issue. *Cambridge Journal of Economics, 48*(1), 1–24. https://doi.org/10.1093/cje/bead057

Cabrita, J., & Boehmer, S. (2016). *Working time developments*. Eurofound.

Campbell, T. T. (2023). The four-day work week: A chronological, systematic review of the academic literature. *Management Review Quarterly, 74*(1), 1791–1807.

Card, D., Cardoso, A. R., & Kline, P. (2016). Bargaining, sorting, and the gender wage gap: Quantifying the impact of firms on the relative pay of women. *The Quarterly Journal of Economics, 131*(2), 633–658.

Chappell, B. (2019). 4-day workweek boosted workers' productivity By 40%, Microsoft Japan Says. *NPR*.

Chung, H. (2022). A social policy case for a four-day week. *Journal of Social Policy, 51*(3), 551–566.

Collewet, M., & Sayermann, J. (2017). Working hours and productivity. *Labour Economics, 47*, 96–106.

Coote, A., Harper, A., & Stirling, A. (2021). *The case for a four-day week*. Wiley.

Cortes, P., & Pan, J. (2019). When time binds: Substitutes for household production, returns to working long hours, and the skilled gender wage gap. *Journal of Labour Economics, 37*(2), 351–398.

Costa, D. L. (2000). The wage and the length of the work day> from the 1890 to 1991. *Journal of Labor Economics, 18*(1), 156–181.

Cuello, H. (2023). *Assessing the validity of four-day week pilots*. JRC working papers series on social classes in the digital age.

Decicienti, F., Grinza, E., & Vannoni, D. (2015). The impact of part-time work on firm total factor productivity: Evidence from Italy. *IZA DP No. 9463*.

Delaney, H. (2018). *Perpetual guardian's 4-day workweek trial: Qualitative research analysis*. The University of Auckland.

Delaney, H., & Casey, C. (2022). The promise of a four-day week? A critical appraisal of a management-led initiative. *Employee Relations: The International Journal, 44*(1), 176–190.

Dembe, A., Erickson, J., Delbos, R., & Banks, S. (2005). The impact of overtime and long work hours on occupational injuries and illnesses: New evidence from the United States. *Occupational and Environmental Medicine, 62*(9), 588–97.

Dias, M. C., Joyce, R., & Parodi, F. (2018). *Wage progression and the gender wage gap: The causal impact of hours of work*. The Institute for Fiscal Studies.

Dong, X. (2005) Long workhours, work scheduling and work-related injuries among construction workers in the United States. *Scand J Work Environ Health, 31(5*, 329–35.

EIGE. (2021). *Gender statistics database*. From https://eige.europa.eu/gender-statistics/dgs/about

Estevão, M., Sá, F., & Petrongolo, B. (2008). The 35-hour workweek in France: Straightjacket or welfare improvement? *Economic Policy, 23*(55), 417–463.

Eurofound. (2024). Working time reduction with a focus on the four-day week: Literature review. *European foundation for the improvement of living and working conditions.*

Farré, L., Felfe, C., González, L., & Scneider, P. (2023) Changing gender norms across. *IZA DP No. 16341.*

Fitzgerald, J. B., Schor, J. B., & Jorgenson, A. K. (2018). Working hours and carbon dioxide emissions in the United States, 2007–2013. *Social Forces, 96*(4), 1851–1874.

Flèche, S., Lepinteur, A., & Powdthavee, N. (2020). Gender norms, fairness and relative working hours within households. *Labour Economics, 65*, 101866.

Garnero, A., Kampelmann, S., & Rycx, F. (2014). Part-time work, wages, and productivity: Evidence from Belgian matched panel data. *Industrial and Labor Relations Review, 67*(3), 926–954.

Giattino, C., Ortiz-Ospina, E., & Roser, M. (2020) *Working hours*. From OurWorldInData.org. https://ourworldindata.org/working-hours

Gilmore, O. (2021). The working week in manufacturing since 1820. In *How was life? Volume II: New perspectives on well-being and global inequality since 1820.* OECD Publishing, Paris, https://doi.org/10.1787/3d96efc5-en

Giuliano, P. (2017). *Gender: An historical perspective*. NBER Working Paper 23635.

Golden, L., Lambert, S., Henly, J., & Wiens-Tuers, B. (2011). Working time in the employment relationship: Perceived control and work-life balance. In K. T. Wilkinson (Ed.), *Edward Elgar handbook on work and employment relations*. Edward Elgar.

Golden, L. (2011). *The effects of working time on productivity and firm performance: A research synthesis paper.* ILO, conditions of work and employment series no. 33.

Goldin, C. (2014). A grand gender convergence: Its last chapter. *American Economic Review, 104*(4), 1091–1110.

Gomes, P. (2021). *Friday is the new saturday*. The History Press Ltd.

Haraldsson, G. D., & Kellam, J. (2021) *Going Public: Iceland's journey to a shorter working week.* Alda and Autonomy.

Harper, A., & Martin, A. (2018). *Achieving a shorter working week in the UK*. The New Economics Foundation.

Heejung, C. (2022). A social policy case for a four-day week. *Journal of Social Policy, 51*(3), 551–566.

Hunnicutt, B. K. (1992). Kellogg's six-hour day: A capitalist vision of liberation through managed work. *The Business History Review, 66*(3), 475–522.

ILO. (2023). *Statistics on working time*. From ILO Stat: https://ilostat.ilo.org/topics/working-time/

Jahal, T., Bardoel, E. A.., & Hopkins, J. (2023), Could the 4-day week work? A scoping review. *Asia Pacific Journal of Human Resources*. https://doi.org/10.1111/1744-7941.12395

Johnson, J. V., & Lipscomb, J. (2006). Long working hours, occupational health and the changing nature of work organization. *American Journal of Industrial Medicine, 49*(11), 921–9.

Kallis, G., Kalush, M., O'Flynn, H., Rossiter, J., & Ashford, N. (2013). Friday off: Reducing working hours in Europe. *Sustainability (Switzerland), 5*(4), 1545–1567. https://doi.org/10.3390/su5041545

Kamerade, D., Wang, S., Burchelll, B., Balderson, S. U., & Coutts, A. (2019). A shorter working week for everyone: How much paid work is needed for T mental health and well-being? *Social Science and Medicine, 241*, 112353.

Keus Van De Poll, M., Nybergh, L., Lornudd, C., Hagberg, J., Bodin, L., Kwak, L., & Bergstrom, G. (2020). Preventing sickness absence among employees with common mental disorders or stress-related symptoms at work: A cluster randomised controlled trial of a problem-solving-based intervention conducted by the occupational health services. *Occupational and Environmental Medicine, 77*(7), 454–461.

Kim, S., McLean, G. N., & Park, S. (2018). The cultural context of long working hours: Workplace experiences in Korea. *New Horizons in Adult Education and Human Resource Development, 30*(2), 36–51.

Knight, K., Rosa, E. A., & Schor, J. B. (2012) *Reducing growth to achieve environmental sustainability: The role of work hours* (Issue 304).

Kodz, K., Lain, D., Strebler, M., Rick, J., Bates, P., & Meager, N. (2023). *Working long hours. A review of the evidence* (vol. 1). Employment Relations Research Series ERRS16 | Department of Trade and Industry.

Künn-Nelen, A., De Grip, A., & Fouarge, D. (2013). Is part-time employment beneficial for firm productivity? *Industrial and Labour Relations Review, 66*(5), 1172–1191.

Landrigan, C., Rothschild, J., Cronin, J., Kaushal, R., Burdick, E., Katz, J., & Czeisler, C. (2004). Effect of reducing interns' work hours on serious medical errors in intensive care units. *The New England Journal of Medicine, 351*(18), 1838–48.

Lepinteur, A. (2019). The shorter workweek and worker wellbeing: Evidence from Portugal and France. *Labour Economics, 58*, 204–220.

Lewis, K., & Stronge, W. (2021). *A shorter working week in Government: A worker-led inquiry into desirability, feasibility and impact.* Autonomy.

Lewis, K., Stronge, W., Kellam, J., & Kikuchi, L. (2023) *The results are in: The UK's four-day week pilot.* Autonomy, 4Day Week, 4 Day Week Global.

Li, W., Ruan, W., Chen, Z., Yi, G., & Wang, D. (2021). A meta-analysis of observational studies including dose–response relationship between long working hours and risk of obesity. *Reviews in Endocrine and Metabolic Disorders, 22*, 837–845.

Marston, G. (2016). Greening the Australian welfare state: Can basic income play a role? In *Basic income in Australia and New Zealand: Perspectives from the neoliberal frontier* (pp. 157–177). New York, Palgrave Macmillan US.

Milner, A., Smith, P., & LaMontagne, A. D. (2015). Working hours and mental health in Australia: Evidence from an Australian population-based cohort, 2001–2012. *Occupational and Environmental Medicine, 72*, 573–579.

Moores, J. E. (1990). A meta-analysis of the effects of compressed work weeks. *Applied HRM Research, 1*(1), 8–14.

Mullens, F., & Ignace, G. (2023). Reducing weekly working hours: Temporal strategies and changes in the organization and experiences of work-results from a qualitative study of 30-hour workweek experiment. *Time and Society, 32*(2), 146–168.

Mullens, F., Verbeylen, J., & Glorieux, I. (2021). Rethinking the workweek: Results from a longitudinal time-use study of a 30-hour workweek experiment. *Journal of Time Use Research, 15*(1), 52–75.

OECD. (2021). *The role of firms in wage inequality: Policy lessons from a large scale cross-country study.* OECD Publishing, Paris. https://doi.org/10.1787/7d9b2208-en

OECD. (2022). *Incidence of FTPT employment.* From OECD Stat: https://stats.oecd.org/Index.aspx?DataSetCode=FTPTC_I

OECD. (2023). *Framework for the evaluation of SME and entrepreneurship policies and programmes 2023.* OECD Studies on SMEs and Entrepreneurship, OECD Publishing.

Pang, A. S. (2020). *Shorter: How smart companies work less, embrace flexibility and boost productivity.* Penguin.

Pencavel, J. (2015). The productivity of working hours. *The Economic Journal, 125*(589), 2052–2076.

Piasna, A., & Spiegelaere, S. D. (2021). Working time reduction, work/life balance and gender equality. *Dynamiques Regionales, 10*, 19–42.

Pullinger, M. (2014). Working time reduction policy in a sustainable economy: Criteria and options for its design. *Ecological Economics, 103*, 11–19. https://doi.org/10.1016/j.ecolecon.2014.04.009

Raposo, P. S., & Ours, J. C. (2010). How working time reduction affects jobs and wages. *Economic Letters, 106*, 61–63.

Rivera, A., Akanbi, M., O'Dwyer, L., & McHugh, M. (2020). Shift work and long work hours and their association with chronic health conditions: A systematic review of systematic reviews with meta-analyses. *PLoS ONE, 15*(4), e0231037.

Rosengren, C. (2019). Performing work: The drama of everyday working life. *Time and Society, 28*(2), 613–633.

Rutherford, S. (2001). Are you going home already? The long hours culture, women managers and patriarchal closure. *Time and Society, 10*(2/3), 259–276.

Sanchis, J. (2023), *Cuatro Días: Trabajar Menos Para Vivir En Un Mundo Mejor.* Barlin Libros.

Schiller, H., Lekander, M., Rajaleid, K., Hellgren, C., Åkerstedt, T., Barck-Holst, P., & Kecklund, G. (2017). The impact of reduced worktime on sleep and perceived stress—a group randomized intervention study using diary data. *Scandinavian Journal of Work, Environment & Health*, 109–116.

Schor, J. B., Fan, W., Kelly, O., Gu, G., Bezdenezhnykh, T., & Bridson-Hubbard, N. (2022). *The four day week: Assessing global trials of reduced work time with no reduction in pay.* Four Day Week Global.

Schwarz, U. V., & Hasson, H. (2011). Employee self-rated productivity and objective organizational production levels: Effects of worksite health interventions involving reduced work hours and physical exercise. *Journal of Occupational and Environmental Medicine, 3*(8), 838–844.

Schwarz, U. V., Lindfors, P., & Lundberg, U. (2008) Health-related effects of worksite interventions involving physical exercise and reduced workhours. *Scandinavian Journal of Work, Environment & Health, 34*(3), 179–188.

Shangguan, R., Devaro, J., & Owan, H. (2021) *Enhancing team productivity through shorter working hours: Evidence from the great recession.* RIETI Discussion Paper Series 21-E-040.

Shepard, E., & Clifton, T. (2000). Are longer hours reducing productivity in manufacturing? *International Journal of Manpower, 21*(7), 540–553.

Sparks, K., Faragher, B., & Cooper, C. (2001). Well-being and occupational health in the 21st century workplace. *Journal of Occupational and Organizational Psychology, 74*(4), 489–509.

Specchia, G. L., & Vandenberghe, V. (2013). Is part-time employment a boon or bane for firm productivity? *Unpublished paper*.

Spencer, D. A. (2022). A four-day working week: Its role in a politics of work. *The Political Quarterly, 93*(3), 401–407.

Spiegelaere, S. D., & Piasna, A. (2017). *The why and how of working time reduction.* European Trade Union Institute.

Stagl, S. (2014). Ecological macroeconomics: Reflections on labour markets. *European Journal of Economics and Economic Policies: Intervention, 11*(2), 171–181. https://doi.org/10.4337/ejeep.2014.02.04

Stronge, W., & Harper, A. (2019). *The shorter working week: A radical and pragmatic proposal.* Autonomy.

US Bureau of Labor Statistics. (2023). *Unemployment rate, FRED Sr. Louis Fed.* From https://fred.stlouisfed.org/series/UNRATE

Veal, A. J. (2023). The 4-day work-week: The new leisure society? *Leisure Studies, 42*(2), 172–187.

Virtanen, M., Jokela, M., & Lallukka, T. (2022). Long working hours and change in body weight: Analysis of individual-participant data from 19 cohort studies. *International Journal of Obesity, 44*, 1368–1375.

Voglino, G., Savatteri, A., Gualano, M., Catozzi, D., Rousset, S., Boietti, E., & Siliquini, R. (2022). How the reduction of working hours could influence health outcomes: A systematic review of published studies. *BMJ Open, 12.*

Weeden, K. A., Cha, Y., & Bucca, M. (2016). Long work hours, part-time work, and trends in the gender gap in pay, the motherhood wage penalty, and the fatherhood wage premium. *RSF: The Russell Sage Foundation Journal of the Social Sciences*, 2(4), 71–102.

Williams, J. (2010). *Reshaping the work-family debate: Why men and class matter*. Harvard University Press.
Wong, K., Chan, A. H., & Ngan, S. (2019). The effect of long working hours and overtime on occupational health: a meta-analysis of evidence from 1998 to 2018. *International Journal Environmental Research and Public Health, 16*(12), 2102.
Worstall, T. (2019). *The problem with the four day work week*. 1–2. https://www.adamsmith.org/blog/the-problem-with-the-four-day-work-week
Zheng, H., Vatsa, P., Ma, W., & Zhou, X. (2023). Working hours and job satisfaction in China: A threshold analysis. *China Economic Review, 77*, 101902.

Part VIII
Thematic Area: The Importance of HRM for Achieving the Sustainable Development Goals (SDGs)

Chapter 14
Voices from Within: How University Workers Perceive HRM

Aléxia Ferreira Gomes, Ana Veloso, Patrício Costa, and Teresa Proença

Abstract Human Resource Management (HRM) in organizations impacts individual and organizational performance and can promote important behaviors at work. The visibility of HRM practices (salience and ease of observation), their utility (perception that they contribute to achieving individual goals), and their understanding (absence of ambiguity) can be important attributes for creating a strong HRM system according to (Bowen and Ostroff in Acad Manage Rev 29:203–221, 2004). In the context of organizations, the implementation of sustainable practices related to aspects of individual development and employee well-being can bring positive results for the individual and the organization (Sypniewska et al. in Int Entrepreneurship Manage J, 2023), with emphasis on engagement (Cleveland and Cavanagh in Human Resource Manage Rev 146–161, 2015) and innovation (Sypniewska et al. in Int Entrepreneurship Manage J, 2023) in the workplace. The objective of this qualitative study was to investigate workers' perceptions of HRM, job insecurity (JI), work engagement (WE) and innovative behavior (IB), seeking singularities and personal and collective meanings of different professional categories in the context of public higher education. Fourteen workers, including technicians and teachers, with and without tenure, from public universities in Portugal and Brazil participated in the study. Data were collected through semi-structured interviews conducted online. Data analysis was performed based on thematic analysis (King in Essential guide to qualitative methods and analysis in organizational research. Sage publications, 2004). The results reveal that workers at Portuguese and Brazilian universities feel that HRM practices influence their WE and IB behaviors and reduce their JI levels. The different perceptions of the existence of the HRM system within the universities studied stand out, revealing the dissatisfaction or lack

A. F. Gomes (✉) · T. Proença
CEF.UP, School of Economics and Management, University of Porto, Porto, Portugal
e-mail: alexiaamfg@hotmail.com

A. Veloso
CICS.NOVA.UMinho, University of Minho, School of Psychology, Braga, Portugal

P. Costa
Center for Psychology at the University of Porto, Faculty of Psychology and Education Sciences, University of Porto, Porto, Portugal

C. Machado (ed.), *Human Resource Management in a Disrupted World*, Lecture Notes on Multidisciplinary Industrial Engineering,
https://doi.org/10.1007/978-3-031-89948-5_15

of perception of support from HRM in the Portuguese context, while in the Brazilian context the role of HRM is highlighted as important and present in other areas related to support for the worker, which highlights the presence of the strength of the HRM system, in its dimensions, visibility, utility and understanding, according to the model by (Bowen and Ostroff in Acad Manage Rev 29:203–221, 2004). These results generate important reflections and demonstrate the need for attention on the part of the managers of these public organizations to invest in this field.

Keywords Human resource management · Job insecurity · Public sector · Thematic analysis · Universities

14.1 Introduction

Human Resource Management (HRM) plays an important role in promoting innovation and work engagement and in reducing levels of job insecurity in organizations. The presence of HRM and the way employees perceive and evaluate it has an impact on their behaviour and attitudes in organizations, such as universities.

The visibility of HRM practices (salience and ease of observation), their utility (perception that they contribute to achieving individual goals), and their understanding (absence of ambiguity) can be important attributes in creating a strong HRM system according to Bowen and Ostroff (2004). In the context of organizations, the implementation of sustainable practices related to aspects of individual development and employee well-being can lead to positive outcomes for the individual and the organization (Sypniewska et al., 2023).

14.1.1 Human Resource Management and Its Relationship with Work Engagement, Innovative Behavior and Job Insecurity

The HRM studied here is considered as an activity present in organizations and a set of interdependent knowledge that, in the search for results, influences the behavior of workers (Neves & Gonçalves, 2009) and is related to the political, economic, cultural and social context in which it is inserted (Keating, 2002). The HRM system, through its practices, can impact the individual performance of workers (Massuno & Jacquinet, 2019) and organizational performance (Aust et al., 2020), as it can promote important behaviors and perceptions at work, which we highlight in this research, the WE, the perceptions of these workers about JI and also the IB.

In the context of sustainable HRM, there are characteristics that distinguish it from conventional HRM, such as respect (as an important stakeholder in the organization) (Proença, 2022), openness (environmental awareness and an outside-in perspective)

and continuity (economic, social and long-term employability approach) (De Prins et al., 2014). The implementation of sustainable practices in the context of organizations related to aspects of individual development and employee well-being can bring positive results for the individual and the organization (Sypniewska et al., 2023), with emphasis on engagement (Cleveland & Cavanagh, 2015) and innovation (Sypniewska et al., 2023) in the workplace. Sustainable HRM practices have been the focus of studies both in innovative behavior (Stankevičiūtė et al., 2020) and in organizational innovation (Zhang, 2023). Along this path, organizations have shown their interest in assuming their social responsibility and contributing to the achievement of the Sustainable Development Goals (SDGs), since this area has a strong connection with sustainability, through the impact on the behaviors and attitudes of individuals, in the promotion of well-being and mental health at work, and in the consumption of available resources, which impacts ecological and social practices (Chamsa & García-Blandón, 2019).

Innovation in universities is an essential factor for the growth and evolution of these organizations. In higher education, these are considered innovative organizations, which operate on the tripod of teaching, research and extension (interaction with the community), dealing directly with science, technology and innovation, and research and development. These characteristics make HRM unique in the educational context (Coda & Ricco, 2010; Jordan, 2005; Katz, 2005), which adapts to meet the demands of this sector. The contribution to the overall competitiveness of these institutions in the areas of research, teaching and interaction with the community is highlighted as the main role of HRM (Alqahtani & Ayentimi, 2021), with the involvement of different actors in the management and administration of processes and practices (Gilbert et al., 2011).

In this study, IB is translated as innovation behavior, which refers to the production or adoption of useful ideas and their implementation (Kanter, 1988). Amabile (1997) shows that innovation can be influenced by individual and environmental factors that interact with each other and are expressed in the behavior of workers. This author points out expertise, creative thinking and intrinsic motivation as individual factors, and management practices, support from colleagues and management and organizational motivation to support innovation as environmental factors.

Studies indicate that intrinsic motivation is associated with increased engagement (Cerasoli et al., 2014) and creativity, which is considered a driver of employee engagement (Ghosh et al., 2020). HRM is also considered a precursor to innovation (Gupta & Singhal, 1993), and HRM's responsibility in the idea generation and implementation phases is highlighted, as it ensures that functions are clear and that the skills necessary for employees to perform their activities are present (Shipton et al., 2006). WE, another important factor related to performance in organizations, can be interpreted as a concept that expresses a relationship between the employee and the work itself, being defined as a cognitive-affective, positive, intense and rewarding state for this employee. In our study, WE was considered according to the definition of Schaufeli et al. (2002), as it has been used in many recent studies (Ahmed et al., 2020; Gürbüz et al., 2024; Unur et al., 2022; Yadav et al., 2022). It includes three dimensions: vigor (high levels of energy and mental resilience, as well as the individual's willingness to invest and persist in the face of difficulties at work), dedication (presence of a sense of significance, inspiration, enthusiasm, pride, and challenge), and absorption (the individual's deep concentration on work, being fully immersed, having the feeling that time passes quickly and demonstrating difficulty in distancing themselves from work).

In higher education institutions, engagement plays an important role in maintaining the quality and survival of an organization (Aboramadan et al., 2020). Studies show that there is a relationship between HRM practices and positive attitudes at work (Karatepe, 2011; Saks, 2006), contributing to team engagement and bringing benefits to the organization. Despite its importance, research related to engagement is still limited in the context of higher education (Aboramadan et al., 2020), and it is important to expand these studies.

The analysis of job insecurity in universities stands out as important in the scenario of economic, social, and cultural changes experienced in current labor relations, and particularly in the face of the pandemic situation. The pandemic has had consequences on labor relations, highlighting the precarious relationships already existing among some categories of workers (Leite, 2020). In view of these changes in labor relations and the economic crises experienced, insecurity is evident in the concern about unemployment and the loss of job characteristics (De Witte, 1999), which is reflected in the increase in the levels of insecurity experienced by workers. This fact highlights the need for an in-depth look at the perception that workers have about job insecurity, as this perception can affect their behaviors, namely IB and WE.

Job insecurity investigated here is defined as a general concern of the worker about the existence of work in the future (De Witte, 1999), and was considered in its two dimensions: quantitative (concern about the continuity of employment) and qualitative (concern about the continuity of the various aspects of work) (Hellgren et al., 1999).

Studies point to the use of job insecurity as an analysis variable in the context of the relationship between workers and organizations and existing employment relationships, which is relevant to their behaviors (Ferreira Gomes et al., 2024). Thus, we emphasize the meaning and interest in investigating insecurity in different work contexts and in the different relationships existing in organizations, not limited to workers with temporary contracts, since precariousness can be present in different categories of workers, self-employed, temporary, part-time, as well as employees with permanent jobs who are dissatisfied with, for example, the prospect of promotion (Findlay & Thompson, 2017). It is also necessary to look at the factors that help to combat the adverse consequences of job insecurity in the different categories of workers in universities.

The analysis of HRM perceptions considered the Bowen and Ostroff (2004) model, which points to the distinctiveness, consistency, and consensus of the practices of an HRM system as factors that, when present, affect the behaviors and attitudes of workers, as they make communication between them and the organization clear, allowing the understanding of what is expected of them (Sanders et al., 2014, 2021). Distinctiveness (the event effect is highly observable) translates into the attributes of visibility (salient and easily observable HRM practices), utility (perception that the HRM system contributes to achieving individual objectives), and understanding of these practices (unambiguous HRM practices) (Ferreira-Oliveira, 2013). This distinctiveness, used here for analysis, together with consistency (the effect is the same in all modalities and over time) and consensus (there is agreement between individual views of the event-effect relationship) form the characteristics of a strong HRM system.

14.2 Objectives and Research Questions

The objective of this study was to explore the relationship between Human Resource Management (HRM) and innovative behavior (IB), work engagement (WE) and job insecurity (JI) in public universities, specifically: (1) the way workers perceive HRM, (2) the influence of this perception on IC, WE and JI behaviors, and (3) the presence of differences in the levels of innovative behavior between the categories of technicians and teachers and between the type of stable and non-stable employment relationship.

The aim is to answer the following research questions:

(IQ1) How do workers perceive the visibility, utility and understanding of HRM at their university?

(IQ2) How do these workers' perceptions of HRM and its practices influence their IB, WE and JI behaviors? and

(IQ3) Are there differences in the levels of innovative behavior among technical and faculty workers and between stable and non-stable workers?

To this end, relevant and diverse organizational actors were chosen, who, according to the roles developed in the work context, would allow us to investigate through interviews, seeking singularities and personal and collective meanings of different professional categories in the context of public higher education.

14.3 Methodology

Studies on HRM with workers from public universities are scarce, so we opted for a qualitative investigation to deepen the problem (Sampieri et al., 2014). This study is classified as cross-sectional, with data collected in a single stage, non-experimental, when there is no intentional manipulation of variables and descriptive, seeking to describe the phenomenon based on the reports of the interviewees (Sampieri et al., 2014).

14.3.1 Participants

A theoretical sample consisting of 14 workers from public sector universities in Brazil and Portugal was selected based on the following inclusion criteria: being a teaching, technical or research worker in public sector universities. The workers were selected in different roles and employment relationships at the university, in order to allow a more comprehensive data collection, with a greater wealth and variety of information and perceptions in the work context.

The sample selection followed the guidelines of Minayo (2017), prioritizing social subjects who held the attributes we aimed to know (different professional categories and contractual relationships), considering HR managers as a relevant category, and at the same time interacting with other categories, with or without a management role, in order to provide complementary information. We also considered Elias's (1984) position that society and group interact, with the interview with someone from a group being at the same time a personal and collective testimony. According to Onwuegbuzie and Leech (2007), in qualitative research, samples should not be considered in terms of quantity, but rather in terms of the scope of the social actors chosen and the reflection of the multiple dimensions of a phenomenon, with their adequacy being determined by the adequacy of the data (O'Reilly & Parker, 2013). Furthermore, attention was paid to the guideline that indicates a minimum of 10 informants in qualitative research (Atran et al., 2005).

Participation was anonymous and voluntary, using a saturation sampling method (Glaser & Strauss, 1967), and the interviews were conducted according to the availability of the participants. To ensure anonymity, the participants were designated P1 to P14, where P stands for Participant and the numbers refer to the total number, not representing the chronological order in which the interviews were conducted. In the study presented, contractual stability was considered as evidence of (in)security at work, and was translated by the type of employment relationship between the worker and the organization. Participants considered "Stable" referred to those who had a "stable" relationship at the Brazilian university (UB) and "open-ended contract" at the Portuguese university (UP), while "Non-Stable" referred to workers with a "probationary internship" and "substitute professor" relationship at the UB, and a "fixed-term contract", "uncertain-term contract" and "research grant" relationship at the UP.

The characteristics of the sample used are presented in Tables 14.1 and 14.2 bellow. In both universities, the participants held the following positions: HR manager (2), school president (2), professor (4), mid-level technician (2) and higher-level technician (4), ages between 33 and 59 years old and tenure ranging from 4 months to 37 years. The majority at both universities were female and in the "Stable" category.

Table 14.1 Sociodemographic and professional characteristics of participants from Portugal

Participant	Function	Gender	Age (years)	Professional category	Type of employment relationship	Seniority at university
P1	HR Manager	Masculine	56	Technical	Stable	23 years
P2	School President	Masculine	57	Professor	Stable	30 years
P3	Professor	Masculine	52	Professor	Stable	28 years
P4	Higher level technician	Feminine	49	Technical	Stable	27 years
P5	Mid-level technician	Feminine	33	Technical	Stable	7 years
P6	Professor	Feminine	43	Professor	Non-stable	2 years
P7	Higher level technician	Feminine	46	Technical	Non-stable	1 years and 6 months

Table 14.2 Sociodemographic and professional characteristics of participants from Brasil

Participant	Function	Gender	Age (years)	Professional category	Type of employment relationship	Seniority at university
P8	HR Manager	Masculine	58	Professor	Stable	37 years
P9	School President	Feminine	59	Professor	Stable	32 years
P10	Professor	Feminine	55	Professor	Stable	22 years
P11	Higher level technician	Masculine	36	Technical	Stable	12 years
P12	Mid-level technician cian	Masculine	30	Technical	Stable	4 years
P13	Professor	Feminine	55	Professor	Non-stable	2 years
P14	Higher level technician	Feminine	41	Technical	Non-stable	4 months

14.3.2 Instruments

An interview guide was adopted (Table 14.3), consisting of 7 questions drawn up based on the literature review, with the addition of an initial question to confirm and collect demographic data, such as functions performed, age, professional category,

Table 14.3 Interview guide questions

Number	Question
1	For you, is there HRM at the university where you work? And how do you evaluate it in terms of utility, visibility and understanding?
2	If you are aware that HRM exists at your university, does this influence your engagement and innovation behaviors?
3	How do you feel about insecurity/security at work?
4	For you, does job security/insecurity influence your engagement and innovation behaviors?
5	In your opinion, can the performance of HRM through its practices employed at your university influence your perception of job security/insecurity? And how can it influence?
6	In your opinion, is there a difference in innovative behavior between teachers and technicians? And between workers with stable and non-stable employment contracts?
7	Is there anything you want to mention or any comments you would like to make?

type of employment relationship and seniority, and ending with a thank you to the participant with space to add any additional aspects.

14.3.3 Data Collection Technique and Procedures

The data were collected in a single stage, from February to March 2023, individually, through semi-structured interviews. The interviews were conducted online, using the Zoom application, and lasted an average of 30 to 45 min. The participants were asked for permission to record audio only, without images, at the beginning of the interview, using a Free and Informed Consent Form (FICF) model, in which information was provided about the objectives of the study, the confidential and voluntary nature of participation, and the guarantee of anonymity and confidentiality of the information provided. The interviews were then fully transcribed, and the data discarded after use.

Initially, authorization to conduct this research was requested at both universities, based on contacts with managers. According to the official university documents for 2020, the total number of employees at UB is 2,279 professors and 3,358 administrative technicians, while at UP, which is smaller, it is 1,292 professors, 704 administrative technicians and 371 researchers.

Participants were selected according to the functions and professional categories of interest for the study and, after individual contact, were invited to participate in the research voluntarily. A preliminary interview was conducted, in which some difficulties in understanding were identified. Consequentely, adjustements were made to the script, which were taken into account for data analysis.

14.3.4 Data Analysis

Thematic analysis (King, 2004, 2023) was used as a method to organize and analyze the data collected in the interviews.

Thematic analysis is a qualitative data analysis method that is considered to be relatively recent in the field of organizational research (King & Brooks, 2018). This technique can be applied to any type of textual data, usually in interviews (King, 2023), and across a range of epistemologies and research questions (Braun & Clarke, 2006). It is considered suitable for investigating diverse groups in the organizational context.In comparison to other data analysis techniques, it offers a degree of flexibility as a salient advantage (Symon & Cassell, 2012).

King (2004) delineates thematic analysis as a qualitative data analysis technique that is based on coding themes and hierarchizing these data in a useful and meaningful way. Themes represent recurring characteristics that distinguish participants in their perceptions and/or experiences (King & Horrocks, 2010), and these will be subjected to the coding process. In this process, codes are identified and attached to the themes, considering the codes as abbreviations or short comments related to the excerpts of the collected text that are observed by the researcher as material of relevance to the research objective (King & Brooks, 2018). It is an ongoing process in which, based on careful reading of the data, new themes can be created or modified (Braun & Clarke, 2006).

There are 7 steps for implementing the technique, which are: (1) Definition of themes a priori; (2) Transcription and reading of the interviews; (3) Initial coding of the data; (4) Construction of the initial template; (5) Development of the template; (6) Interpretation and writing of the final template; (7) Verification of the quality of the data analysis King (2023). Although the author presents these steps as a suggestion, he reinforces that the template is a tool that facilitates data interpretation, and that researchers should never consider it as a final model The possibility of revising it is always present, and it is the researcher's responsibility to consider the relationship between the amount of time and effort spent on the analysis and the enrichment of the data, in order to find a good model (King, 2023).

Following the guidelines in the literature on procedures for developing the chosen technique, themes were defined a priori according to the study variables and their dimensions, HRM (visibility, utility, understanding), WE (vigor, dedication and absorption), JI (quantitative and qualitative) and IB present in the questions in the script, for later refinement after data exploration. We sought coherence in the number of pre-defined themes in our analysis, with the aim of allowing the exploration of possible more pertinent issues in the reports, listing 4 themes, presented below in the results (Table 14.4), taking into account that many pre-defined themes can limit the analysis (King, 2004).

After completing the full transcriptions of the interviews and carefully reading them, coding continued, identifying the parts of the reports and including them in the defined themes, based on the relationship with the variables studied and the relevance to the study objectives. The coding was performed manually using colored markers.

Table 14.4 Initial and final template

Initial template	Final template
1. Perception of HRM at the University 1.1. HRM 1.1.1. Existence of an HRM system 1.1.2. Utility 1.1.3. Visibility 1.1.4. Understanding 1.2. Identification of practices 1.2.1. Welcoming new employees 1.2.2. Recruitment and selection of new employees 1.2.3. Performance evaluation 1.2.4. Training 1.2.5. Administrative procedures in Human Resources Management (holidays, absences, continuity and termination of employment contracts, etc.) 1.2.6. Communication (information sharing) 1.2.7. Teamwork	1. Perception of HRM at the University 1.1. HRM 1.1.1. Existence of an HRM system 1.1.2. Utility 1.1.3. Visibility 1.1.4. Understanding 1.2. Identification of practices 1.2.1. Welcoming new employees 1.2.2. Recruitment and selection of new employees 1.2.3. Performance evaluation 1.2.4. Training 1.2.5. Administrative procedures in Human Resources Management (holidays, absences, continuity and termination of employment contracts, etc.) 1.2.6. Practices of external control bodies to the university
2. Impact of HRM on JI perception and WE and IB behaviors 2.1. JI 2.1.1. Quantitative 2.1.2. Qualitative 2.2. WE 2.2.1. Vigor 2.2.2. Dedication 2.2.3. Absorption 2.3. IB	2. Impact of HRM on JI perception and WE and IB behaviors 2.1. JI 2.1.1. Quantitative 2.1.2. Qualitative 2.2. WE 2.2.1. Vigor 2.2.2. Dedication 2.2.3. Absorption 2.2.4. Intrinsic motivation 2.3. IB 2.3.1. Intrinsic motivation 2.4. Impact on individual and organizational performance

(continued)

Table 14.4 (continued)

Initial template	Final template
3. Perception of JI and the impact on WE and IB behaviors 3.1. JI 3.1.1. Quantitative 3.1.2. Qualitative 3.2. WE 3.2.1. Vigor 3.2.2. Dedication 3.2.3. Absorption 3.3. IB	3. Perception of JI and the impact on WE and IB behaviors 3.1. JI 3.1.1. Quantitative 3.1.2. Qualitative 3.2. WE 3.2.1. Vigor 3.2.2. Dedication 3.2.3. Absorption 3.2.4 Motivation 3.3. IB 3.3.1 Risk 3.3.2 Creativity
4. Differences in IB between the groups “Professional category” and “Employment relationship” 4.1. Professional category 4.1.1. Technical 4.1.2. Professor 4.2. Employment relationship 4.2.1. Stable 4.2.2 Non-stable	4. Differences in IB between the groups “Professional category” and “Employment relationship” 4.1. Professional category 4.1.1. Technical 4.1.2. Professor 4.1.3. Resources provided by the university 4.2. Employment relationship 4.2.1. Stable 4.2.1.1. Accommodation 4.2.1.2 Leadership Type 4.2.2 Non-stable 4.2.2.1. JI

Caption: HRM: human resource management; WE: work engagement; JI: job insecurity; IB: innovative behavior

New themes were modified or created throughout the process. An initial template was created and subsequently, after analyzing all the responses, the codes created were reviewed, obtaining the final template (see Table 14.4).

The saturation point in this study was considered to have been reached after the 14th interview. Studies in the literature report the use of the saturation point after the 10th interview in research (Atran et al., 2005), other authors suggest that the saturation point usually occurs by the 12th interview and argue that central categories tend to appear by the 6th interview (Guest et al., 2006; Hoffman et al., 2016; Thiry-Cherques, 2009).

During the process of constructing the template, an external expert selected based on knowledge of the technique and the topic of study participated in order to perform peer reflexivity and quality verification of the technique used. Initially, the material was coded in its entirety by the author and the categories were defined. Subsequently, 10% of the material was coded independently by the specialist. At the end of this stage, the identified coding inconsistencies and divergences were discussed among the specialists, resolving the author's doubts and reaching a mutual agreement.

14.4 Results and Discussion

14.4.1 Thematic Analysis

The initial and final templates obtained after data analysis are presented below in Table 14.4.

14.4.2 Themes

14.4.2.1 Perception of HRM at the University

To answer (IQ1) How do employees perceive the visibility, utility and understanding of HRM at their university?, Theme 1 "Perception of HRM at the university" was proposed. The responses to question 1 were analyzed (see Table 14.3).

Different reports were found among participants in the different countries. At (UP), the perception is of a mechanistic, non-existent and decentralized function:

"(…) we are not managed by people, we are being managed by algorithms",[…] of regulations, and therefore there is no […] people management. I think it is non-existent…" (P3).

"(…) at the university where I work we have a division, a human resources department, but part of these services are decentralized by faculties…" (P4).

At UB, the data collected recognizes the existence of HRM due to its presence in the organizational structure:

"(…) there is indeed people management, there is an entire sector here, (…) and this sector is well structured, […] there is the training division […] it is very active" (P13).

The perception of HRM as a predominantly administrative system more involved with the administration of laws is present in the majority of UP workers, evidencing the absence of a system active in other issues related to human resources management. people, although some workers are aware of their actions.

"(…) I'm not sure there is people management, quite frankly, I think there is vacancy management, in the sense that the university is fundamentally concerned with the need to hire people" (P2).

"(…) I think there is essentially administrative management, there is not really strategic management…" (P3).

The workers' perception of the presence of an active HRM system within the organization is more apparent among workers at UB, which was verified in most of the reports that highlighted the different areas developed by management.

"(…) there is indeed people management and the structure is a Pro-Rector of People Management that involves the area of payment, the area of quality of life, the area of development, the area of training, the area of medical expertise…" (P8).

In search of greater depth on the perception of the existence of HRM in organizations, we explored the dimensions of HRM, visibility, utility and understanding, according to the Bowen and Ostroff model (2004), seeking to understand the perception of workers about each of these dimensions.

The majority of workers at UB perceive the "Visibility" of HRM based on the presence of several practices, which confirms the presence of prominent and easily observable practices in these institutions.

"(…) we work with quality of life, with development, with training, with the whole part of functional progression, payment, medical expertise, […] it ends up having a very high visibility and also a very high usefulness…" (P8).

"(…) it is well publicized and we receive notices of several actions that take place in the area of people management […] it has been very visible, I believe that our university has worked on a very strong part of both external and internal communication, I think this has grown in recent years" (P11).

Other workers, these from UP, do not perceive the existence of visibility of HRM in their institution, which is consistent with the fact that they complain about the non-existence of HRM, or when they do perceive it, they reinforce its administrative and mechanistic nature.

"(…) no, it is not visible, it is not perceived, and in fact there is great discouragement in this regard" (P1).

"(…) Yes, visible, […] in terms of hiring, salary progression, salary policies, absenteeism, regularization of absences, etc., everything goes through the human resources sector, so they have visibility at this level, […] they are the central services" (P4).

In the "Utility" dimension, workers demonstrate the perception that the HRM system contributes to the achievement of their individual objectives, which can be verified in the use of the practices offered by the institutions and in the support they feel. The response pattern in all interviews is that of the visibility of HRM dimension that we presented previously. UB refers transversally in the interviews to the utility of HRM, but UP does not.

"(…) it's because almost all of the university's 6,000 employees use this people management service and if it weren't for this, logically the employees wouldn't be able to progress, they wouldn't be able to do medical assessments, they wouldn't be able to join our quality of life program, so I think its usefulness is very broad, broad and also complex, right? But people management is useful" (P8).

"(…) yes, it's very useful, right, because it provides support to the employee when needed" (P12).

The importance of an HRM system that operates through unambiguous practices was highlighted by the UB, with the "Understanding" dimension of the system being considered an impactful factor in the perception of the system by employees.

"(…) not all practices are very understandable, […] the further away you are from people management, […] even physically, often the more difficult it is to understand some possibilities, some benefits that are not so clear" (P11).

It is interesting to note that the understanding of HRM practices and their impact exists at the UP when the participants previously mentioned that HRM did not exist.

"(…) when the processes and procedures are clear and understood, therefore the perception that people have of the support they receive from human resources is positive. When they are not always clear, are not understood or are not properly explained, often they are not really known, they are new situations, the perception that one has at the level of human resources management can be a little more biased" (P4).

According to the workers' reports, it is suggested that there is a need for a complementary look at the dimensions of HRM, which can be seen in their recognition of the need to make HRM practices visible and explicit within the organization in order to achieve the utility and understanding of these practices. This understanding perceived by the workers reinforces the need to deepen the dimensions of the B&O Model verified in the previous study, where the study by de la Rosa-Navarro et al. (2020) is highlighted, in which some of the dimensions were added.

"(…) it would be useful if it were made explicit, because it would most likely make us aware of all the rules of the game, […] if there is not something unsaid here, something hidden, something that seems to me to be unhealthy" (P2).

Also regarding the workers' perception of the performance of HRM in universities and its dimensions, the "Identification of the practices" highlighted in the interviews by the participants was carried out. "Welcoming new employees" was observed as a practice present at UB, but was not highlighted by workers at UP.

"(…) it provides full assistance in terms of the employees' adaptation, […] in this contractual aspect as well, […] it welcomes them, holds the inauguration ceremony, […] everything is very organized" (P13).

Another practice observed was "Recruitment and selection of new employees", with this report only being verified at UP, which can be justified by the fact that at UB admissions are only made through public competition, while at the Portuguese university other forms of entry may be present.

"(…) guest assistants, which is where there is most staff recruitment, there is no active hiring, it is much more passive, in terms of career development it is […] very much guided by administrative and bureaucratic restrictions and […] budgetary restrictions of the university itself, that is, there is not much point in planning a career in the sense that it is dependent, given that there are quotas for moving up in position" (P2).

"(…) at the hiring level, what we do is propose to our superiors, those who are authorized […] all the processes of ranking, selection of people are done by the services…" (P4).

The practices of "Performance evaluation" and "Training" stand out, which were reported as the main practices observed by workers in the performance of HRM in both institutions, with performance evaluation being the most prominent at UP.

"(…) there is this training offer by the university that is very important for this innovation effort…" (P4).

"(…) in a public university there must always be this, this management, […] there is a performance evaluation that is done annually, whether or not it is in the trial period" (P6).

The workers mentioned "Practices of the university's external control bodies" as important at universities, particularly at UB.

"(…) the control bodies in Brazil, the Federal Court of Auditors, the Federal Comptroller General, they are very active in the area of people management" (P8).

The practices related to "Administrative procedures in Human Resources Management (vacations, absences, continuity and termination of employment contracts, etc.)" and "Communication (information sharing)" did not present relevant reports.

14.4.2.2 Impact of HRM on the Perception of JI and on the Behaviors of WE and IB

To answer (IQ2) How do these workers' perceptions of HRM and its practices influence their IB, WE behaviors and JI levels?, theme 2 "Impact of HRM on the perception of JI and on the behaviors of WE and IB" was proposed. The answers to questions 2, 3 and 5 of the interview were analyzed in particular (see Table 14.3).

In the analysis of the impact of HRM on "JI", workers from both universities point to HRM as an important factor in reducing levels of insecurity. These results confirm the important role of HRM in reducing levels of insecurity, helping to reduce unwanted behaviors among workers (Kerse et al., 2018) and contributing to the superior performance of organizations (Gould-Williams, 2003).

"(…) the existence of some type of practice ends up having this, this influence on my perception, […] to enhance security depends on the type of practices, but generally they interfere with the feeling of security" (P6).

"(…) this engagement of a management sector with the sectors will have a positive impact in the sense that it will provide more security, training, understanding, (…) I leave ignorance behind, I understand better how a system works, I will feel more secure, I need to know in order to understand" (P10).

The most prominent and easily observable HRM practices (HRM_Visibility) were recognized as an important factor in reducing concerns about job loss and "Quantitative" insecurity. It is worth noting that these reports are more common among UP workers, who, despite not realizing this dimension is present in their organization, recognize the importance of reducing JI levels.

"(...) perhaps a more explicit policy on human resources reduces insecurity, it can either decrease or increase it, [...] because obviously by making this explicit it can make some people finally realize the goal they have to achieve [...] is for me to know exactly what the consequences of my behavior are" (P2).

"(...) from the moment we saw that human resources management is also committed to stabilizing our situation, we also ended up feeling safer, right, more supported." (P7).

Workers confirm the perception that the HRM system contributes to achieving individual objectives (HRM_Utility) and unambiguous practices (HRM_Understanding) reduce concerns about the continuity of various aspects of work, "Qualitative" insecurity. The reports are mostly from UB interviewees, which is consistent with the perception of these dimensions previously exposed.

"(...) I believe so [...] it gives you security in terms of how to develop your work, [...] to do better because you are also better guided" (P13).

"(...) I realize through the personnel development policies that they, that the university is investing in its staff, [...] and this provides the assurance that management wants people to remain in their positions and develop and become more satisfied and be able to have a slightly better life, including financially, as a result" (P14).

It is important to note that the testimonies regarding the impact of HRM in its different components (visibility, utility and understanding) on the perception of insecurity in its Quantitative and Qualitative dimensions are mentioned respectively by respondents from UP and UB, thus remaining consistent with the recognition of HRM that we mentioned previously and with the greater existence of temporary contracts at UP.

In the interviews, workers realize that practices such as those related to external control bodies, which must be complied with by universities, can be a source of JI_Qual (continuity of the various aspects of work).

"(...) the control bodies in Brazil, the Federal Court of Auditors, the Federal Comptroller General, they work heavily in the area of people management, which is also the area with the highest cost, [...] and these control practices often have an impact on the employee's pockets, [...] in the correction of some actions carried out by some employees, and this is still a practice that bothers and can affect engagement, [...] also the employee's safety in some aspect" (P8).

HRM was also identified as a driver of engagement and innovation behaviors in universities, suggesting that higher education workers, as they perceive the performance of HRM in their organization, will demonstrate more engagement behaviors, with higher levels of enthusiasm and energy, and also high levels of innovation (Hewett et al., 2018; Wang et al., 2020). The results indicate that these levels of WE and IB behaviors can be enhanced as HRM acts to reduce JI levels, corroborating the studies by Kerse et al. (2018) and Gould-Williams (2003), which indicate, respectively, that the performance of HRM in the levels of job insecurity helps to reduce unwanted behaviors of workers and contributes to the superior performance of organizations.

There is a greater perception of this impact in the reports of workers at UB, compared to UP. Although, in the latter, there are reports on the recognition of the importance of HRM in promoting WE behaviors, it is noteworthy that these workers do not mention the effect in practice, as shown in the following reports.

"(...) I think that people management practices can influence, yes, [...] mainly by giving me security, [...] guarantees that I can innovate, [...] do different things with reduced risks, [...] feeling support from this area will make me be more engaged and innovate more" (P11).

"(...) I believe that having strong people management allows people to be more engaged and also more innovative in the workplace. [...] I believe that the greater the perception, the more the person tends to respond positively, I do believe so" (P14).

In the analysis of "WE" in its dimensions, it is clear that the perception of the impact of HRM in the "Vigor" dimension (WE_Vigor) is present in the workers' reports at the UP. The workers demonstrates higher levels of energy and mental resilience, and a willingness to invest and persist in the face of work difficulties. The "Dedication" dimension (WE_Dedication) was also highlighted, with the performance of HRM being considered fundamental for these workers to feel inspired, enthusiastic and to show feelings of pride and challenge in the face of work. No clear reference was found in the interviews to the "Absorption" dimension (WE_Absorption).

"(...) It is clear that when we feel supported, cherished, encouraged, I mean, we give everything, there is no doubt about that" (P1).

"(...) it helps us to be more committed on a day-to-day basis and therefore to carry out the processes more rigorously, with more vigor, with more attention to some aspects of the law that we probably did not know about and without their support it would be very difficult to get there" (P4).

Employees at both universities perceive HRM as an important tool for the development of "IB", corroborating studies that point to the important role of HRM in promoting innovative behaviors (Jiménez-Jiménez & Sanz-Valle, 2008; Rodrigues & Veloso, 2013; Seeck & Diehl, 2017; Rasool et al., 2019; Stankevičiūtė et al., 2020), with emphasis on the HRM_Utility component.

"(...) in fact, there is this training offer by the university that is very important for this innovation effort" (P4).

"(...) I can do certain activities, I can have more freedom, flexibility to propose changes to innovate, without any risk, [...] something along those lines, are examples of actions that could be taken by the HR management area" (P11).

Among the workers at both universities, some revealed that their engagement and innovation behaviors are not directly related to the HRM practices that exist at the university, pointing to "Intrinsic motivation" as the main factor, however, the importance of the university as a provider of resources for these behaviors to be applied was highlighted.

"(…) by nature I am a person dedicated to work, right, whether the working conditions are good or not, […] so there is this intrinsic motivation, but at the same time the university has, provides here some resources that are also fundamental to increase it" (P6).

"(…) being aware that there is a HR management system may make us in a somewhat more comfortable situation, but this does not condition the role and the exercise of the role itself. I don't think it affects me directly, maybe more indirectly" (P7).

"(…) I'm an innovative girl, so it's because of my decision to go to the gym, right? Because of my life principles, maturity, and responsibility, I chose the gym as my mission" (P13).

"(…) I notice the movement in me, but not directly because of people management, […] the general movement of people who are very engaged in their work, each one here engaged in their role, right, and with the intention of proposing improvements, right, of proposing innovation, proposing improvements in the work" (P14).

The positive "Impact on individual and organizational performance" of an HRM system, based on its practices, was highlighted by workers, recognizing HRM as one of the fundamental elements in improving individual (Esteves & Caetano, 2010; Hewett et al., 2018; Massuno & Jacquinet, 2019; Wang et al., 2015) and organizational (Aust et al., 2020) performance. The results confirm that having this perception of HRM performance generates positive attitudes at work (Karatepe, 2011).

We can appreciate in our results the perception of the impact of HRM on organizational results, namely in the IB, even though it has emerged in a generalist way regarding its expression, namely in the UB. Due to its nature and relevance, we added a new subtheme to the initial template. "(…) people management certainly has a huge impact, […] it impacts the organizational culture, […] on how the university will function, the better structured it is, the more we will be able to see results" (P11).

"(…) yes, it has a positive impact, I especially value the training part…" (P13).

"(…) And I myself noticed the management movement in terms of training, […] it gives motivation" (P14).

This impact of HRM on individual and organizational performance was also highlighted by workers, based on the importance of clear communication between these workers and the organization, providing an understanding of what is expected of them (Sanders et al., 2014) (Sanders et al., 2021), a factor considered in the Bowen and Ostroff (2004) model as necessary for building a strong HRM system.

"(…) people knowing what to do, who to look for and how they should behave throughout their lives works, so yes, I consider it very important to have a people management area" (P11).

14.4.2.3 Perception of JI and Its Impact on WE and IB Behaviors

Still aiming to answer (IQ2) How do these workers' perceptions of HRM and its practices influence their IB, WE and JI levels of behaviors?, theme 3 was proposed: "Perception of JI and its impact on WE and IB behaviors". This aspect interested us because the participants are part of a public university whose contractual relationship is perceived as secure. In Portugal and Brazil, the participants reported a low level of job insecurity. The responses to questions 3 and 4 were analyzed in particular (see Table 14.3).

"(…) I don't feel insecure, I don't feel any kind of insecurity, […] but I realize that this is a factor […] that weighs heavily on, let's say, people's involvement in their careers…" (P2).

"(…) totally safe, […] I have job security, right, because of the length of service I have, I am entitled to a full pension, right? […] so for me it is totally safe" (P9).

"(…) I would say that in a Brazilian federal public university practically no one will have the perception, the feeling of job insecurity, since we have job security, right, it is very difficult for someone to be fired or dismissed under the legislation we have in Brazil, so I feel zero insecurity" (P11).

Workers reveal levels of perception of "Qualitative" insecurity JI_Qual (continuity of the various aspects of work) when compared to "Quantitative" insecurity JI_Quant (continuity of employment), where the majority, in both universities, do not report this perception. These results corroborate studies that indicate that in the public service, where there is less risk in relation to employment, the perception of JI_Quant was lower, when compared to JI_Qual (Chirumbolo et al., 2020).

"(…) I don't think I'm at risk of losing my job, so I think I have a stable contract. The instability is much greater from the point of view of remuneration, for example, […] we had a real loss in the value of what we receive, so the contract remains but the remuneration is lower, and here it generates instability from the point of view of remuneration and the associated salary" (P3).

"(…) at this moment, what is a more structural aspect and that has to do with the financial situation at the university, we all know that it is not the best and naturally this ends up being an issue that also causes some insecurity here, about how this can affect the resources that we will have available, the working conditions that we will have available […] all of this ends up affecting" (P6).

"(…) the issue of security is only in relation to the value of the salary […] there is a salary policy, it is an economic issue of the country […] in the general context of workers it is quite stable, it is quite secure" (P9).

"(…) zero quantitative, right? Yes, qualitative, in the sense that, […] not having the resources to develop my research… this gives me tremendous insecurity, […] now it has been decided that the unhealthy conditions will be eliminated, the unhealthy conditions are R$2,000 off my salary, this is also an insecurity" (P10).

Regarding the impact of the perception of JI on "WE" and "IB" behaviors, workers at both universities confirm the importance of safety for increasing engagement (Altinay et al., 2019; Unur et al., 2022) and innovation (De Spiegelaere et al., 2014; Niesen et al., 2018; Probst et al., 2020), while the perception of insecurity can reduce these levels (Wang et al., 2015). The "Dedication" dimension of WE stands out as being impacted by JI in the workers' reports, with no reports being found in the other dimensions "Vigor" and "Absorption".

"(…) there is naturally some insecurity due to the fact that our contract is experimental… because I know that I have to prove that I deserve the position, I end up getting more involved and dedicating myself more to ensure that I will actually keep the position" (P6).

"(…) certainty, at least the engagement part, which I see more as commitment,… we feel more committed when I am at ease regarding my security, both in terms of keeping my job and in terms of being able to carry out my activities knowing that they will not be affected" (P8).

"(…) of course, a secure person is calm and dedicated, they do not worry about other issues, right? They do not waste energy thinking that they will not be continued, so they have to think about plan B, right, and not dedicate themselves so much to their plan A" (P14).

Workers at UP refer to "Motivation", while those at UB refer to "Creativity", as factors present in the relationship between JI and WE and IB behaviors.

"(…) for me it is essential because I have a family and I have bills to pay, it gives me some guarantees and also keeps me motivated to continue." (P4).

"(…) it makes me feel more confident in creating, innovating, taking risks, because as I said, we are privileged, we don't even have a boss, (…) and this has a lot to do with my security, quantitatively" (P10).

The UB workers highlight "Risk" as a factor related to IB. For them, low levels of JI can enhance IB behaviors, allowing these workers to take risks and consequently be more innovative in their work.

"(…) and in relation to innovative behavior, yes, it changes completely, right, because we know that innovation is a way of working that involves risks […] When I feel that I have the possibility of taking risks and making mistakes without losing anything, I will definitely innovate more" (P11).

14.4.2.4 Differences in IB Between the "Professional Category" and "Employment Relationship" Groups

To answer (IQ3) Are there perceived differences in the levels of innovative behavior between technical and teaching staff and between stable and non-stable staff?, theme 4 "Differences in IC between the "Professional category" and "Employment relationship" groups" was proposed. The responses to question 6 were analyzed in particular (see Table 14.3).

"(…) they are very different missions and in fact the functional contents of each career or staff group are very different" (P1).

"(…) I cannot compare between technicians and teachers because I see, in my opinion, a decision for innovation in both, and in some yes, and in some no (P13).

The majority of workers at both universities recognize that there are more conditions and "Resources made available by the university" for innovation in the teaching category, when compared to technicians.

"(…) In the case of teachers, there may be a greater scope, there may be more room for innovation and not so much in the role of higher technician" (P7).

"(...) both are qualified, but the conditions are more favorable for teachers... research projects, extension projects, so those who coordinate the projects, those who can make major changes and innovate are still [...] the teachers." (P9).

"(...) teachers have the use of didactics in the classroom, right, where they have freedom to innovate, [...] They have this possibility of making this innovation, of directing teaching, studying, according to their will. Technicians have a slightly more technical job, right, of making things happen. I believe that teachers have more, a wider range of activities, so, greater freedom to act and to innovate" (P14).

In the analysis of the group regarding the type of "Employment relationship", no significant differences were found in the IB behaviors between workers with a "Stable" type of relationship and workers with a "Non-Stable" type of relationship. In the participants' reports, most of them from both universities, it can be observed that both stable and non-stable workers emphasize that the type of relationship is not a relevant factor for their IB behaviors.

"(...) I don't think that's it, I don't think about it, I mean when I get more or less involved and let's be clear, I'm a very involved type always" (P2).

"(...) well, I think that it depends on the characteristics of the person's personality, right, and the environment in which they are, right, [...] I believe that the question of being stable or not stable doesn't even have much influence" (P4).

"(...) what motivates me is not just the fact that I'm on a trial period, it's certainly not that" (P6).

"(...) engagement, I think my personal perception doesn't influence me, I work. I will work the same way with or without the risk of losing a possible role, for example," (P11).

"(...) I think it depends a lot on how the person knows themselves, how they understand themselves as a person, how they understand their mission. [...] so commitment, for me, is independent of something temporary or definitive" (P13).

"(...) it's not a factor of being on probation, it doesn't directly influence how much I dedicate myself, I believe I'm dedicating myself for other reasons, because I'm involved, because the work environment is very welcoming and stimulating," (P14).

With regard to workers with stable contracts, reports were observed that this group tends not to demonstrate innovative behaviors, pointing to "Accommodation" as a limiting factor to innovation, which was verified in both universities. Some worker, at UP, points to the "Type of leadership" in this group of workers as an important factor for motivation.

"(...) I would say that probably those who have stable contractual relationships are more accommodating and those who don't work more so that they can later have a more stable contract" (P4).

"(...) what we observe, the difference between those who are stable and those who aren't stable would be the accommodation, right, sometimes they become accommodating, right [...] those who would be able to innovate are those who are already secure and stable. Now, when he is stable, he can both innovate and settle down" (P9).

"(...) now, despite having a more secure contract [...], I also have a completely different type of management than I had [...] I am much better now, without a doubt, in terms of everything, in terms of the work environment, in terms of motivation" (P5).

Another group of workers considers that being in a non-permanent contract can impact higher levels of WE and IB behaviors, since in this group levels of "JI" are present and can influence these behaviors. The reports below highlight the UP, as it has a greater presence of temporary contracts, when compared to UB.

"(...) I think that non-permanent workers are more innovative, sometimes because they want to show more work and, therefore, when they later have a contract, they tend to become more... to do what they are supposed to do, but therefore take much, much less risks" (P3).

"(...) because I know I have to prove that I deserve the position, I end up getting more involved and dedicating myself more to ensure that I actually stay in the position" (P6).

"(...) those who are in a more unstable situation may always have this need to give a little bit more [...] I think those who are unstable end up innovating more than those who are stable" (P7).

"(...) I understand that people who are not yet in a stable position have a desire to innovate, even if they are newcomers, they bring ideas from outside and have a greater tendency to present innovation than those who have been there for a long time," (P14).

14.4.3 Comparison Between Countries

The analysis of the data collected in the interviews made it possible to compare some important points in the perception of workers in different countries, which were presented throughout the results and discussion of this study. Below, with the aim of sparking reflections and making understanding clearer, we present a summary of the main differences and similarities observed in the analyses carried out.

In Portugal, the existence of an HRM system in universities stands out as mechanistic, non-existent and decentralized, being a system that is more heavily administrative and governed by laws, with an absence of strategic management and support for workers. HRM was highlighted only in the understanding dimension, although there is a perception that HRM does not exist in this country or that it is only perceived as administrative. The main HRM practices stressed in the organization were: recruitment and selection of new employees, performance evaluation, training, and administrative procedures in HRM (holidays, absences, continuity and termination of employment contracts, etc.). The impact of HRM on JI is recognized, particularly in the quantitative dimension, since it has a greater number of workers with temporary contracts. Furthermore, the impact of HRM on WE behaviors was mentioned as important, with emphasis on the dimensions of vigor and dedication, and on IB, although this is not effectively perceived in practice. Innovation was also recognized in this impact.

In Brazil, HRM is perceived as active and present in its organizational structure and is presented with the indication of several areas of action in people management. HRM was highlighted in its three dimensions analyzed, visibility, utility and understanding. The main HRM practices emphasized in the organization were: welcoming new employees, performance evaluation, training, administrative procedures in human resources management (vacations, absences, continuity and transfer of employment contracts, etc.), and practices of external control bodies at the university. The impact of HRM in its dimensions, in JI, was recognized, mainly in the qualitative dimension, which coincides with the fact that in Brazil the majority of contracts are permanent. The positive impact of HRM on WE and IB behaviors stands out in this group, with the more general impact on individual and organizational performance also being recognized. The main similarities observed concern the low levels of JI verbalized by the participants, which when it occurs refers to the qualitative dimension, as well as the report on the importance of safety for increasing WE and IB. For both countries, there are no significant differences in IB in relation to the professional category, and both highlight that there is greater availability of resources for the development of IB for teachers, when compared to technicians. Regarding the employment relationship, similarly, there are no differences in the perception between stable and non-stable employees, considering that the type of relationship is not a determining factor for the development of these behaviors.

14.5 Conclusions, Limitations and Future Studies

This study revealed the perceptions of workers about HRM, in its dimensions (visibility, utility, understanding),within the organizations in which they work, in the different countries, and concerning the influence of this HRM and its practices on their WE and IB behaviors, as well as on the levels of JI. The study also revealed the perception of workers regarding the differences in the levels of IB between the different categories (technicians and teachers) and between employment relationships (stable and non-stable).

In general, two contrasting realities can be seen regarding HRM in these organizations, one mechanistic and non-existent, and the other present and active in several areas. At UP, the interviewees do not recognize the presence of HRM, and when they do, it is viwed as merely administrative, reflecting the presence of a traditional model present in the evolution of HRM (Gomez, 1999) at this university. In contrast, UB workers recognize the role of HRM beyond administrative issues and highlight its stronger role in other areas related to support and assistance to workers.

In fact, universities, as part of the public administration, are subject to the bureaucracy present in this context, which is reflected in all sectors of the institution, including the HRM sector. This consists of several bureaucratic rules and procedures in the administrative routine, which can harm the organization's progress (Pereira &

Silva, 2016). It is suggested that this perception of lack of support observed in the UP may arise from the fact that these workers do not feel supported by HRM in other areas related to caring for people, such as the worker's quality of life, reporting performance evaluation as one of the main practices observed in HRM performance, which is mandatory and used merely for career progression.

The importance of the dimensions of HRM, visibility, utility and understanding, was highlighted at the UB, which is reflected in the perception of the effects of HRM in this organization. The lack of visibility of the HRM system, or the recognition of a merely administrative system, conditions the effects of this system at the UP, also leading to a lack of perception of its utility, which also occurs with understanding. The need for a complementary look at the dimensions of HRM to enhance its effects in organizations becomes evident. The results of this study demonstrate the recognition of the impact of HRM on WE and IB behaviors (Ahmed et al., 2020; Urbini et al., 2020) on the levels of JI (Gould-Williams, 2003) and the strong relationship of these variables in public universities (Yadav et al., 2022). We highlight here engagement, a concept linked to work itself (Schaufeli et al., 2002), which is present in workers, but justified by intrinsic motivation, and not specifically by the performance of HRM in these organizations, which highlights a lack of policies and practices that encourage the development of this variable in these organizations.

The verification of higher levels of qualitative JI highlights the subjective and perceptive nature of this construct (Brockner et al., 1992), demonstrating that even in work relationships considered safe, such as in public universities, this feeling can be present (Chirumbolo et al., 2020). In universities, the quantitative and qualitative dimensions are not equally present, which we understand to be justified by the difference in the number of types of employment relationships of workers, impacting a greater perception of quantitative insecurity in UP compared to Brazil.

The importance of safety for increasing WE and IB behaviors was highlighted by workers (Altinay et al., 2019; De Spiegelaere et al., 2014; Niesen et al., 2018; Probst et al., 2020; Unur et al., 2022), reinforcing the need for an in-depth look at this variable from HRM.

It was possible to observe the absence of differences in the professional category and employment relationship groups in relation to WE and IB behaviors at universities. This suggests that employment relationship is not a relevant factor for the development of these behaviors, although it was evidenced by workers that there is greater availability of resources for innovation for the teaching category compared to technicians.

Conducting qualitative research is a challenge. Sometimes it faces some prejudices from those who are more familiar with quantitative research, and who are surprised by the rigor criteria used in qualitative research (Braun & Clarke, 2006). In order to demonstrate this methodological rigor, we sought to follow guidelines found in the literature that indicate the existing criteria for conducting good qualitative research, both in data collection and analysis, explaining information regarding sample size and selection, as well as determining the theoretical saturation point in the analysis of the collected data (Atran et al., 2005; Minayo, 2017; Onwuegbuzie & Leech, 2007; Parker, 2004). The use of thematic analysis as a technique for data analysis proved to be challenging when applied to a different context, such as universities, and contributes to the growth of the use of the technique in the organizational area, where there are still few reported studies. In this study and in relation to thematic analysis, we sought to apply the rigor criteria identified by Braun and Clarke (2006, p. 96), such as rigor in transcription, attention to data analysis during coding, analyzing the data by interpreting it and not just describing it, and dedicating enough time so that this technique could be applied without rushing.

This study has some limitations, the main one being the impossibility of generalizing the data from this research to all organizations or workers, restricting the analysis of the phenomenon to public universities, excluding the private sector, and to workers in higher education. Another point to highlight was the use of a sample considered small and the exclusive reliance on interviews as the primary data collection method. This may be limiting, but which we believe does not harm the research considering the careful observation of the saturation point in the technique.

We also emphasize that considering the theoretical saturation point as a reference for the cessation of qualitative data collection can be seen as a technique with limitations, due to the difficulty of accuracy regarding new information that may still emerge. Therefore, the possibility of errors and divergences in determining the point must be considered, although we followed the appropriate guidelines in the procedure, which does not undermine the reliability of the technique (Falqueto et al., 2018).

Finally, these results make us understand and reflect on the importance of having a strong HRM system (Bowen & Ostroff, 2004), with visibility, utility and understanding of the practices employed in universities, to improve individual and organizational performance. Such improvement can be achieved through behaviors of engagement and innovation, which only become possible with the existence of HRM that goes beyond administrative issues, with investment focused on the care, support and assistance of employees. Investing in practices that promote people's involvement is essential to achieving these objectives. In this sense, HRM, by developing positive behaviors and attitudes in the workplace, through the promotion of employee well-being and development, can be an important tool in contributing to the achievement of the SDGs. Regarding theoretical and practical implications, we believe that research into the relationships between HRM and WE and IB behaviors, as well as JI, with an emphasis on the study of the dimensions of visibility, utility and understanding, can contribute to raising awareness of the role of HRM in higher education organizations, particularly public universities, and to the need for investment in practices to support and assist workers in these organizations.

In light of these results, we suggest that new studies be conducted in other universities, in order to cover other categories of workers, and also to extend the study to the private sector, enabling future comparisons with the study presented here.

References

Aboramadan, M., Albashiti, B., Alharazin, H., & Dahleez, K. A. (2020). Human resources management practices and organizational commitment in higher education. *International Journal of Educational Management, 34*(1), 154–174. https://doi.org/10.1108/IJEM-04-2019-0160

Ahmed, U., Kura, K. M., Umrani, W. A., & Pahi, M. H. (2020). Modeling the link between developmental human resource practices and work engagement: The moderation role of service climate. *Global Business Review, 21*(1), 31–53. https://doi.org/10.1177/0972150919837813

Alqahtani, M., & Ayentimi, D. T. (2021). The devolution of HR practices in Saudi Arabian public universities: Exploring tensions and challenges. *Asia Pacific Management Review, 26*(2), 86–94. https://doi.org/10.1016/j.apmrv.2020.08.005

Altinay, L., Dai, Y.-D., Chang, J., Lee, C.-H., Zhuang, W.-L., & Liu, Y.-C. (2019). How to facilitate hotel employees' work engagement: The roles of leader-member exchange, role over-load and job security. *International Journal of Contemporary Hospitality Management, 31*(3), 1525–1542. https://doi.org/10.1108/IJCHM-10-2017-0613

Amabile, T. M. (1997). Motivating creativity in organizations: On doing what you love and loving what you do. *California Management Review, 40*(1), 39–58. https://doi.org/10.2307/41165921

Atran, S., Medin, D. L., & Ross, N. O. (2005). The cultural mind: Environmental decision making and cultural modeling within and across populations. *Psychological Review, 112*(4), 744–776. https://doi.org/10.1037/0033-295X.112.4.744

Aust, I., Matthews, B., & Müller-Camen, M. (2020). Common good HRM: A paradigm shift in sustainable HRM? *Human Re-source Management Review, 30*(3). https://doi.org/10.1016/j.hrmr.2019.100705

Bowen, D. E., & Ostroff, C. (2004). Understanding HRM-firm performance linkages: The role of the "Strength" of the HRM system. *The Academy of Management Review, 29*, 203–221. https://doi.org/10.5465/AMR.2004.12736076

Braun, V., & Clarke, V. (2006). Using thematic analysis in psychology. *Qualitative Research in Psychology, 3*(2), 77–101. https://doi.org/10.1191/1478088706qp063oa

Brockner, J., Grover, S., Reed, T. F., & Dewitt, R. L. (1992). Layoffs, job insecurity, and survivors' work effort: Evidence of an inverted-U relationship. *Academy of Management Journal, 35*(2), 413–425. https://doi.org/10.2307/256380

Cerasoli, C. P., Nicklin, J. M., & Ford, M. T. (2014). Intrinsic motivation and extrinsic incentives jointly predict performance: A 40-year meta-analysis. *Psychological Bulletin, 140*(4), 980–1008. https://doi.org/10.1037/a0035661

Chamsa, N., & García-Blandón, J. (2019). On the importance of sustainable human resource management for the adoption of sustainable development goals. *Resources, Conservation & Recycling, 141*, 109–122.

Chirumbolo, A., Callea, A., & Urbini, F. (2020). Job insecurity and performance in public and private sectors: A moderated measurement model. *Journal of Organizational Effectiveness: People and Performance, 7*(2), 237–253. https://doi.org/10.1108/JOEPP-02-2020-0021

Cleveland, J. N., & Cavanagh, T. M. (2015). The future of HR is HR: Respect for humanity at work. *Human Resource Management Review*, 146–161.

Coda, R., & Ricco, M. F. (2010). Behavioral styles of science, technology and innovation (S, T&I) professionals. *Journal of Administration Contemporary, 14*(3). https://doi.org/10.1590/S1415-65552010000300008

De la Rosa-Navarro, D., Díaz-Fernández, M., & Lopez-Cabrales, A. (2020). Disentangling the strength of the HRM system: Effects on employee reactions. *Employee Relations: THe International Journal, 42*(2), 281–299. https://doi.org/10.1108/ER-12-2018-0322

De Prins, P., Van Beirendonck, L., De Vos, A., & Segers, J. (2014). Sustainable HRM: Bridging theory and practice through the 'Respect Openness Continuity (ROC)'-model. *Socioeconomics Studies, 25*(4), 263–284. https://doi.org/10.1688/mrev-2014-04-Prins

De Spiegelaere, S., Van Gyes, G., De Witte, H., Niesen, W., & Van Hootegem, G. (2014). On the relation of job insecurity, job autonomy, innovative work behavior and the mediating effect of work engagement. *Creativity and Innovation Management, 23*(3), 318–330. https://doi.org/10.1111/caim.12079

De Witte, H. (1999). Job insecurity and psychological well-being: Review of the literature and exploration of some unresolved issues. *European Journal of Work and Organizational Psychology, 8*(2), 155–177. https://doi.org/10.1080/135943299398302

Elias, N. (1984). *The society of individuals*. Jorge Zahar Publishing House.

Esteves, T. P., & Caetano, A. (2010). Human resource management practices and organizational results: Studies, theoretical and methodological controversies. *Journal of Psychology, Organizations and Work, 10*(2), 159–176. http://pepsic.bvsalud.org/scielo.php?script=sci_arttext&pid=S1984-66572010000200012&lng=pt&tlng=pt

Falqueto, J. M. Z., Valmir, E. H., & Farias, J. S. (2018). Theoretical saturation in qualitative research: Report of an application experience in a study in the area of administration. *Journal of Administrative Sciences, 20*(52), 40–53. https://doi.org/10.5007/2175-8077.2018V20n52p40

Ferreira Gomes, A., Veloso, A., & Costa, P. (2024). Psychometric analysis and validation of the job insecurity scale in universities. *Psychology: Theory and Research, 40*, e40509. https://doi.org/10.1590/0102.3772e40509.en

Ferreira-Oliveira, A. T. M. (2013). *Perceptions of the human resource management system and the role of organizational trust*. [Doctoral thesis in psychology, University of Minho]. Repository. https://hdl.handle.net/1822/7703

Findlay, P., & Thompson, P. (2017). Contemporary work: Its meanings and demands. *Journal of Industrial Relations, 59*(2). https://doi.org/10.1177/0022185617667225

Ghosh, D., Sekiguchi, T., & Fujimoto, Y. (2020). Psychological detachment: A creativity perspective on the link between intrinsic motivation and employee engagement. *Personnel Review*. https://doi.org/10.1108/PR-12-2018-0480

Gilbert, C., De Winne, S., & Sels, L. (2011). The influence of line managers and HR department on employees' affective commitment. *The International Journal of Human Resource Management, 22*(8). https://doi.org/10.1080/09585192.2011.565646

Glaser, B., & Strauss, A. (1967). *The discovery of grounded theory: Strategies for qualitative research*. Aldine Publishing Company.

Gould-Williams, J. (2003). The importance of HR practices and workplace trust in achieving superior performance: A study of public-sector organizations. *The International Journal of Human Resource Management, 14*(1), 28–54. https://doi.org/10.1080/09585190210158501

Guest, G., Bunce, A., & Johnson, L. (2006). How many interviews are enough? An experiment with data saturation and variability. *Field Methods, 18*(1), 59–82. https://doi.org/10.1177/1525822x05279903

Gupta, A. K., & Singhal, A. (1993). Managing human resources for innovation and creativity. *Research-Technology Management*. https://doi.org/10.1080/08956308.1993.11670902

Gürbüz, S., Schaufeli, W. B., Freese, C., & Brouwers, E. P. M. (2024). Fueling creativity: HR practices, work engagement, personality, and autonomy. *The International Journal of Human Resource Management, 35*(22), 3770–3799. https://doi.org/10.1080/09585192.2024.2429125

Hellgren, J., Sverke, M., & Isaksson, K. (1999). A two-dimensional approach to job insecurity: Consequences for employee attitudes and well-being. *European Journal of Work and Organizational Psychology, 8*(2), 179–195. https://doi.org/10.1080/135943299398311

Hewett, R., Shantz, A., Mundy, J., & Alfes, K. (2018). Attribution theories in human resource management research: A review and research agenda. *The International Journal of Human Resource Management, 29*(1), 87–126. https://doi.org/10.1080/09585192.2017.1380062

Hoffmann, V. E., Belussi, F., Martínez-Fernández, M. T., & Reyes Jr., E. (2016). United we stand, divided we fall? Clustered firms' relationships after the 2008. *Entrepreneurship & Regional Development*, 1–24. https://doi.org/10.1080/08985626.2017.1343869

Jiménez-Jiménez, D., & Sanz-Valle, R. (2008). Could HRM support organizational innovation? *International Journal of Human Resource Management, 19*(7), 1208–1221. https://doi.org/10.1080/09585190802109952

Jordan, G. B. (2005). What matters to R&B workers? *Research Technology Management, 48*(3), 23–32. https://doi.org/10.1080/08956308.2005.11657312

Kanter, R. M. (1988). Three tiers for innovation research. *Communication Research, 15*(5), 509–523. https://doi.org/10.1177/009365088015005001

Karatepe, O. M. (2011). The effects of coworker and perceived organizational support on hotel employee outcomes. *Journal of Hospitality & Tourism Research, 36*(4), 495–516. https://doi.org/10.1177/1096348011413592

Katz, R. (2005). Motivating technical professionals today. *Research Technology Management, 48*(6), 29–27.

Keating, J. (2002). The nature of the human resource management function. In A. Caetano, & J. Vala (Eds.), *Human resource management: Contexts, processes and techniques* (2nd ed., pp. 103–116). RH Editora.

Kerse, G., Kocak, D., & Ozdemir, S. (2018). Does the perception of job insecurity bring emotional exhaustion? The relationship between job insecurity, affective commitment and emotional exhaustion. *Business and Economics Research Journal, 9*(3), 651–663. https://doi.org/10.20409/berj.2018.129

King, N. (2004). Using templates in the thematic analysis of testing. In G. Symon & C. Cassell (Eds.), *Essential guide to qualitative methods and analysis in organizational research* (pp. 256–270). Sage publications.

King, N., & Brooks, J. (2018). Thematic analysis in organizational re-search. In C. Cassell, A. Cunliffe, & G. Grandy (Eds.), *The SAGE handbook of qualitative business and management research methods* (pp. 219–236). Sage publications.

King, N., & Horrocks, C. (2010). *Interviews in qualitative research.* Sage publications.

King, N. (2023) *What is template analysis?*. University of Huddersfield. https://research.hud.ac.uk/research-subjects/human-health/template-analysis/

Leite, K. C. (2020). The (un)expected pandemic and its implications for the world of work. *Psychology & Society*. https://doi.org/10.1590/1807-0310/2020v32240215

Massuno, A., & Jacquinet, M. M. (2019). Human resources management in one of the Angolan higher education institutions. *Multidisciplinary Association for Scientific Research (AMIC) Angolan Journal of Sciences, 1*(1), 70–94. https://doi.org/10.5281/zenodo.3837139

Minayo, M. C. de S. (2017). Sampling and saturation in qualitative research: consensus and controversies. *Revista Pesquisa Qualitativa, 5*(7), 1–12. https://editora.sepq.org.br/rpq/article/view/82

Neves, J., & Gonçalves, S. (2009). Research in human resource management in Portugal: Results and trends. *Revista Portuguesa e Brasileira de Gestão, 8*(4). https://bibliotecadigital.fgv.br/ojs/index.php/rbpg/article/view/78842

Niesen, W., Van Hootegem, A., Vander Elst, T., Battistelli, A., & De Witte, H. (2018). Job insecurity and innovative work be-haviour: A psychological contract perspective. *Psychologica Belgica, 57*, 174–189. https://doi.org/10.5334/pb.381

O'Reilly, M., & Parker, N. (2013). 'Unsatisfactory Saturation': A critical exploration of the notion of saturated sample sizes in qualitative research. *Qualitative Research, 13*(2), 190–197. https://doi.org/10.1177/1468794112446106

Onwuegbuzie, A. J., & Leech, N. L. (2007). Sampling designs in qualitative research: making the sampling process more public. *The Qualitative Report, 12*(2), 238–254. https://doi.org/10.46743/2160-3715/2007.1636

Parker, I. (2004). Criteria for qualitative research in psychology. *Qualitative Research in Psychology, 1*(2), 95–106. https://doi.org/10.1191/1478088704qp010oa

Pereira, L., & Silva, S. (2016). Bureaucracy in the Human Resources Sector of an Autarchy. *Journal of Administration, Society and Innovation, 2*(2), 118–131. https://doi.org/10.20401/rasi.2.2.86

Probst, T., Chizh, A., Hu, S., Jiang, L., & Austin, C. (2020). Explaining the relationship between job insecurity and creativity: A test of cognitive and affective mediators. *Career Development International, 25*, 247–270. https://doi.org/10.1108/CDI-04-2018-0118

Proença, T. (2022). Sustainability and the Role of HRM. *Business Administration Magazine, 62*(5). https://doi.org/10.1590/S0034-759020220513

Rasool, S., Samma, M., Anjum, A., & Khan, T. (2019). Relationship between modern human resource management practices and organizational innovation: Empirical investigation from banking sector of China. *International Transaction Journal of Engineering, Management, & Applied Sciences & Technologies, 10*(19), 1–13. https://doi.org/10.14456/ITJEMAST.2019.265

Rodrigues, A. F., & Veloso, A. (2013). Contributions of human resources management to organizational creativity and innovation. *Revista Psicologia Organizações e Trabalho, 13*(3), 293–308.

Saks, A. M. (2006). Antecedents and consequences of employee engagement. *Journal of Managerial Psychology, 27*(1), 600–619. https://doi.org/10.1108/02683940610690169

Sampieri, R. H., Collado, C. F., & Lucio, P. B. (2014). *Research methodology*. McGraw-Hill Publishing

Sanders, K., Shipton, H., & Gomes, J. (2014). Guest editors' introduction: Is the HRM process important? Past, current, and future challenges. *Human Resource Management, 53*(4), 489–503. https://doi.org/10.1002/hrm.21644

Sanders, K., Guest, D., & Rodrigues, R. (2021). The role of HR attributions in the HRM—outcome relationship: Introduction to the special issue. *Human Resource Management Journal, 31*(2). https://doi.org/10.1111/1748-8583.12358

Schaufeli, W. B., Salanova, M., González-romá, V., & Bakker, A. B. (2002). The measurement of engagement and burnout: A two sample confirmatory factor analytic approach. *Journal of Happiness Studies, 3*(1), 71–92. https://doi.org/10.1023/a:1015630930326

Seeck, H., & Diehl, M.-R. (2017). A literature review on HRM and innovation—taking stock and future directions. *The International Journal of Human Resource Management, 28*(6), 913–944. https://doi.org/10.1080/09585192.2016.1143862

Shipton, H., West, M. A., Dawson, J. F., & Birdi, K. (2006). HRM as a predictor of innovation. *Human Resource Management Journal, 16*(1). https://doi.org/10.1111/j.1748-8583.2006.00002.x

Stankevičiūtė, Ž, Staniškienė, E., & Ciganė, U. (2020). Sustainable HRM as a driver for innovative work behavior: Do respect, openness, and continuity matter? *The Case of Lithuania. Sustainability, 12*(14), 5511. https://doi.org/10.3390/su12145511

Symon, G., & Cassell, C. (2012). *Qualitative organizational research: Core methods and current challenges*. Sage publications.

Sypniewska, B., Baran, M., & Klos, M. (2023). Work engagement and employee satisfaction in the practice of sustainable human resource management—based on the study of Polish employees. *International Entrepreneurship and Management Journal*. https://doi.org/10.1007/s11365-023-00834-9

Thiry-Cherques, H. R. (2009). Saturation in qualitative research: Empirical sizing estimate. *Brazilian Journal of Marketing Research (PMKT), 3*, 20–27.

Unur, M., Atai, G., Capkiner, E., & Arasli, H. (2022). Can safety leadership be an antidote in the COVID-19 fear of job insecurity and the work engagement relationship in the Norwegian service industry? The moderated-mediation model. *Sustainability, 14*(19), 12203. https://doi.org/10.3390/su141912203

Urbini, F., Presti, A. L., Chirumbolo, A., & Cllea, A. (2020). Two is worse than one. The mediating role of precariousness of life in the association between qualitative job insecurity and distress among Italian temps. *Electronic Journal of Applied Statistical Analysis, 13*(3). https://doi.org/10.1285/i20705948v13n3p634

Wang, H.-J., Lu, C.-Q., & Siu, O. (2015). Job insecurity and job performance: The moderating role of organizational justice and the mediating role of work engagement. *The Journal of Applied Psychology, 100*, 1249–1258. https://doi.org/10.1037/a0038330

Wang, C., Cheng, Z., Yue, X. G., & McAleer, M. (2020). Risk management of COVID-19 by universities in China. *Journal of Risk and Financial Management, 13*(2), 36–41. https://doi.org/10.3390/jrfm13020036

Yadav, M., Bhakar, S., & Upadhyay, Y. (2022). Impact of HRM practices on employee engagement and employee performance with special reference to academicians and work professionals. *International Journal of Health Sciences*, *6*(S3), 5443–5456. https://doi.org/10.53730/ijhs.v6nS3.7144

Zhang, B. (2023). human resource management (HRM) practices and sustainable organizational innovation: A staged exploration of the role of HRM based on a dynamic environment. *Highlights in Business, Economics and Management*, *3*, 157–169. https://doi.org/10.54097/hbem.v3i.4730

Chapter 15
HRM in the Small Family Business: Case Studies in the Interior of Portugal

Maria Amélia Marques and Jorge Rodrigues

Abstract The main objective of this study was to understand human resource management practices in two small family businesses in the interior of Portugal. We aimed to explore how the structure of family businesses influences their governance models and human resource management practices. Additionally, we wanted to understand the challenges they face regarding succession planning. To meet our objective, we employed a case study approach (Yin in Case study research: design and methods. Sage, 1994) in two small family businesses located in the interior of Portugal. We conducted semi-structured interviews with two managing partners from one company and an entrepreneur from another, accompanied by a documentary analysis. The data collected was analysed using categorical content analysis (Bardin in Edições 70, 2011). Our findings align with family business studies: both companies have low formalization of practices and an arbitrary model, centralized around the entrepreneurs' discretionary power (Pichault and Shoenaers in Appl Psychol 52:120–143, 2003). The entrepreneurs appear to have stewardship characteristics (Breton-Miller and Miller in Acad Manage Learn Educ 14:386–399, 2015), such as the capability to adapt and to learn by doing, and insights about the business, but are conservative in their management (Marques in Empresas familiares da região norte mapeamento, retratos e testemunhos, Universidade do Minho (CICS.NOVA), 2018).The differences between the companies HRM practices, and governance models, show that these are shaped by their sector of activity, legal ownership structure, and family business structures (Guerreiro in Pequenas E Médias Empresas Familiares. Tipos de relações das famílias com as empresas. Associação Portuguesa de Sociologia. Práticas e processos de mudança social. III Congresso Português de Sociologia, 1996). In the public limited company, which is a family business with a phratry structure and an open autocracy (Gomez in Référentiel pour une gouvernance raisonnable des entreprises. Institut Français de Gouvernement des Entreprises, 2015) governance model, with market culture (Cameron and Quinn, Diagnosing and changing

M. A. Marques (✉)
School of Business Administration, Polytechnic Institute of Setubal, Setúbal, Portugal
e-mail: amelia.marques@esce.ips.pt

J. Rodrigues
Lisbon Accounting and Business School, Polytechnic Institute of Lisbon, Lisboa, Portugal

C. Machado (ed.), *Human Resource Management in a Disrupted World*, Lecture Notes on Multidisciplinary Industrial Engineering,
https://doi.org/10.1007/978-3-031-89948-5_16

organizational culture. Jossey-Bass, 2006), there appears to be a tendency for more formalisation of practices than in the private limited company with a family business marital structure, which is a closed autocracy (Gomez in Référentiel pour une gouvernance raisonnable des entreprises. Institut Français de Gouvernement des Entreprises, 2015), and has characteristics of a clan culture (Cameron and Quinn, Diagnosing and changing organizational culture. Jossey-Bass, 2006). The differences in family business structures and governance models also reflect the challenges they encounter in succession planning. Although neither model has formal succession planning, the situations differ between open and closed autocracies. In an open autocracy, the founder and chairman face limitations in their choice of successor, while in a closed autocracy, the entrepreneur has the freedom to choose their successor.

Keywords Human resource management practices · Governance models · Family business · Succession · SME's

15.1 Introduction

As the European Commission (2024) reported, family businesses constitute over 60% of all companies in Europe and around 70% of companies in Portugal in 2021 (Association of Portuguese Family Businesses, 2021). This recognition has brought family businesses to the forefront of institutional and academic discourse, stimulating political and scholarly debates, and increasing scientific production in the different social sciences (Rodrigues & Marques, 2019; Marques & Rodrigues, 2023) However, there have been several challenges in defining and addressing this topic, as well as in systematising the information, stemming from the concept of family, and consequently, family business, assumed by the diverse fields of social sciences (Rodrigues & Marques, 2019). In an investigation by the European Commission (2009) into family businesses around 90 definitions were identified, meaning that some countries have formally more than one definition. Matching the entrepreneurial landscape in Portugal, most of these businesses are micro and small, operated by entrepreneurs trying to navigate personal, professional, and family projects.

Family businesses consist of intricate emotional, financial, informational, and political relationships among members, significantly influencing their corporate governance (Rodrigues, 2019a) and human resource management practices (Marques, 2018; Marques & Rodrigues, 2023). Family businesses tend to prefer to follow intuitive procedures (Daspit et al., 2016), and usually have less formalised practices, often adopting more arbitrary models (Guerreiro, 1996; Kok et al., 2003; Marques, 2010; Pichault & Schoenars, 2003). The latter appears to be enhanced in micro and small firms where authority or power is concentrated in the entrepreneur.

The main objective of this study was to understand human resource management practices in two small family businesses in the interior of Portugal. Considering that Human Resource Management Practices are influenced by both internal and external contextual factors (DiMaggio & Powell, 1983; Marques, 2010), and recognising that family businesses encounter unique challenges in today's socioeconomic environment, we have defined the following specific objectives: (a) to characterise the company; (b) to characterise the business and governance model; (c) to characterise human resources management practices; (d) to characterise the challenges of succession.

The methodology selected for this study was qualitative (Creswell, 2009; Fortin, 2009), specifically using a case study approach (Yin, 1994). Data collection involved semi-structured interviews with two managing partners from one company and an entrepreneur from another, complemented by a documentary analysis. The gathered data was analysed through categorical content analysis (Bardin, 2011).

This chapter is divided into four sections. The first section reviews the literature on family businesses, including topics such as governance, human resource management practices, and succession. The second section outlines the objectives and methodology used in the study. The third section analyses and discusses the results, leading to the final conclusions presented in the last section.

15.2 Theoretical Framework

Family businesses play a crucial role in Europe's socio-economic landscape. Their significance underscores the necessity for more systematic research into their dynamics and practices. In this chapter, we will examine the concept of family business, governance models, the specifics of human resource management practices, and the challenges associated with succession.

15.2.1 Family Business: Concept and Types

The family business can be seen as a network of particular and organic links between the people of a business family, which influence the company's governance (Rodrigues, 2019a; Rodrigues & Marques, 2021). This network of connections has four dimensions—emotional, financial, informational and political—which differ from family to family and translate into different degrees of influence on the governance of the family business. The division of powers attributed to the board of directors (the governing body) and the directors (functional managers) in the family business should make it possible to reconcile concerns such as the continuity of ownership of the capital in the business family, the legitimacy of this continuity and economic viability. Various flexible control mechanisms are thus used to resolve conflicts in different ways (Hirigoyen, 2002). The uniqueness of family businesses stems from their vision, intentions, and behaviour, which involve the family and the capabilities and ambitions of family members to advance the company's development (Porfírio et al., 2020). Concerning the type of relationship between families and companies, Guerreiro (1996) characterised them as follows: (a) clan: capital is owned by a significant number of family members, belonging to several generations and with extended kinship, consanguinity and alliance ties, and generally has no non-family partners involved; (b) lineage: unlike the previous model, this one involves ownership and management being strictly vertical, with assets and power passed down through the founder's direct descendants, excluding non-family partners; (c) nuclear ownership involves only parents and children in managing companies, including those founded by the entrepreneur or their parents; (d) marital: ownership is shared by the couple, with management primarily centred on the entrepreneur. The wife typically participates to varying degrees, often handling logistical and administrative support functions; (e) phratry: this company is founded by siblings seeking autonomy and facing job market challenges. These individuals typically have prior work experience and aim to leverage their skills in new ventures. While family involvement is mainly from the brothers, women may occasionally collaborate. These entrepreneurs take on both managerial and operational roles; (f) individual: this is the company, which is linked to just one entrepreneur, with no other family members involved in the activity or ownership.

The family understood as a dynamic system, comprises individuals whose backgrounds, motivations, and interests may not always be aligned (Bernhoeft & Gallo, 2003). The intricate dynamics of family businesses necessitate holistic and multidisciplinary approaches that consider various dimensions and articulate different logic. These can range from the logic of economic capital (associated with the capitalist bourgeoisie) or land as a form of capital (related to the peasantry) to the logic of educational capital and its conversion into professional and educational opportunities (Guerreiro, 1996). In a European Commission (2009) study, approximately 90 definitions of family businesses were identified, indicating that some countries had more than one formal definition. Wherefrom the European Commission (2009) defined a "Common European definition of a family business"[1] we will adopt in this study. Donnelley (1964) only classifies as family businesses those that have been in the family for more than one generation and when there is reciprocal intergenerational influence. Whereas Bernhoeft (1989) mentions the family business as a founder's ideal that worked. Lodi (1998) reinforces Donnelley's concept of the second generation by emphasising that the family business only emerges when the founder paves the way for the future generation or successors. In the founder's phase, a company tends

[1] Criteria for a family business: any firm or company, regardless of its size (EC, 2009):

1. "The majority of decision-making rights are in the possession of the natural person(s) who established the firm, or in the possession of the natural person(s) who has/have acquired the share capital of the firm, or in the possession of their spouses, parents, child, or children's direct heirs.

2. The majority of decision-making rights are indirect or direct.

3. At least one representative of the family or kin is formally involved in the governance of the firm.

4. Listed companies meet the definition of family enterprise if the person who established or acquired the firm (share capital) r their families or descendants possess 25 per cent of the decision-making rights mandated by their share capital." (pp.4).

to be more 'personal', while in the second generation, it becomes more 'family-oriented' (Hoon et al., 2019). The concepts of Lodi (1998) and Rodrigues and Marques (2019) are similar when they distinguish between family businesses and entrepreneurial families, with a governance model and/or professionalised management model from the second generation onwards analysis. Nordqvist and Melin (2010), and Nordqvist et al. (2013), describe the second generation of family businesses as the stage where the company transitions into an entrepreneurial phase. They argue that family business succession should be viewed from an entrepreneurial process perspective. According to Nordqvist et al. (2013), the foremost business opportunity during the succession process lies within the family business itself. During this transition, the family may invite new members, either from within or outside the family, who bring fresh management skills without necessarily increasing the capital. This influx of talent can professionalise the management of the business.

In Portugal, Marques (2018) coordinated a study that mapped family businesses in the northern region of Portugal, where the following characteristics stand out: the vast majority were micro and small businesses, and almost 95.0% were owned by the family or a family member, managed by the founder. Entrepreneurs were predominantly male, with an average age of around 47 and a secondary education qualification. Female entrepreneurs were younger (41) and more qualified. Micro-enterprises predominated (70.7%), with female labour being less significant in micro-enterprises compared to larger companies. The average age of the companies studied was less than 20 years, with first and second generations coexisting. Regarding their legal status, the majority were private limited companies. Most were self-financed, not depending on bank loans, and did not have sophisticated formal governance structures (Marques, 2018).

15.2.2 Governance Models

Corporate governance models depend on how each power—sovereign, executive, supervisory—is exercised: strongly or weakly (Rodrigues, 2019b). Thus, organisational governance models, like political systems, are defined according to the density of the distribution of powers. Gomez (2015) identifies six configurations: (a) Closed autocracy, when all three powers are exercised simultaneously by the same actor; (b) Open autocracy, when a sovereign power is partially held by minority shareholders who do not exercise the other powers; (c) Shareholder domination: when shareholders have a strong influence but do not directly exercise executive power; (d) Management domination: when shareholders have little influence and the executive branch enjoys strong autonomy in practice; (e) Board domination: when shareholders have little influence, the executive branch has little autonomy in practice, because the board of directors has appropriated part of the effective power; (f) Organisational democracy: when the three powers are perfectly separated and are exercised by different actors.

The shareholder structure exhibits a concentrated and robust configuration characteristic of closed organisational autocracy, where both executive and sovereign powers are vested in the same individual or group (Rodrigues, 2019b). In this setup, executive power, represented by the managers, governs the capital and supervisory bodies. Consequently, the supervisory authority is weakened, primarily due to the absence of independent supervisory agents, whose role is largely limited to mere representation (Rodrigues, 2019b). This arrangement is commonly found in organisations directly managed by the entrepreneur or in property companies with a concentrated and closed shareholder framework (Rodrigues, 2019b). In contrast, the open autocracy configuration also retains executive and sovereign powers within the same individual or group (Rodrigues, 2019b). Nevertheless, in this case, although executive power is predominantly held by majority shareholders, minority shareholders are always present. Their involvement significantly impacts the organisation's governance system, as the oversight authority can serve as a platform for either dialogue or conflict. Minority shareholders can vary in influence, exhibiting either limited engagement or a strong desire to participate in governance (Rodrigues, 2019b). Their role in management distinguishes open autocracy from closed forms. This configuration typically includes organisations led by the entrepreneur alongside family minority shareholders (who are excluded from day-to-day management) or financial minority shareholders (such as investment funds), as well as publicly listed entities with low capital dispersion.

In the configuration characterised by shareholder domination, the separation of executive power and supervisory power is evident (Rodrigues, 2019b). Dominant shareholders possess the sovereign power and the supervisory authority, while professional managers are entrusted with executive power, even in the presence of minority shareholders. The supervisory board plays a crucial role in this system, acting as the arena where the dominant shareholders exercise control over the managers (executive power) (Rodrigues, 2019b). Minority shareholders can be actively involved, which may lead to conflicts with the dominant shareholders (Rodrigues, 2019b). This configuration is typically observed in heritage organisations where the entrepreneurial family does not directly manage the organisation.

In management-dominated configurations, sovereign power is weakened due to a fragmented shareholder base that struggles to unite effectively during general meetings (Rodrigues, 2019b). In this context, professional directors oversee corporate governance by approving strategies and appointing managers, contrasting with autocratic models where power is heavily centralised. Consequently, oversight power remains weak, characteristic of patrimonial organisations with dominant shareholders, such as public companies.

In scenarios where the board of directors wields significant authority, influential supervisors may undermine executive power, resulting in a situation where the board effectively makes all decisions and executive power becomes subordinate to their expectations (Rodrigues, 2019b). This differs from shareholder dominance, where control lies with major shareholders. Instead, influential supervisors—often external critics of a weak executive—exert power, particularly in organisations with minority board appointees or during uncertain succession periods. Such dynamics are common in associations where executive discretion is limited.

15.2.3 Human Resources Management in Family Businesses

Despite the diversity among family businesses, human resource management (HRM) in these firms is often less professionalised, less formalised (Kok et al., 2003; Marques, 2010; Rodrigues, 2019a), and more arbitrary (Pichault & Schoenaers, 2003). Family firms with a high level of family influence are less likely to establish an HRM department or adopt high-performance work systems (HPWS) than non-family firms (Hoon et al., 2019). Furthermore, the long-term orientation or hiring practices (Kim et al., 2024) and characteristics associated with "human capital development models" (Lepak & Snell, 1999) in family firms do not necessarily lead to greater investment in HRM or professional training (Basco et al., 2023). Additionally, attracting talent appears to be hindered by the roles occupied by family members within the company (Basco et al., 2023).

Studies suggest that in family businesses, everything or almost everything seems to be centred on the founder-entrepreneur (Kelly et al., 2000). Additionally, family businesses have greater risk aversion than other companies (Basco et al., 2023). As Marques (2018) pointed out, most family businesses in northern Portugal were micro and small enterprises and almost 95.0% were owned by the family or a family member and managed by the founder. Most entrepreneurs in the study were male, with an average age of approximately 47 and a secondary education qualification. In contrast, female entrepreneurs were younger, averaging 41 years old, and generally held higher qualifications. Micro-enterprises represented 70.7% of the businesses, but the involvement of female workers was notably lower in these smaller enterprises compared to larger companies. The average age of the companies examined was under 20 years, with both first and second generations of ownership coexisting. In terms of legal structure, private limited companies were the most common. Most businesses were self-financed or relied on bank loans and did not have sophisticated formal governance structures. Even in larger companies or when not all the partners were family members, there seemed to be a tendency in this type of company towards an 'open autocracy', that is, an executive power and a sovereign power being held by the same person or group of people (Rodrigues, 2019b). Although companies also look for young people with technical-professional qualifications, suited to the business and higher academic qualifications, they tend to look for workers with professional experience, regardless of age (Marques, 2018).

Family firms often face challenges in attracting employees due to perceptions of lower compensation and limited opportunities for training and promotion (Vardaman et al., 2024). The two key factors that attract non-family employees are (Vardaman et al., 2024): corporate social responsibility and long-term orientation. The former is related to the family firm's reputation and the latter to job security. In their study on family firms, Vardaman et al. (2024) hypothesised that recruiting non-family employees through referrals can help lessen agency conflicts and act as an intermediary factor in the relationship between the essence of the family firm and its overall performance. It's important to note that referred employees generally exhibit higher productivity, and tend to remain with the company longer, but often experience slower salary growth when compared to non-referred employees (Brown et al., 2012).

From an agency perspective, Nelson and Constantinidis (2017) found that family members are unlikely to make investments if benefits are not directly appropriable by the firm. Although family firms are better financial stewards of their companies, they perform worse as organisational stewards than nonfamily firms (Nelson & Constantinidis, 2017). Specifically, they offer lower compensation, invest less in employee training, and exhibit higher voluntary turnover and lower labour productivity (Nelson & Constantinidis, 2017). The family businesses that perform best are those whose owners and managers embrace profound ideals and talents of stewardship, and these experiences begin in early childhood and depend on the learning and education environment provided by their family (Breton-Miller & Miller, 2015). Miller (1996) designed a typology of six types of learning that foster stewardship, which he divides into two kinds of modes of thought and action: methodical, namely: analytic, experimental, structural, and emergent: synthetic, interactional, and institutional. DiMaggio and Powell (1983) addressed decision-making and institutional learning using the concept of isomorphism. Their "new institutionalism approach" (DiMaggio & Powell, 1983) explains why and how organisations resemble similar organisations in their environment to obtain legitimacy in their markets by isomorphism: competitive and institutional. Competitive isomorphism is linked to a system of rationality that emphasises market competition, niche change, and fitness measures. In contrast, institutional isomorphism is an alternative perspective that focuses on three mechanisms of decision-making in organisations: coercive mechanisms (change stems from political influence and the problem of legitimacy), mimetic mechanisms (change results from standard responses to uncertainty, copying practices of other organisations), and normative mechanisms (change stems from professionalisation). Breton-Miller and Miller (2015) emphasise the following key characteristics for successful stewardship: (1) moral values and belief in ethical behaviour toward stakeholders; (2) personal discipline and willingness to work hard; (3) ample social and political skills to encourage interactive learning and manage conflicts; (4) capability to adapt and to learn by doing, that is, to do experimental learning; (5) having technical and business knowledge, and information to run a business sustainably; (6) having insight reflection, that is, synthetic learning: the ability to understand how all the elements of an enterprise and all the interests of its stakeholders fit together; how they, align and complement one another.

15.2.4 Succession in Family Business

One of the challenges in family businesses is succession. While succession is often described in the literature as a formalised procedure, in family businesses it tends to be approached more intuitively, like other decision-making processes and practices, in these contexts (Daspit et al., 2016). Succession is typically a problematic issue for family businesses because it involves relational and emotional aspects and economic and legal issues (Decker et al., 2016; Handler, 1994). Based on Decker et al. (2016) five-stage succession model, the main challenges family businesses face are: (1) preparing and socialising the next generation; (2) choosing the successor; (3) transferring the ownership and leadership; (4) managing the incumbent's withdrawal (conflict between predecessor and successor); (5) post-succession stage. Noteworthy to highlight that the decision of the next generation to participate in the family business is essential for successful intra-family succession. Gender is a key issue in family business succession (Decker et al., 2016), despite the "family business succession research as a whole historically is a theoretical in terms of gender or it embeds a liberal feminist view of gender when gender thinking appears" (Nelson & Constantinidis, 2017, pp. 230). While female heirs may be interested in the family business, they are often less likely to pursue succession if the firm is in a male-dominated industry (Schröder et al., 2011, as cited by Decker et al., 2016). Also, if a successor is unavailable, owner-managers will be less likely to engage in succession planning. The socialisation of successors remains a blind spot in the literature (Decker et al., 2016), depending mostly on how the founder passes on his values (García-Álvarez et al., 2002). García-Álvarez et al. (2002) define two types of socialisations: family socialisation and business socialisation. The former is related to family values and includes formal education of the family members, that is, potential successors. The latter is oriented toward the potential successors and begins when they work full-time in the business.

15.3 Objectives and Methodology

The study's main objective was to understand human resource management practices (HRMPs) in two small family businesses in the Interior of Portugal. Considering that Human Resource Management Practices (HRMPs) are influenced by both internal and external contextual factors (DiMaggio & Powell, 1983; Marques, 2010), and recognising that family businesses encounter unique challenges in today's socioeconomic environment, having defined the following specific objectives: (a) to characterise the company; (b) to characterise the business and governance model; (c) to characterise human resources management practices; (d) to characterise the challenges of succession. Based on insights from previous studies (Daspit et al., 2016; Marques, 2018; Marques & Rodrigues, 2023), we intended to explore how the structure of family businesses, along with the educational background of their entrepreneurs and family members, influences their governance models and human resource management practices. Additionally, we wanted to understand the challenges they face regarding succession planning.

The selected methodology for this study was qualitative research (Creswell, 2009; Fortin, 2009), employing a case study approach (Yin, 1994). Using the EU's (2009) definition and considering Guerreiro's characterisation of family businesses, we selected two small companies with different structures for our case studies. Both consider themselves successful and have registered development over the past five years. Also, the willingness of each company to participate in our research, which depended highly on trust and previous relationships, strongly influenced our choice. Notwithstanding their willingness, access to other participants was not enabled, and documentary information was limited.

Data collection involved semi-structured interviews with two managing partners (Chairman and his brother) from one company and an entrepreneur (Chairman/husband) from another, complemented by a documentary analysis (company website and documents). The interviews were about 90 min. The gathered data was analysed through categorical content analysis (Bardin, 2011).

15.4 Analysis and Discussion of Results

15.4.1 Characterisation of the Companies

Both companies operate in the trade sector but have different focuses. Company A, located in Alto Alentejo, sells beverages to the HORECA market (hotels, restaurants, and cafes). Company B, based in Beira Baixa, primarily sells home textiles and decorations. The companies also differ in their family ownership structures (Guerreiro, 1996): Company A predominantly presents characteristics of a phratry, while Company B shows marital characteristics.

Both are first-generation companies; however, when they were founded more than 35 years ago, they had formally different legal ownership structures. Company A was founded in 1975 by five unrelated partners as a private limited company. Three of them had returned from the Colonial War and decided to take over the management of the company they worked for. The Managing Partner/Chairman had previous experience in accountancy and assumed the role of Manager.

> Those were difficult and troubled times here in the Alentejo, but we had a completely different attitude towards our former boss. We rented the premises from him and the takeover was a peaceful process. Then we carried over to negotiating with suppliers (…) We managed to convince all the suppliers to grant us all the representations. A series of large manufacturers that ended up being the backbone of our business. (Managing Partner/Chairman).

In 2007 due to "the size achieved, the revenue, and the need to secure the future" (Chairman), they transitioned from a private limited company to a public limited company, with the founders retaining their capital and reinforcing their position as a wholesale alcoholic beverages distributor. The company only became a family firm, by definition of the EU (2009), in 2012 when one of the founding partners sold his share to the brother of the current Chairman of the Board of Directors (21.4%). Subsequently, the company experienced the loss of one of the founding partners due to death, with his heirs remaining as shareholders (28.6%). Additionally, another partner chose to retire and sold part of his share back to the company, while donating a 10% share to an employee who then joined the Board of Directors. The family holds a 50% stake, while three partners on the Board jointly own 64.3% of the company. Today, the capital structure is the result of a dynamic of share sales by shareholders and share buybacks by the company, which currently stands at: 28.6% (Chairman), 21; 4 Brother) and 14.3% (Other member of the Board). The Chairman holds the largest single share.

Company B was founded in 1989 by the entrepreneur as an individual enterprise. The entrepreneur's endeavour represents what García-Álvarez et al. (2002) describe as family socialisation, as he was raised in a family business in the same sector. Initially, he sought to forge his path by pursuing higher education and gaining experience in other companies. However, despite these efforts toward independence, he ultimately felt drawn to start his own business. When his family suggested that he open a branch focused on home textiles, he seized the opportunity. In 2010, the company transitioned to a private limited company, with his wife holding 50% of the shares. Thus, becoming a family business with the characteristics of a "marital structure" (Guerreiro, 1996).

Both companies are small and, despite their assumed recent development, wish to remain small. Company A's decision to remain small in size is for the sake of efficiency: "We don't have the structure for it" (Chairman), reinforced by his brother "We're not interested in growing geographically or in terms of our portfolio. It's the average order that has to be bigger". The entrepreneur of Company B seems to reveal the need for a better work-life balance or sense of achievement: "We could grow more, but there is more to life than this." Company A has 49 employees, whereas Company B has 25, and both engage in seasonal hiring. And, both companies are geographically dispersed.

As assumed in small and family firms, the formalisation is low (Kok et al., 2003; Marques, 2010). Both companies have a website with description of their products, but Company A has more formalised artefacts (Schein, 1983) than Company B. Its website details its history, and its slogan: 'Passion for what we do is reflected in our services'. The company has a written document, similar to a presentation, that was created in collaboration with a consultant. This document outlines the company's mission, goals, objectives, and action plan. It was developed a few years ago when the company needed insights into its situation. Upon reflection, the company' stated its mission 'To be recognised by customers as their best supplier' and values: 'To satisfy the HORECA drinks market with profitability and excellent service through 'Integrity, Rigour, Responsibility, Quality, and Excellence". The company's formalised objectives are very much centred on "Maintaining the growth" and "Maintaining the internal structure". The latter corroborating the decision to keep small. Another objective is to keep "Improving working conditions". For this and the coming years, it has set itself the goals of "Increasing the effectiveness of collecting payments, improving internal communication and retaining staff" (Company's Document). These artefacts are appointed towards a prescribed market culture (Cameron & Quinn, 2006).

Company A has three operational centres covering 19 municipalities, some in the neighbouring province of Ribatejo. The company's formalised slogan is 'Passion for what we do is reflected in our services', and its communication is strongly customer-orientated, as 'It has a portfolio of products suited to the needs of the market and a team of professionals trained to respond effectively to the needs of its customers.' (company website).

Company B has 6 stores in 3 municipalities in Beira Baixa. The company has no formal slogan, but its website enhances the following: "[..] is a brand that represents a group of shops dedicated to selling decorative items. (..) The largest textile decoration shop in the country! More than 5,000 products of great variety and the highest quality." Although they are not formalised in writing, the core values of the company, according to the entrepreneur, are: "Close family: when a collaborator has a problem we have a problem; Honesty; Customer proximity, Good relationship between the company and clients." (Entrepreneur) Currently, the decoration segment accounts for 40% of the company's total activity, compared to only 20% in 2010:

> With, let's say, changing the architecture of the flats and houses. All this space in much larger areas has forced people here. They had to use more professionalised services. And so, we took advantage of this area, because it's an area with more added value. It's personalised, so it allows us to do tailored-made work. We can achieve completely different profit margins. (Entrepreneur)

These artefacts of the company, along with its size and the aforementioned organisational dynamics of a family business, lead us to hypothesise the characteristics of clan culture (Cameron & Quinn, 2006).

15.4.2 Organisational Structure and Corporate Governance Models

None of the companies have formal organisational charts. Both companies have simple organisational structures and characteristics of entrepreneurial configurations (Mintzberg, 1995), being strongly centralised on the strategic apex or founder. However, there are differences between the two associated with the structure of family business (Guerreiro, 1996): one being a phratry, with non-executive shareholders (the heirs, which hold 28, 6%), and the other a marital structure. The founders of both companies are older than the average age of the Portuguese entrepreneurs in Marques's (2018) studies, and the founder of Company A has a lower education level, which we hypothesise may be associated with his age and or education context in Portugal, and social background (started working at an early age).

Company A is centralised in the Chairman of the Board of Directors, the founder, who is 71[2] years old, has an elementary school education level (6th grade), and received professional training in accounting along with some behavioural training from suppliers. He is, in his own words, "the man who represents and has a vision

[2] Noteworthy, in Portugal compulsory schooling for six years, targeting the entire schoolable population up to the age of 14 was only enacted in 1964 with the Decree-Law no. 45810, of 9 July.

for the company" (Chairman). The other managing partner, who is his brother, is 62 years old. He has a secondary school education and has spent almost his entire career in a multinational company, gaining extensive professional experience in logistics. The other managing partner, who is his brother, is 62 years old, has a secondary school education, and has spent almost his entire career in a multinational company, gaining extensive professional experience in logistics. Although he said: "I've always known and worked for the company, even as a teenager" (Brother), the Chairman stated that "He joined at a time when he left the other company, and we needed him. He brought a lot of experience to us. I persuaded him to take over my partner's position" (Chairman CA). The brother's experience outside the family business reinforces Guerreiro's (1996) definition of phratry. The brother oversees logistics in the company and manages one of the warehouses. The non-familial managing partner, a former employee, who was granted an initial 10% share, is in her forties. She has a secondary education level and continues to perform administrative duties. In practice, she, the daughter of the Chairman, who also has a secondary education level, and another administrative employee are the company's staff support (Mintzberg, 1995), because they are responsible for all the administrative work. The Company's operating core (Mintzberg, 1995) consists of the warehouses' supervisors or "managers", the sales teams, and the distribution teams (truck drivers and assistants): "We created logistics warehouses dependent on the head office, a warehouse manager, an administrative for bureaucratic matters and drivers for distribution". Chairman). The company exhibits characteristics of an open autocracy model (Gomez, 2015), centered around the Chairman with supervision from the shareholders.

As a family business with a marital structure, Company B is centralised in the entrepreneur. Although he emphasises that he and his wife have complementary roles, he also admits that the "final decision rests with him." He is responsible for making strategic decisions and handling the main purchases:

> I take care of the financial part, the negotiation, banking, insurance. Tapestry products. Technicians go through me a bit, upholstery. She does most of the shopping for lingerie and pyjamas, which is exclusively hers. Fabrics she and I sometimes make these purchases together when they are in very large quantities. When we negotiate or import certain products, then we both work together. (Entrepreneur)

The entrepreneur, 62 years old, and his wife have a secondary school education level. As support staff, Company B has a Bachelor with a degree in Management, the only person in the company with a higher education degree, who is the entrepreneur's right hand, and responsible for managing and organising payments, and all the accounting, since bookkeeping is outsourced: "She's the person who organises all the management, and accounting. Bookkeeping, it's not done here, it's an external company, but she organises everything that has to do with....let's say, the management of payments." (Entrepreneur).

The operating core consists of a store manager for all six stores, a trustworthy senior employee, that is:

> The person in charge in our absence, and even in when we are present, is the person who, in a way, makes many decisions without going through us, because she already knows the

whole way of working with product distribution, product repetition, and she ends up being the person who is there, even if we are there. (Entrepreneur)

The operating core consists of the saleswomen, who also work in the warehouse, and the seamstresses. Company B, a private limited company with a marital structure centered on the entrepreneur, seems to display key characteristics of a closed autocracy model (Gomez, 2015).

15.4.3 Human Resources Management Practices

Neither company has a human resource management department or personnel assigned exclusively to this function, which supports the observation made by Hoon et al. (2019) about the influence of the family, more specifically the entrepreneur, in the family business. In Company A, the Chairman and his brother (Managing Partner), both members of the Board of Directors, are responsible for Human Resource Management (HRM). They are assisted by the staff, which includes a member of the Board (the non-familial member).

Company A has 49 employees, predominantly male (more than 90%), with an average age of 40 years, with permanent employment contracts, and high seniority (the majority with more than 10 years). These 49 employees are the core workers (Atkinson, 1984) of the company because in peak seasons the company hires "peripherical workers" with short-term employment contracts.

The company has grown around groups of warehouses: Each warehouse has 1 warehouse manager, 1 administrative manager, sales staff, drivers and drivers' assistants. Only in the last two years has the number of workers remained fixed throughout the year due to the company's growth. In general, hiring is seasonal, that is, the company only hires the maximum number of workers during the peak season (summer). (Chairman)

Following their "market culture" (Cameron & Quinn, 2006) and the nature of their activities, Company A's HRM practices focus on maximizing results and retaining talent. Although the formalisation of practices, and procedures, is low, the key practices for Company A are: recruitment and selection, and performance appraisal. Training and development are often deprioritized, which is common in many family businesses (Nelson & Constantinidis, 2017). Most training is on the job and/or occurs primarily due to legal compliance, that is, to coercive isomorphism (DiMaggio & Powell, 1983). Thus, HRMPs are both administrative and technical due to talent retention needs.

Recruitment, corroborating Hoon et al. (2019) and Marques (2018), is through referrals, favouring people with previous experience regardless of age:

We were always looking to recruit people with experience in this area, through personal knowledge. Always through personal knowledge. In the first instance to head the warehouse. We met someone for [...] ... who had already headed a warehouse, and had left the company and had his own business. He was someone who knew the overall shape of the business and how to manage the warehouse and other people. And that's how we went about recruiting. (Chairman CA)

As far as selection is concerned, the company has a policy of not hiring candidates who reside outside the municipality, even for driver positions. They believe that the advantage they provide is that employees can always return home to sleep after completing their routes.

Company A implements a management by objectives performance appraisal system. This practice aligns with the requirements and objectives defined by their main supplier, resulting in an approach influenced by mimetic isomorphism and competitive isomorphism (DiMaggio & Powell, 1983). In the last two years, the company has changed its performance appraisal to ensure better equity and results between employees and achieve the goal of improving the effectiveness of collecting payments. This included "Paying bonuses to the distribution employees related to sales' (Brother). The company felt that the drivers and helpers deserved to receive bonuses because they are the ones who make the payment collections and by doing so they are also facilitating sales. 'They let the salesperson be a salesperson… Be closer to the customer" (Brother).

Corroborating other studies (Bosco et al., 2023; Kim et al., 2024), long-term hiring does not mean more investment in training. It only hires drivers with the necessary qualifications, hence the difficulty in hiring staff in peak seasons. Although the company has faced problems in attracting professional truck drivers, it does not consider financing their licence as part of training:

> The state doesn't invest in training professionals. The state invests in higher education and doesn't realise that a driving licence costs €2,500.00 and people don't have the means to qualify to drive. On top of that, they still need to have the CAM (driver's certificate) and the CQM (driver's qualification licence), which is another thousand euros. (Brother).

Regarding compensation and rewards, employees receive a fixed salary, which the company considers competitive, a food allowance, and bonuses. The information gathered about this practice was limited, as interviewees were not very forthcoming. Despite having a performance appraisal system linked to bonuses, the information we gathered indicates a tendency toward arbitrariness in recognizing compensation based on role, seniority, and performance. This observation supports the findings of Pichault and Schoenaers (2003).

The company has no formal communication and participation mechanisms. They hold "Informal daily meetings at the warehouse" (Brother) to discuss the daily activities. The delivery routes are not fixed, but the teams may have certain preferences that are taken into account. Occasionally, conflicts arise regarding the types of vans or trucks that drivers are assigned to drive, as drivers tend to prefer the newer vehicles. These conflicts are typically resolved by the brother, a Member of the Board of Directors, if necessary. However, as a rule, the Board of Directors meets monthly to discuss the company's "most strategic issues: at commercial, financial and logistics level " (Chairman). Considering the dynamics of the organisation, we hypothesize that Company A primarily follows an arbitrary HRM model (Pichault & Schoenaers, 2003), which relies more on managerial discretion and personal relationships than on formal, transparent rules. Nevertheless, the Chairman and his brother reveal some of Breton-Miller and Miller (2015) characteristics of successful stewardship, such as extensive technical knowledge of the business and insight and capacity to reflect on the business,.

Despite its long-term hiring, Company A´s HRM practices lead us to hypothesise that its system has not only characteristics of an internal development model (Lepak & Snell, 1999), that is, of commitment, but also of an acquisition model (Lepak & Snell, 1999). Despite the differences between the two models in their conception of the value and uniqueness of human capital (Lepak & Snell, 1999), the first model perceives human capital as both high in value and unique, while the second model sees human capital as low in value but high in uniqueness. Our hypothesis focuses on the hiring of experienced employees who possess skills that are not specific to the organisation, leading to a subsequent low investment in training. This scenario creates what is known as a "symbiotic employment relationship" (Lepak & Snell, 1999). Company B has 25 employees, 24 females and one male (the entrepreneur). The average age of the employees is around 40, the majority are married, and all hold permanent employment contracts. The seniority within the company exceeds 15 years. Four of the employees are seamstresses. Additionally, the company hires employees with short-term employment contracts during peak seasons, such as Christmas and summer holidays.

Following their "clan culture", HRM practices, like all other procedures in the company, are not formalised. All the practices are centralised in the entrepreneur with the support of his wife and the store manager. Company B's key HRM practices reinforce the clan culture and promote strong socialisation processes, intuitively implementing a "value model" (Pichault & Schoenars, 2003). Company B combines characteristics of an arbitrary model, based on the description of the entrepreneur and a value model, based on reinforcing the clan culture; the key HRM practices in this company are recruitment and selection, onboarding, training and development, and the mechanisms of communication and participation. Concerning recruitment and selection, most of the recruitment is through referrals (corroborating Hoon et al., 2019), and most of the employees recruited were former clients. The company was able to "assess them in advance" (Entrepreneur), which meant that the candidates were pre-socialised to the company's values. The entrepreneur conducts the selection process through interviews. Onboarding is a key practice in the company that reflects its value model. The onboarding of a new employee occurs in two stages over the course of two weeks: the first stage takes place in the warehouse, followed by the second stage in the store, both of which are supervised by the store manager. Training and development is another practice that is well-developed in the company. Although there is no formal training plan, the company outsources training from their Commercial Association, Employment and Training Institute and suppliers. Their main fields of training are: "There are modules adapted to each specific area, but basically we have window dressing, customer service and decoration." (Entrepreneur). Once a year there is also a guided visit to a manufacturer: "It's a visit that has a fun aspect, but also an educational one. We explore the production process." (Entrepreneur). They also visit textile fairs with some employees. All the training is for the saleswoman; seamstresses are not included because their work is standardised: "She has to sew 10 cm each time, and it's programmed." (Entrepreneur).

Outsourced training is assessed by the provider, according to the entrepreneur. We presume it is a satisfaction evaluation (Kirkpatrick, 1983).

The company has no formalised performance appraisal system. The reason was that "We tried once. We had a system by objectives. It didn't work. ... We want people to focus on the service and not on the objective. ...It created some rivalry and even created some unease." (Entrepreneur). They decided to create an alternative: "At the time, we decided to give bonuses in terms of, let's say, variable remuneration." (Entrepreneur). The bonuses are given to the employees: "That are more accessible, targeting both high and low spending customers" (Entrepreneur). The employees are assessed by him, his wife and the store manager, but the bonus and its value are only known to the individual. Thus, this is an example of an arbitrary model common in family businesses, as Pichault and Schoenaers (2003) asserted.

As to compensations, the salary is updated according to the national minimum wage, which means that the employees' salaries depend on their seniority and the bonuses. Benefits include food allowance, a meal preparation space, and "Free bottled water" (Entrepreneur). Some employees also have access to transport allowance through petrol payments. The latter is also attributed to the same evaluation methods, which are arbitrary.

Apart from the onboarding and training, the Company has several participation mechanisms that foster the clan culture and seem to reflect what Breton-Miller and Miller (2015) characterise as successful stewardship, such as insight and knowledge about the business and learning by doing. These activities include outdoor training and picnics. In contrast to Company A, Company B's human resource management (HRM) model exhibits characteristics of a development human capital model (Lepak & Snell, 1999). We hypothesise this based on the emphasis on employee socialisation and training. Additionally, it is reinforced by career advancement in terms of salary.

Neither company A nor B have career planning. Both are flat organisations, coherent with their size and configurations.

15.4.4 Succession

None of the companies have a formal succession plan, but both companies are at a crossroads in succession planning, but for different reasons: Company A does not have a suitable successor in the family, whereas Company B appears to be socialising the potential successor. This is typically a problematic issue for family businesses because it involves relational, emotional, and economic and legal issues (Decker et al., 2016; Handler, 1994). In Company A, the Chairman's daughter is an employee, and he acknowledges that "'She knows everything about the company, but she's not yet ready to take on leadership roles'. (Chairman). His brother, a Member of the Board of Directors, agrees with him. The problem is that they disagree on the solution. The Chairman, a founding member of the company, has a deep emotional attachment to it. He believes that it is necessary to hire an external firm to study the succession transition, as his daughter is not yet ready to take over. In contrast, his brother has a more rational approach and believes they should focus on growing the company to sell it. Unlike the Chairman's daughter, the brother's son has a university degree and is building a career abroad. He would be a potential successor, but he is not interested or aligned with the family's business (Handler, 1994).

In Company B, the entrepreneur is at the stage of preparing and socialising the next generation (Decker et al., 2016) and slowly withdrawing from the "long hours at the company to enjoy life" (Entrepreneur). The couple, an entrepreneur and his wife have two offspring: a son who is an architect and a daughter who is a nurse. Both studied and live abroad. According to the entrepreneur, his son is showing interest in coming home and perhaps taking on some projects:

> And my son, I even have some hope for him. .... he works for a reputable architecture and design company, but he sees that his colleagues who have been there for 10 years are still earning more or less the same. For 10 years, they are still earning more or less the same. Moreover, he said, 'Me, I can see myself in five or six years earning what he's earning now, which is very little'. And you have a house with a swimming pool with these things. You've got so and so, going on holiday, and travelling here and there. I don't know. And I realise that being here, I can't achieve anything. In between the lines it might seem like a sign. (Entrepreneur)

The son´s dialogue, referred by the entrepreneur points to the values of family socialisation mentioned by García-Álvarez et al. (2002), but the following statements are pointing more to business socialisation:

> Yes, we shared a bit and he said to me the other day, 'Ah, look, I'd even be willing to do your marketing now. Image creation and all that. I can work online and stuff. That is interesting. I like that approach, don't you? Because integrated them. And, I always did that to see if they got a bit motivated, but it did not seem like much to me. There's a lot of interest now. So I welcome that. I would like him to stay in business, I'd like him to stay in this business. (Entrepreneur)

Thus, the successor has been chosen, although there is no formal plan in action. The entrepreneur stated that his daughter has never been interested in the family business. As to the withdrawal, it is not only the sense of retirement but of better work-life balance:

> As of January, we will no longer be coming to the company on Mondays. In other words. It is a decision because the company manages to survive perfectly well with its people…at our age, will also try to make decisions to reduce our working hours. (Entrepreneur)

Reinforcing the lack of a succession plan and the related uncertainties is the entrepreneur also contemplates he might sell the business if the right offer arises, although he would prefer not to.

15.5 Final Considerations

Family businesses are vital economy economic drivers (EU, 2009; Nordqvist et al., 2013). They are complex and dynamic organisations of intricate emotional, financial, informational, and political relationships (Rodrigues, 2019a), influencing their corporate governance and human resource management practices (Marques, 2018; Marques & Rodrigues, 2023). They often prefer to follow intuitive procedures (Daspit et al., 2016) and have less formalised practices, and adopt more arbitrary models (Pichault & Schoenars, 2003). This tendency is especially enhanced in micro and small firms, where authority or power is concentrated in the hands of the entrepreneur.

This study´s main objective was to understand human resource management practices in two small family businesses in the Interior of Portugal. Based on the assumption that human resource management is shaped by internal and external factors (DiMaggio & Powell, 1983; Marques, 2010), we aimed to explore how the structure of family businesses, along with the educational background of their entrepreneurs and family members, influences their governance models and human resource management practices. Additionally, we wanted to understand the challenges they face regarding succession planning.

To meet our objective, we employed a case study approach (Yin, 1994) in two small family businesses located in the interior of Portugal. We conducted semi-structured interviews with two managing partners from one company and an entrepreneur from another, accompanied by a documentary analysis. The data collected was analysed using categorical content analysis (Bardin, 2011).

Our findings support most family business studies regarding data access and the results collected. Despite the willingness to participate, access to other organisational actors or participants and documentary data was limited. Both companies have low formalisation of practices and procedures and characteristics of an arbitrary model (Pichault & Schoenaers, 2003), centralised and based on the discretionary power of the entrepreneurs. The entrepreneurs appear to have stewardship characteristics (Breton-Miller & Miller, 2015), such as the capability to adapt and to learn by doing, insight and capacity to reflect on the business, along with technical and business knowledge, but they are very conservative in their management. Thus, they decided to keep it small in size to maximize their efficiency and development and centralized in the "strategic apex" (Mintzberg, 1995).

There were, however, differences between the companies shaped by their sector of activity, legal ownership structure, and family business structures (Guerreiro, 1996). In the public limited company, which is a family business with a phratry structure with an open autocracy (Gomez, 2015) governance model, and market culture (Cameron & Quinn, 2006), there appears to be a tendency for more formalisation of practices than in the private limited company with a family business marital structure, which is a closed autocracy (Gomez, 2015), and a clan culture (Cameron & Quinn, 2006). In the open autocracy, although the company is centered on the founder, the oversight from the shareholders seems to have influenced the formalisation of the company's mission and values and held it accountable. In line with the latter and revealing the capability to adapt, most practices were also introduced due to institutional isomorphism (DiMaggio & Powell, 1983) by mimetic mechanisms. Whereas, in a closed autocracy with a marital structure, there is less formalization and more significant characteristics of an arbitrary model (Pichault & Schoenaers, 2003). The company is centred on the entrepreneur, with support from his wife, and all the practices are based on socialisation and commitment. Thus, reinforcing the clan culture, with characteristics of both an arbitrary model, typical of these businesses, and a value model (Pichault & Schoenaers, 2003). These more sophisticated practices, although introduced intuitively (Daspit et al., 2016), also emphasise the entrepreneur's stewardship skills, such as his insights about the business.

The differences in family business structures and governance models also reflect the challenges they encounter in succession planning. Although neither model has formal succession planning, the situations differ between open and closed autocracies. In an open autocracy, the founder and chairman face limitations in their choice of successor, while in a closed autocracy, the entrepreneur has the freedom to choose their successor. In an open autocracy, the company faces the challenge of having no potential successor, whereas, in a closed autocracy, the company has a potential successor that is being prepared for the family business (business socialisation) (García-Álvarez et al., 2002).

Studying family businesses presents unique challenges, and this research faced several limitations that we hope to address in future studies. Specifically, we aim to understand better the dynamics among family members within the family business and to gain insight into employees' perceptions of practices within these firms. Although formal documentation is limited, having access to personnel data—such as sociodemographic information, seniority, and job descriptions—would have also been beneficial.

References

Associação das Empresas Familiares Portuguesas. (2021). Associação das Empresas Familiares – Associação das Empresas Familiares. https://www.empresasfamiliares.pt/

Atkinson, J. (1984). Manpower strategies for flexible organizations. *Personnel Management, 16*(8), 28–31.

Bardin, L. (2011). Análise de conteúdo. *Edições, 70.*

Basco, R., Bassetti, T., Maso, L. D., & Lattanzi, N. (2023). Why and when do family firms invest less in talent management? The suppressor effect of risk aversion. *Journal of Management and Governance, 27*, 101–130. https://doi.org/10.1007/s10997-021-09599-1

Bernhoeft, R. (1989). Empresa familiar: sucessão profissionalizada ou sobrevivência comprometida (2nd ed.). Nobel.

Bernhoeft, R., & Gallo, M. A. (2003). Governança na Empresa Familiar (2nd ed.). Editora Campus.

Breton-Miller, I. L., & Miller, D. (2015). Learning stewardship in family firms: For family, by family, across the life cycle. *Academy of Management Learning & Education, 14*(3), 386–399. https://doi.org/10.5465/amle.2014.0131

Brown, M., Setren, E., & Topa, G. (2012). Do informal referrals lead to better matches? Evidence from a firm's employee referral system. *Journal of Labor Economics, 34*(1), 161–209.

Cameron, K. S., & Quinn, R. E. (2006). *Diagnosing and changing organizational culture.* Jossey-Bass.
Creswell, J. W. (2009). Research design: Qualitative, quantitative and mixed methods approaches (3ª ed.). SAGE.
Daspit, J. J., Holt, D. T., Chrisman, J. J., & Long, R. G. (2016). Examining family firm succession from a social exchange perspective a multiphase. *Multistakeholder Review, 29*, 44–64. https://doi.org/10.1177/0894486515599688
Decker, C., Heinrichs, K., Rau, S., & Jaskiewicz, P. (2016). What do we know about succession in family businesses? Mapping current knowledge and future territory. In F. Kellermanns & F. Hoy (Eds.), *The Routledge companion to family business* (pp. 15–44). Routledge.
Dimaggio, P. J., & Powell, W. W. (1983). The iron cage revisited: Institutional isomorphism and collective rationality in organizational fields. In *The new institutionalism in organizational analysis* (pp. 63–82). The University of Chicago Press.
Donnelley, R. G. (1964). The family business. *Harvard Business Review, 42*(4), 93–105.
European Commission. (2009). Final report of the expert group. Overview of the family-business-relevant issues: research, networks, policy measures and existing studies. Enterprise and Industry Directorate-General, Ref. Ares (2015) 2102740 - 20/05/2015.
Fortin, M. (2009). *Fundamentos e Etapas do Processo de Investigação.* Lusodidacta.
García-Álvarez, E., López-Sintas, J., & Saldaña Gonzalvo, P. (2002). Socialization patterns of successors in first- to second-generation family businesses. *Family Business Review, 15*(3), 189–203. https://doi.org/10.1111/j.1741-6248.2002.00189.x
Gomez, P. Y. (2015). *Référentiel pour une gouvernance raisonnable des entreprises* (2nd d.). Institut Français de Gouvernement des Entreprises. https://www.researchgate.net/publication/278685414_Referentiel_pour_une_gouvernance_raisonnable_des_entreprises
Guerreiro, M. D. (1996). Pequenas E Médias Empresas Familiares. Tipos de relações das famílias com as empresas. Associação Portuguesa de Sociologia. *Práticas e processos de mudança social. III Congresso Português de Sociologia.* https://aps.pt/wp-content/uploads/2017/08/DPR492ed4c6b71c1_1.pdf
Handler, W. C. (1994). Succession in family business: A review of the research. *Family Business Review, 7*(2), 133–157. https://doi.org/10.1111/j.1741-6248.1994.00133
Hirigoyen, G. (2002). Droit et finance: De l´ignorance à la gouvernance. *Revue Du Financier, 133*, 1–25.
Hoon, C., Hack, A., & Kellermanns, F. W. (2019). Advancing knowledge on human resource management in family firms: An introduction and integrative framework. *German Journal of Human Resource Management, 33*(3), 147–166. https://doi.org/10.1177/2397002219847883
Kelly, L. M., Athanassiou, N., & Crittenden, W. F. (2000). Founder centrality and strategic behavior in the family-owned firm. *Entrepreneurship Theory and Practice, 25*(2), 27–42. https://doi.org/10.1177/104225870002500202
Kim, T., Medina-Craven, M. N., Marler, L. E., Long, R. G., & Fugate, M. (2024). Attracting nonfamily applicants to the family firm: A signalling theory approach. *Journal of Small Business Management.* https://doi.org/10.1080/00472778.2024.2378906
Kirkpatrick, D. L. (1983). Four steps to measuring training effectiveness. *Personnel Administrator, 28*(11), 19–25.
Kok, J. M. P., Uhlaner, L. M., & Thurik, A. R. (2003). Human resource management with small firms, facts and explanations. *Erim Report Series Research In Management.* http://papers.ssrn.com/sol3/papers.cfm?abstract_id=411655
Lepak, D. P. E., & Snell, S. A. (1999). The human resource architecture: Toward a theory of human capital allocation and development. *Academy of Management Review, 24*(1), 31–48.
Lodi, J. B. (1998). *A empresa familiar* (5ª ed). Pioneira.
Marques, M. A., & Rodrigues, J. (2023). A gestão de recursos humanos numa empresa familiar do Alto Alentejo no setor da distribuição alimentar: um estudo exploratório. Atas do XII Congresso Português de Sociologia. Trabalho, Organizações e Profissões.(XII-APS-16772). APS. https://aps.pt/atas-xii-congresso/

Marques, M. A. A. (2010). Modelos organizacionais e práticas de gestão de recursos humanos - um estudo multicaso. [Tese de Doutoramento. Universidade Técnica de Lisboa. Instituto Superior de Economia e Gestão]. http://hdl.handle.net/10400.5/2250

Marques, A. P. (Coord.) (2018). *Empresas familiares da região norte mapeamento, retratos e testemunhos*. Universidade do Minho (CICS.NOVA).

Miller, D. (1996). A preliminary typology of organizational learning: Synthesizing the literature. *Journal of Management, 22*(3), 485. https://doi.org/10.1177/014920639602200305

Mintzberg, H. (1995). *Estruturas e Dinâmicas Organizacionais*. Publicações Quixote.

Neckebrouck, J., Schulze, W., & Zellweger, T. (2018). Are family firms good employers? *Academy of Management Journal, 61*(2), 553–585. https://doi.org/10.5465/amj.2016.0765

Nelson, T., & Constantinidis, C. (2017). Sex and gender in family business succession research: A review and forward agenda from a social construction perspective. *Family Business Review, 30*(3), 219–241. https://doi.org/10.1177/0894486517715390

Nordqvist, M., Wennberg, K., Bau, M., & Hellerstedt, K. (2013). An entrepreneurial process perspective on succession in family firms. *Small Business Economics, 40*, 1087–1122.

Nordqvist, M., & Melin, L. (2010). Entrepreneurial families and family firms. *Entrepreneurship & Regional Development, 22*(3–4), 211–239. https://doi.org/10.1080/08985621003726119

Pichault, F., & Schoenaers, F. (2003). HRM practices in a process of organisational change: A contextualist perspective. *Applied Psychology, 52*, 120–143. https://doi.org/10.1111/1464-0597.00127

Porfírio, J. A., Felício, J. A., & Carrilho, T. (2020). Family business succession: Analysis of the drivers of success based on entrepreneurship theory. *Journal of Business Research., 115*, 250–257. https://doi.org/10.1016/j.jbusres.2019.11.054

Rodrigues, J. (2019a). *Família Empresária Estrutura, recursos e poder*. Editora RH.

Rodrigues, J. (2019b). *Introdução à governabilidade organizacional*. Associação dos Professores do Concelho de Almada.

Rodrigues, J., & Marques, M. A. (2021). Governabilidade na Família Empresária: Principais dimensões. IPL.

Rodrigues, J., & Marques, M. A. A. (2019). Family firms and family business: A conceptual approach about the ambiguities, paradoxes and uniqueness of family businesses. *European Journal of Social Sciences Studies, 4*(5), 150–172. https://doi.org/10.46827/ejsss.v0i0.698

Schein E. H. (1983). *Organizational culture: A dynamic model*. WP#1412–83 Alfred P. Sloan School of Management. https://dspace.mit.edu/bitstream/handle/1721.1/48689/organizationalcu00sche.pdf?seq

Vardaman, J. M., Tabor, W. E., Hargrove, D. C., & Zhou, F. (2024). Reducing agency costs through recruitment: Staffing referrals and family business success. *Journal of Family Business Management, 14*(1), 103–119. https://doi.org/10.1108/JFBM-05-2023-0063

Yin, R. K. (1994).*Case study research: Design and methods* (2.ª ed.), Sage.

Index

A
Academia, 3, 13–15
Adapting, 313, 314, 320, 324, 326
Adopting, 333
Advantages, 198, 202, 205, 206, 210, 211
Arguments, 333, 334, 337, 349
Artificial intelligence, 47–58, 60–63, 68, 71, 74–77
Assistant role, 95
Attraction, 32, 36–38, 40–42
Authentic Leadership (AL), 164, 165
Authoritarian Leadership (ATL), 165, 166
Automation of routine tasks, 93

B
BANI world, 8
Behaviors of WE and IB, 373
Benefits, 25–27, 30–33, 36
Beyond, 195, 197, 207, 208
Build-up, 39
Burnout, 275, 277, 278, 280, 282–287, 289, 290, 292–297

C
Candidate profile, 200, 209
Candidate screening, 93, 99, 108
Candidate selection, 89–91, 93, 94, 97, 99, 102, 104, 107, 108
Career transitions, 243–248, 250, 251, 253, 254, 257–264, 266, 268–270
Certified public accountants, 141, 142
Challenges, 22, 24, 26, 27, 31, 35
Change, 3, 4, 7–15, 113–116, 125, 130, 163, 164, 166, 169, 172–181, 183, 184
Change management, 163, 173
Characteristics, 23, 24, 29, 31
Coaching, 215–218, 220, 221, 225–227, 236, 237
Collaborative technologies, 69
Communication, 23, 25, 27–30, 32, 34, 37, 39–41
Concept, 390–393, 396
Construct clarity, 5
Construction sector, 317
Continuity, 3, 4, 6–10
Corporate governance models, 393, 400
Countries, 370, 380, 381
Country of origin, 313
COVID-19 pandemic, 3, 8, 9, 236, 275, 278, 280, 281
Cranet database, 3, 10
Creating interesting, 38
Critical analysis, 333
Culture's influence, 181

D
Data, 47, 49, 52–57, 59–77
Decline, 40
Despotic leadership, 167
Destination country, 311, 313–315, 317, 320, 325, 326
Destructive Leadership (DL), 167, 168, 182
Development of individuals, 69
Different periods in time, 333, 334
Digital nomadism, 21–29, 32, 41

C. Machado (ed.), *Human Resource Management in a Disrupted World*, Lecture Notes on Multidisciplinary Industrial Engineering, https://doi.org/10.1007/978-3-031-89948-5

Digital nomads, 21–29, 31–42
Disadvantages, 198, 205, 206, 209, 210, 211
Disadvantages of technology, 205, 206, 211
Disrupted times, 4, 14
Disrupted world, 4, 7–10, 14
Drawbacks, 26, 27

E

Emerging categories, 248, 250, 251, 268, 269
Employee monitoring, 32, 37, 40
Engagement, 275, 277, 278, 280–282, 284–288, 290–297
E-recruitment, 196, 197
Ethical Leadership (EL), 165, 168, 169, 183
Evaluation, 113–130
Evaluation criteria, 117, 122–124, 128, 130, 131
Executive coaching, 215, 217, 218, 227, 237
Expatriate workers, 309, 314, 317, 318, 324, 327
Expatriate workers scale, 318
Expatriation, 311–314, 317, 321, 322, 324, 325, 327
Expatriation process, 311–314

F

Factor analysis, 288, 289, 318, 319
Family business, 389–393, 395–397, 399–402, 405–408
Field of study, 3, 5, 6, 13, 14
First chatGPT prompt, 96, 99, 100
Focus group, 47, 49, 64–67, 70–72
Focus group study, 201, 211
Four continents, 21, 30
4-day workweek, 333, 334, 337, 339, 340, 344–347
4-day workweek arguments, 333, 334, 337, 349
4-day workweek pilots, 333, 334, 338, 340, 343, 345–347, 349, 350

G

Gender imbalances, 333, 334, 337, 341, 349
General satisfaction with life, 284
Generative artificial intelligence, 89–91, 93–96, 99, 102, 107–109
Governance models, 389–391, 393, 397, 400, 407, 408
Group I, 221
Group II, 221, 225

H

Hiring, 21, 26, 27, 36–39, 41
HRM strategy, 215, 218
Human Resource Management (HRM), 3–10, 12–15, 21, 22, 28, 31, 34, 89, 108, 113–115, 120, 129–131, 196, 201, 211, 215–218, 220, 231–233, 235–237, 359, 360, 363, 369, 389–391, 395, 397, 402, 405, 407
Human resource management practices, 389–391, 397, 402, 404, 407
Human resources, 21, 22, 28–31, 34, 36–39
Human resources management level, 34

I

I4.0 technologies, 70
Impact, 49, 51, 52, 54, 57, 59, 63, 66, 70, 113, 115, 119, 122, 124–126, 128, 129
Impact of HRM, 373–376, 380–382
Impact on adaptation, 309, 314
Impact on WE and IB behaviors, 369, 376
Implications, 26, 113, 125, 128, 130, 323, 324
Importance, 115, 116, 123, 125, 127
Improvement, 338, 341, 345, 347
Income growth, 34, 37, 40
Increase of the productivity, 338
Individual analysis, 251
Industry 4.0, 47–49, 51–55, 57–59, 61–64, 66–78
Influence, 141–145, 147, 149–153, 155
Influence at work, 141, 142, 144, 145, 147, 149–155
Innovative behaviour, 359–361, 363, 366, 369, 375, 378, 379
Insights, 35, 42, 260, 261, 269
Instrument, 219, 221, 236
Integrated, 216, 217, 236, 237
Integrating the technology, 203
Integration, 22, 31, 34, 39, 41
Interior of Portugal, 389, 390, 397, 407

Interventation, 215, 216, 218, 237, 324, 327
Intervention, 324, 327

J

Job advertising, 197, 207
Job insecurity, 359, 360, 362, 363, 369, 375, 377
Job satisfaction, 113, 115, 116, 126, 127, 134, 141–145, 147, 150–152, 154

K

Knowledge management, 47–50, 53, 56, 61–64, 66–68, 71–77
Knowledge transfer, 50, 71, 74, 76, 177
Kotter's model, 175
Kurt Lewin's CATS, 173, 176

L

Leadership, 163–171, 173, 174, 178, 180–183
Leadership's impact, 163, 164, 170, 181
Leave, 35, 40, 41
Life satisfaction, 275, 277–279, 284, 285, 287, 289–295, 297

M

Main arguments, 334
Managing digital nomads, 31, 32, 36, 40–42
Maturity, 40
Meaningful work, 141–147, 151–154
Mediating, 150, 152
Mental health, 215, 217, 227
Moderating, 142–144, 146, 155
Motivation, 275–286, 288–297
Motivation for expatriation, 317, 321, 322, 324, 325

N

Nadler and Tushman congruence model, 174
Narrative approach, 243, 245, 246, 248
Narrative elements, 248–251, 261, 268, 269
Navigating complexity, 163
Negative, 122, 124, 126–128, 132, 135
Non-technological Portuguese SME, 113

O

Obstacles, 69, 71
Offsetting, 205
Onboarding, 32, 37, 39
Online Recruitment (OR), 195–201, 204, 206–210
Openness to Change (OTC), 164, 181, 184
Operational framework, 21, 31, 37, 38, 42
Organiational structure, 400
Organisational values, 195, 208
Organisations, 22, 26–28, 31, 41
Organizational Change (OC), 163, 164, 166, 172, 173, 181, 183, 184
Organizational culture, 163, 164, 169, 175–179, 181, 182–184
Organizational level, 32, 33
Organizational strategies, 309
Outset, 195, 203

P

Participative Leadership (PL), 166
People management strategy, 21
Perception, 113, 115, 117, 121, 123, 124, 126, 127
Perception of HRM, 368, 370, 371, 376
Perception of JI, 369, 373, 376, 377
Performance evaluation, 113–116, 118–122, 125–128
Performance evaluation methods, 118
Performance evaluation sources, 117
Performance Evaluation System (PES), 113, 114, 116, 117, 120, 121, 124, 125, 127–131
Perspectives, 312
PE satisfaction, 115, 116, 126, 128
PES functioning, 122, 131
PES impact, 113, 124
Portuguese context, 195, 196, 199, 200, 212
Portuguese population, 315, 316
Portuguese recruiters, 195, 211
Positive, 116, 119, 122, 124, 126, 128, 129, 132, 134
Positive exiting procedures, 35, 41
Predictors of adaptation, 323
Premises, 116, 124
Process, 115–118, 120–126, 131
Profile matching, 93
Program, 3–6
Promotion of general well-being, 340
Public sector, 364

R

Reasons, 277, 279–281, 283, 287
Recommendations, 113, 125, 127–129

Recruiters, 195–201, 203–206, 208–211
Recruitment, 26, 27, 36–38, 195–201, 203, 204, 206–210, 212
Recruitment and Selection (R&S), 195–201, 204–212
Reducing working time, 333, 334, 341, 343
Reduction of environmental impact, 344
Relationship, 141–147, 150–154, 245, 250, 256, 258, 260, 268, 361–367, 369, 377–379, 381–383
Remote employee lifecycle, 21, 31, 37, 38, 42
Remote work experiences, 295
Remote working, 32, 33
Repeat, 40
Research methods, 245
Resistance to change, 164, 173, 175, 179
Rethink HRM, 3, 4, 14
Rise, 10, 12–14
Role, 276, 277, 281, 282, 295
Roles of stress, 142, 144

S

Selection, 195–199, 203, 207
Selection process, 197, 207
Self-determination theory, 275, 277, 279, 288, 293
Servant leadership, 169, 170, 182, 183
Shaping, 276, 277, 281
Simulations of interviews, 93
Slightly, 3
Small and Medium Enterprises (SME's), 390
Small family business, 389, 390, 397, 407
Socialisation, 25, 37, 40
Social media, 195, 197, 199–201, 207, 208
Society, 23
Sociocultural adaptation, 309, 316, 318, 321–324, 326
Stace and Dunphy's model, 174
Staff assistance, 34, 39
Stages, 311
Stakeholder, 3, 6, 8, 12–15
Strategic human resource management, 5
Stress, 141–147, 150–155
Subcultures' influence, 179
Subversive view, 3
Success factors, 312, 314
Succession, 389–391, 393, 394, 396, 397, 405, 407, 408
Succession in family business, 396
System role, 95

T

Team level, 35
Technological change, 11
Technological innovations, 68
Technology, 195–201, 203–208, 210–212
Telework, 275–289, 292–297
Teleworking, 275–282, 284–287, 289, 290, 292–297
Telework motivations, 286, 288, 289
Telework well-being, 275–280, 283–285, 289, 293, 297
Thematic analysis, 359, 367, 370, 382, 383
The role of leadership, 71, 74, 75, 77
Transactional Leadership (TCL), 171
Transformational Leadership (TL), 170, 171, 181
Types, 391, 395–397, 403
Types of knowledge, 396

U

Unchanged, 195, 209
Universities, 359–364, 366, 372–375, 377–383
User role, 95

V

Validation, 244, 269, 296
Van de Ven and Poole's Model, 175
VUCA world, 3, 7, 8

W

Well-being, 275–280, 282–285, 289, 292–294, 296, 297
Work, 3–7, 9–15
Work engagement, 275, 281, 286–288, 293, 294, 359, 360, 363, 369
Workers, 359–364, 366, 371–383
Work-Family Conflicts (WFC), 141–147, 149–155
Working-hours management, 39
Working time, 333–337, 339, 341, 343, 345, 350
Workplace, 215–218
Work-related satisfaction, 126, 127
Work time reduction, 346

The manufacturer's authorised representative in the EU is Springer Nature Customer Service Centre GmbH, Europaplatz 3, 69115 Heidelberg, Germany. If you have any concerns regarding our products, please contact ProductSafety@springernature.com

Printed and bound by CPI Group (UK) Ltd, Croydon, CR0 4YY
23/09/2026
02221529-0003